VOLUME SEVENTY EIGHT

ADVANCES IN BOTANICAL RESEARCH

Genomes and Evolution of Charophytes, Bryophytes, Lycophytes and Ferns

ADVANCES IN BOTANICAL RESEARCH

Series Editors

VOLUME SEVENTY EIGHT

ADVANCES IN BOTANICAL RESEARCH

Genomes and Evolution of Charophytes, Bryophytes, Lycophytes and Ferns

Volume Editor

STEFAN A. RENSING

University of Marburg, Germany

ELSEVIER

AMSTERDAM • BOSTON • HEIDELBERG • LONDON
NEW YORK • OXFORD • PARIS • SAN DIEGO
SAN FRANCISCO • SINGAPORE • SYDNEY • TOKYO

Academic Press is an imprint of Elsevier

Academic Press is an imprint of Elsevier
125 London Wall, London EC2Y 5AS, UK
The Boulevard, Langford Lane, Kidlington, Oxford OX5 1GB, UK
50 Hampshire Street, 5th Floor, Cambridge, MA 02139, USA
525 B Street, Suite 1800, San Diego, CA 92101-4495, USA

First edition 2016

ISBN: 978-0-12-801102-7
ISSN: 0065-2296

For information on all Academic Press publications
visit our website at http://store.elsevier.com/

CONTENTS

CONTRIBUTORS

A.A. Carrell
Duke University, Durham, NC, United States; Oak Ridge National Laboratory, Oak Ridge, TN, United States

A.C. Cuming
University of Leeds, Leeds, United Kingdom

C.F. Delwiche
University of Maryland, College Park, MD, United States

J.P. Der
California State University, Fullerton, CA, United States

N. Devos
Duke University, Durham, NC, United States

V. Knoop
Universität Bonn, Bonn, Germany

D. Lang
University of Freiburg, Freiburg, Germany

S.F. McDaniel
University of Florida, Gainesville, FL, United States

P.-F. Perroud
University of Marburg, Marburg, Germany

R. Reski
University of Freiburg, Freiburg, Germany; FRIAS – Freiburg Institute for Advanced Studies, Freiburg, Germany; BIOSS – Centre for Biological Signalling Studies, Freiburg, Germany

K. Sakakibara
Kanazawa University, Kanazawa, Japan

M. Schallenberg-Rüdinger
Universität Bonn, Bonn, Germany; University of Marburg, Marburg, Germany

J. Schmutz
HudsonAlpha Institute of Biotechnology, Huntsville, AL, United States

E.B. Sessa
University of Florida, Gainesville, FL, United States

A.J. Shaw
Duke University, Durham, NC, United States

S. Shu
Department of Energy Joint Genome Institute, Walnut Creek, CA, United States

P. Szövényi
University of Zurich, Zurich, Switzerland; Swiss Institute of Bioinformatics, Lausanne, Switzerland; ELTE, Budapest, Hungary

K.K. Ullrich
University of Marburg, Marburg, Germany

N. van Gessel
University of Freiburg, Freiburg, Germany

D.J. Weston
Oak Ridge National Laboratory, Oak Ridge, TN, United States

PREFACE

Genome sequences of nonvertebrate genomes have greatly informed our understanding of animal evolution. In similar fashion, the sequencing of the genomes of the moss *Physcomitrella patens* and of the lycophyte *Selaginella moellendorffii* have helped to unriddle plant evolution by means of comparative genomics and evolutionary developmental (evo-devo) studies. However, many extant land plant lineages of non-seed plants still did not have a representative with a sequenced genome until recently, or are work in progress. The focus of this book is on these so far neglected lineages, namely charophyte algae that are the sister group of the land plants, the paraphyletic bryophytes (hornworts, mosses and liverworts), as well as the lycophytes (club mosses) and ferns.

In the first two chapters of this book, development and evolution of land plants are reviewed. While Chapter 1 by Keiko Sakakibara focuses on non-seed plant evo-devo, Chapter 2 by Volker Knoop and Mareike Schallenberg-Rüdinger highlights a very peculiar feature of land plants, organellar RNA editing.

Chapters 3–6 deal with sequenced genomes. The first non-seed plant genome, *P. patens*, is reviewed in Chapter 3 (written by Daniel Lang, Nico van Gessel, Kristian Ullrich and Ralf Reski), followed by the genomes of two other mosses that are work in progress, the purple moss *Ceratodon purpureus* (reviewed by Stuart McDaniel, Andrew Cuming, Peter Szovenyi and Pierre-Francois Perroud) and the peat moss, *Sphagnum* (by Jon Shaw, David Weston and colleagues). The hornworts are covered in Chapter 6 by Peter Szovenyi dealing with *Anthoceros agrestis*. While the actual genome drafts have not yet been published, Chapters 4–6 review what is known from transcriptomic data and provide a glimpse at how these genomes will soon inform our understanding of land plant evolution.

The two final chapters look primarily into transcriptomic data. Chapter 7 by Joshua Der and Emily Sessa deals with lycophytes and ferns and will, after reviewing transcriptomic data, outline which genomes are work in progress. What we know about charophyte algae based on their transcriptomes (and one genome) is reviewed in Chapter 8 by Charles Delwiche.

The first genomes of conifers have recently been published, closing an important gap. Hopefully, the other lineages of gymnosperms (*Ginkgo*, cycads) will also be covered in the near future. Two projects are now tackling the large genomes of the ferns, *Azolla* and *Ceratopteris*. Several projects aim at unravelling more genomes of mosses (eg, *Takakia*) and liverworts (eg, *Jungermannia*), as well as of several lineages of charophyte algae (eg, *Chara*). The genome of the liverwort *Marchantia polymorpha* has been sequenced but not yet published. The next decade will see most of the major gaps in land plant phylogeny closed in terms of sequenced genomes of representatives. This book reviews what we know today about how land plants evolved, and lays the ground for many more exciting findings to come.

Stefan A. Rensing

ACKNOWLEDGEMENTS

I would like to thank Christopher Grosche, Pierre-Francois Perroud and Kristian Ullrich for their help with editing this book.

SECTION ONE

Evolution

CHAPTER ONE

Technological Innovations Give Rise to a New Era of Plant Evolutionary Developmental Biology

K. Sakakibara
Kanazawa University, Kanazawa, Japan
E-mail: bara@staff.kanazawa-u.ac.jp

Contents

Abstract

Land plants evolved from freshwater algaelike ancestors approximately 480 million years ago. Land plants developed many new morphological features during the evolution including the origin of a multicellular diploid sporophyte, a sporophytic apical meristem that produces complex body architecture, stomata and the production of lateral organs (leaves), vascular tissue, roots, seeds and flowers. Genetic changes drove the development of these morphological features and recent technological innovations in genomics and genetic modification technologies have allowed us to investigate these genetic changes. Here, I describe recent research on several transcription factors that contributed to the morphological evolution of land plants, including the homeodomain proteins KNOX, BELL and WOX, as well as transcription factors of the bHLH, NAC and FLO/LFY families. I also discuss the conservation of plant hormone signalling, which acts to coordinate the multicellular body plan.

Advances in Botanical Research, Volume 78
ISSN 0065-2296
http://dx.doi.org/10.1016/bs.abr.2016.01.001

1. INTRODUCTION

The origin and diversification of land plants precipitated a tremendous increase in diversity on Earth. Terrestrialization of metazoans depended on the subsequent diversification and spread of land plants, which changed the biosphere and established a habitable global environment. Molecular phylogenetic information indicates that land plants (embryophytes) originated from a freshwater green alga (Karol, McCourt, Cimino, & Delwiche, 2001; Lewis & McCourt, 2004; Wickett et al., 2014). The closest relatives of land plants, charophycean algae, exhibit a haplontic life history, in which a multicellular haploid organism produces gametes that fuse to form a diploid cell that undergoes meiosis to yield spores. By contrast, land plants possess a haplodiplontic life cycle, the so-called alternation of generations, which involves alternation of two morphologically distinct bodies, the sporophyte and gametophyte. The evolution of body plan in land plants involved fundamental changes in the forms of both the haploid gametophyte and the diploid sporophyte. These changes, which were inherited by descendant land plants, allowed the presence of two body plans in different generations with the same genome.

Alternation of generations evolved several times in different groups of algae (Friedman, 2013), but evolved only once in the streptophytes, which include the land plants and charophycean green algae. Two major theories have been proposed to explain the origin of the alternation of generations in land plants: the homologous (or transformation) (Pringsheim, 1878) and antithetic (or interpolation) theories (Bower, 1890). The homologous theory posits that the ancestor of land plants exhibited alternating isomorphic generations and that the morphological difference in each generation gradually arose (Kenrick & Crane, 1997b). By contrast, the antithetic theory, which is more widely accepted, suggests that the multicellular diploid sporophyte body of the embryo evolved through a delay in zygotic meiosis and an intercalation of extensive mitotic cell divisions between fertilization and the production of spores via meiosis (Bower, 1890; Graham, 1993).

A number of researchers supposed that early land plants were haploid gametophyte dominant, similar to extant bryophytes (Bower, 1890; Kenrick & Crane, 1997a). The evolution of a diploid sporophyte-dominant life cycle in vascular plants was accompanied by dramatic changes in body plan that allowed for the development of larger and more complex body plans in the diploid sporophyte generation.

Fundamental innovations of the land plant body plan include the origin of a multicellular diploid sporophyte (zygote retention), cell specialization capacity in both gametophyte and sporophyte generations, the placental transfer cell, a junction of gametophyte and sporophyte tissues, stomata, a pore differentiating in sporophyte epidermis for transpiration and gas exchange were established. Also subsequent innovations were achieved such as a sporophytic apical meristem with the capability of branching, the production of lateral organs (leaves), specialized nutrient and water-conducting cells (vascular tissue), the origin of roots in vascular plant lineage, and the origin of seeds and subsequently flowers in the seed plant lineage (Friedman, Moore, & Purugganan, 2004; Graham, Cook, & Busse, 2000; Mishler & Churchill, 1985).

Much research has aimed to identify the genetic changes that drive such fundamental morphological innovations. Initial research identified several genes in nonflowering plants that are homologues of developmental genes in flowering plants, and researchers have examined the phylogeny among these genes in diverse taxa and their roles during evolution, based on sequence similarities and expression patterns (Floyd & Bowman, 2007; Nishiyama, 2007; Rensing et al., 2008). In ten years after human genome project, sequencing techniques have advanced dramatically with the development of next-generation sequencing. Genome sequencing projects have been conducted in most major taxa, including gymnosperms (Birol et al., 2013; Nystedt et al., 2013), lycophytes (Banks et al., 2011), bryophytes (Rensing et al., 2008), charophytes (Hori et al., 2014), chlorophytes (Derelle et al., 2006; Merchant et al., 2007; Prochnik et al., 2010), red algae (Collén et al., 2013; Matsuzaki et al., 2004; Nakamura et al., 2013; Schönknecht et al., 2013), Glaucophyta (Price et al., 2012) and brown algae (Cock et al., 2010) (Fig. 1). Genomics has accelerated the identification of genes and allowed researchers to identify nonflowering plant homologues of developmental genes in flowering plants, and to detect gene losses and acquisitions. Comparative genomics offers a powerful approach to address major questions in plant evolutionary biology, such as identifying the developmental genetic changes that gave rise to important plant innovations. This approach has revealed a certain level of conservation between land plants and charophycean green algal lineages related to land plants at the genome level (Hori et al., 2014). Ultimately, the aim is to describe morphological evolution in terms of gene acquisitions and losses (Banks et al., 2011; Lang et al., 2010). Phylogenetic analyses of developmental genes can give

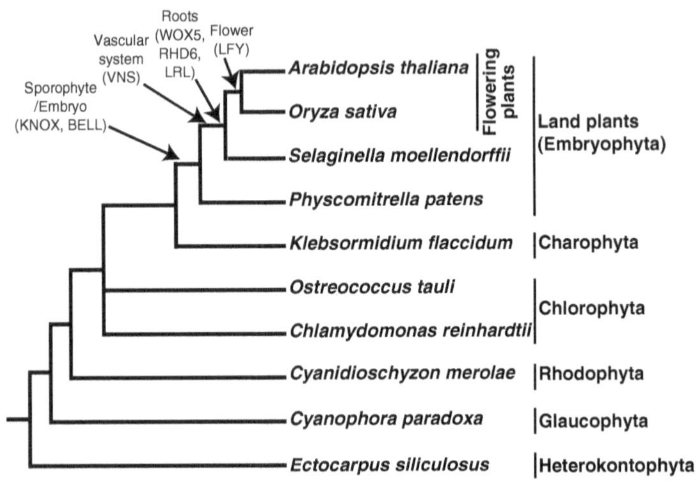

Figure 1 Phylogenetic relationship among species used for comparative analysis. *Arrows* indicate the timing of acquisition of morphological innovations and genes related to the morphological innovation are shown in the parentheses.

rise to hypotheses regarding the morphological evolution of land plants, but testing these hypotheses requires functional analyses.

In addition to the expression data, functional analysis would reveal the conservation and change of gene function. Genetic modification techniques in most major taxa, especially ferns (Muthukumar, Joyce, Elless, & Stewart, 2013; Plackett, Huang, Sanders, & Langdale, 2014), bryophytes (Ishizaki, Chiyoda, Yamato, & Kohchi, 2008; Schaefer & Zrÿd, 1997) and charophycean green algae (Abe et al., 2011; Sørensen et al., 2014), have emerged from late 1990's and provide powerful tools for new perspectives in evolutionary developmental research. These techniques provide opportunities to examine the functions of genes hypothesized to be involved in the evolution of morphological characteristics.

In this chapter, I discuss the evolution of major developmental innovations by land plants in both genomic and molecular developmental genetic contexts. I provide several examples of studies that focus on transcription factors that were fundamental to the evolution of land plants. Changes in the function of these transcription factors can yield aberrant morphological features. Consequently, gene duplication events that release one gene copy from selective pressure are thought to be a common mechanism by which transcription factors acquire new activities (Conant & Wolfe, 2008; Rensing, 2014). I discuss the evolution of developmental genes that contributed to morphological innovation in land plants, with an emphasis on transcription factors, including members of the KNOX (knotted1-like homeobox), BELL, WOX (WUSCHEL-related homeobox), and bHLH (basic helix-loop-helix), FLO/LFY (Floricaula/Leafy), and NAC (NAM (no apical meristem); ATAF (Arabidopsis transcription activation factor); CUC (cup-shaped cotyledon)) gene families. In addition, I discuss the evolution of regulatory networks related to these transcription factors. Subsequently, I introduce the conservation of plant hormone signalling, which is essential for multicellular body plan development. At the end, I discuss the future of plant evolutionary developmental research.

2. TRANSCRIPTION FACTORS

2.1 TALE-Type Homeobox Genes, KNOX and BELL: Diploid Developmental Genes

In bryophytes, the predominant generation of the life cycle is the free-living gametophyte, while the sporophyte is dependent on the gametophyte for its nutrition; by contrast, in vascular plants, the sporophytic generation is predominant. Thus, the sporophyte enlarged and became independent during the course of land plant evolution (Graham et al., 2000). It was hypothesized that the evolution of genes encoding two transcription factors, KNOX and BELL, mediated the emergence of the multicellular diploid generation of land plants (Lee, Lin, Joo, & Goodenough, 2008).

Homeodomain containing transcription factors already existed and diversified into at least two subfamilies, TALE (three-amino acid length extension) and non-TALE, in the last common eukaryote ancestor (Derelle, Lopez, Le Guyader, & Manuel, 2007). KNOX and BELL are both TALE homeodomain proteins. The genes encoding KNOX and BELL share an ancient origin and have been isolated from diverse green plants and the red alga *Cyanidioschyzon merolae* (Table 1; Figs 2 and 3; Mukherjee, Brocchieri, & Bürglin, 2009). In the unicellular chlorophyte *Chlamydomonas reinhardtii*, the diploid generation is regulated by the KNOX–BELL pathway (Lee et al., 2008). *Chlamydomonas reinhardtii* has two gamete mating types, 'plus' and 'minus', which express two homeobox genes, *GSP1* (*Gamete-specific plus1*), encoding a BELL-related protein, and *GSM1* (*Gamete-specific minus1*), encoding a KNOX-related protein, respectively. These two homeodomain proteins are cytoplasmic in each gamete, but upon gamete fusion form a heterodimer and translocate into the nucleus

Table 1 Classification and Number of Transcription Factors Retrieved From Plant Genomes

		Eudicots	Monocots	Lycopodiophyta	Bryophyta	Charophycean alga	Unicellular Green Alga		Red Alga	Brown alga	
Super Family	**Family (subfamily)**	***Arabidopsis thaliana***	***Oryza sativa***	***Selaginella moellendorffii***	***Physcomitrella patens***	***Klebsormidium flaccidum***	***Chlamydomonas reinhardtii***	***Ostreococcus tauri***	***Cyanidioschyzon merolae***	***Ectocarpus siliculosus***	**Resource**
	WOX	15	14	9	3	1	–	1	–	–	Deveaux et al. (2008), in this study
TALE	KNOX	8	13	5	5	2	1	1	1	–	Mukherjee et al. (2009), in this study
	(KNOX1)	4	9	3	3	1	–	–	–	–	Mukherjee et al. (2009)
	(KNOX2)	4	4	2	2	1	–	–	–	–	Mukherjee et al. (2009)
	BELL	13	14	2	4	1	1	2	2	–	Mukherjee et al. (2009), in this study
bHLH	bHLH	158	173	103	98	?	3	1	1	4	Pires and Dolan et al. (2010), Rayko, Maumus, Maheswari, Jabbari and Bowler (2010)
NAC	VNS	13	12	4	8	–	–	–	–	–	Nakano, Yamaguchi, Endo, Rejab and Ohtani (2015)
	LFY	1	1	1	2	1	–	–	–	–	Sayou et al. (2014)

? indicates that explicit report was not found.

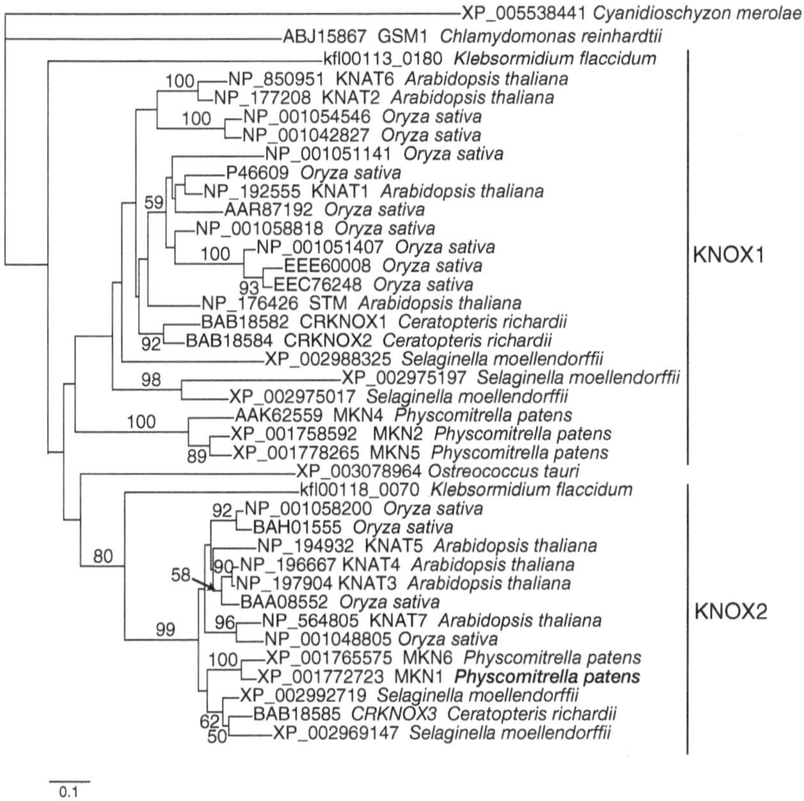

Figure 2 A maximum likelihood phylogenetic tree of the KNOX homeobox genes constructed using conserved 96 amino acid sequences in representative green plants and the red alga *Cyanidioschyzon merolae*. Bootstrap probabilities of more than 50% are shown on the branches. The horizontal branch length is proportional to the estimated evolutionary distance. The *bar* represents 0.1 substitutions per site. The *brackets* to the right indicate the subfamilies of the KNOX gene family.

to initiate zygotic development. Thus, the KNOX–BELL pathway regulates the haploid-to-diploid transition in chlorophytes.

In contrast to chlorophytes, land plant genomes contain two subfamilies of KNOX genes, class 1 (KNOX1) and class 2 (KNOX2), resulting from a gene duplication in the lineage leading to land plants (Fig. 2; Mukherjee et al., 2009). KNOX1 genes regulate sporophytic (diploid) meristematic genes (Hay & Tsiantis, 2010) and KNOX2 genes function in the haploid-to-diploid transition in mosses (Sakakibara et al., 2013) and in tissue differentiation of flowering plant sporophytes (Furumizu, Alvarez, Sakakibara, & Bowman, 2015).

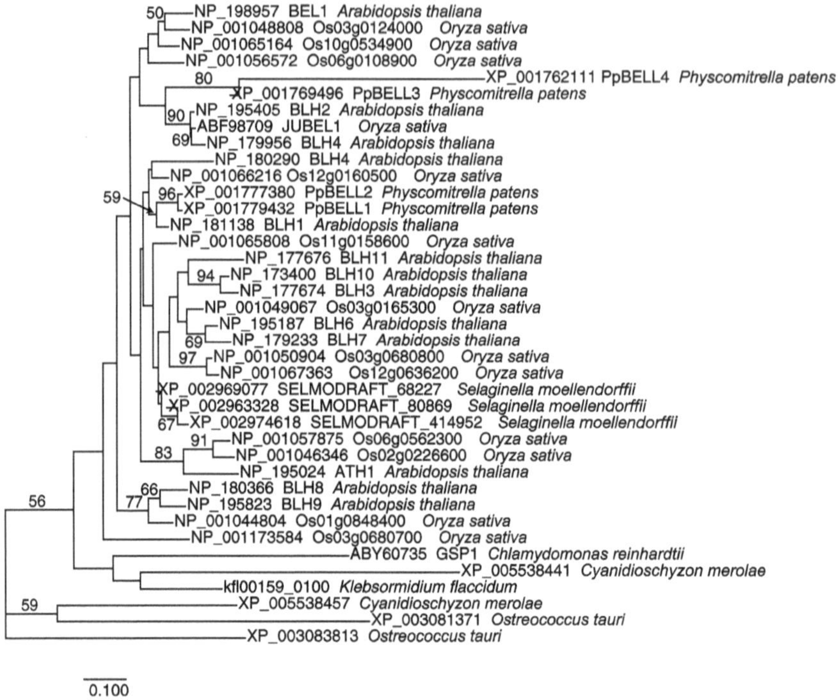

Figure 3 A maximum likelihood phylogenetic tree of the BELL homeobox genes constructed using conserved 50 amino acid sequences in representative green plants and the red alga *Cyanidioschyzon merolae*. Bootstrap probabilities of more than 50% are shown on the branches. The horizontal branch length is proportional to the estimated evolutionary distance. The *bar* represents 0.1 substitutions per site.

Flowering plants form and maintain various meristems in the sporophyte body, including a shoot apical meristem, root meristem and vascular meristem. As a result, flowering plants can produce a complex body structure in the sporophyte generation. Genetic analyses conducted in *Arabidopsis thaliana* have elucidated the gene regulatory network underlying the formation and maintenance of the shoot apical meristem. An *A. thaliana* gene isolated by mutant analysis, *shoot meristemless* (*STM*), is required for shoot apical meristem formation during embryogenesis. *STM* encodes a TALE homeodomain protein (Long, Moan, Medford, & Barton, 1996) similar to the one affected in the maize leaf mutant *knotted1* (*kn1*). Maize *kn1* was the first plant homeobox gene to be reported (Vollbrecht, Veit, Sinha, & Hake, 1991). After the discovery of *kn1*, numerous related genes were reported from diverse plant species and named KNOX genes. KNOX1 genes are characteristically expressed in the shoot apical meristem. In flowering plant species with simple leaves like *A. thaliana*, KNOX1 genes are expressed only in the shoot apical meristem and not in lateral organ primordia. The expression ceases in the regions that are destined to form the next leaf, prior to any visible differentiation (Hake et al., 2004). In flowering plant species with complex leaves like tomato and the *A. thaliana* relative *Cardamine hirsuta*, KNOX1 gene expression is reactivated during leaf development to facilitate leaflet formation (Hay & Tsiantis, 2010). Thus, KNOX1 genes play a critical role in regulating the meristematic properties of cells in flowering plant sporophytes (Hay & Tsiantis, 2010). In flowering plants, a complex consisting of ARP (Asymmetric Leaves1/Rough Sheath2/Phantastica), which contains an MYB domain, AS2 (Asymmetric Leaves2), which harbours a LOB (lateral organ boundaries) domain, and YAB (YABBY) protein represses KNOX1 gene expression in lateral organ primordia and defines the domain between an indeterminate shoot apex and a determinate lateral organ (Byrne et al., 2000; Hake et al., 2004; Kumaran, Bowman, & Sundaresan, 2002).

In the moss *Physcomitrella patens*, a model system of basal land plant lineages, the haploid generation dominates over the diploid generation. Gametophytes, the haploid plant body, produce shootlike structures with an indeterminate apical cell responsible for repetitive organ production, whereas sporophytes, the diploid body, do not have indeterminate growth potential and terminate with the formation of a sporangium. Sakakibara, Nishiyama, Deguchi, & Hasebe (2008) examined the expression and function of all three KNOX1 orthologs in *P. patens* and concluded that *P. patens* KNOX1 genes are expressed in the apical cell and meristematic region of the sporophyte and modulate determinate growth in the diploid generation, but do not function in the haploid indeterminate meristem. This result indicated that the genetic networks governing the indeterminate meristem in land plants are divergent, and the KNOX1-regulated networks governing the diploid indeterminate meristem likely evolved *de novo* in the flowering plant lineage (Sakakibara et al., 2008).

Studies of the gene regulatory network governing KNOX1 repression revealed that AS1 orthologs are not found in *P. patens*, and AS2 and YAB orthologs are found only in seed plants, whereas the *P. patens* genome contains members of both the MYB and LOB gene families (Banks et al., 2011; Lang et al., 2010). The establishment of complex regulatory networks that govern KNOX1 expression in seed plant lineages might have enabled the formation of indeterminate apical meristems that undergo repetitive organ production in the flowering plant sporophyte (Sakakibara et al., 2008).

The disruption of KNOX2 genes in *P. patens* causes the formation of gametophyte-like structures in the sporophytes without meiosis (apospory), indicating that KNOX2 genes are required for repression of the gametophyte developmental program in diploid sporophytes (Sakakibara et al., 2013). Thus, KNOX2 regulates the alternation of generations. Furumizu et al. (2015) revealed the function of KNOX2 in flowering plants using multiple loss-of-function KNOX2 mutants. *Arabidopsis thaliana* KNOX2 genes function with distinct BELL genes to modulate organ differentiation in all above-ground sporophyte organs. Furthermore, KNOX2 genes act antagonistically with KNOX1 genes in the above-ground parts of the sporophyte body (Furumizu et al., 2015).

The KNOX1 and KNOX2 subfamilies arose after the divergence of chlorophytes and streptophytes and the duplication facilitated the evolution of more complex gene regulatory networks with respective critical roles in the diploid phase of the alternation of generations. Gene duplication also had instrumental roles in establishing the multicellular diploid generation in land plants and the complex above-ground sporophyte body in flowering plants.

A recent phylogenetic analysis that includes KNOX genes in the charophycean algae *Klebsormidium flaccidum* indicates that KNOX1 and KNOX2 were duplicated in the streptophyte lineage (Fig. 2), even though charophycean algae do not exhibit alternation of generations. Functional analysis of KNOX genes using transformable charophycean algae will provide insight into the origin of KNOX1 and KNOX2 function in the alternation of generations (Abe et al., 2011; Sørensen et al., 2014).

2.2 WOX: The Origin of Stem Cell Genes

In addition to the KNOX–BELL gene regulatory pathway, which maintains indeterminacy of meristematic tissues in the shoot apical meristem, *WUSCHEL* (*WUS*) and its related homeobox genes maintain stem cell populations in the seed plant sporophyte body. Originally, the loss-of-function *wus* mutant was described as lacking stem cells in the shoot apical meristem and exhibiting termination of the shoot apical meristem of flowering plant sporophytes (Laux, Mayer, Berger, & Jurgens, 1996). The *WUS* gene encodes a plant-specific homeodomain transcription factor and belongs to a large gene family, named WOX. The *A. thaliana* genome encodes 15 WOX genes classified into three subfamilies, WUS-like, WOX9-like and WOX13-like (Fig. 4; Haecker et al., 2004; van der Graaff, Laux, & Rensing, 2009). Most members of the WUS subfamily are expressed in different

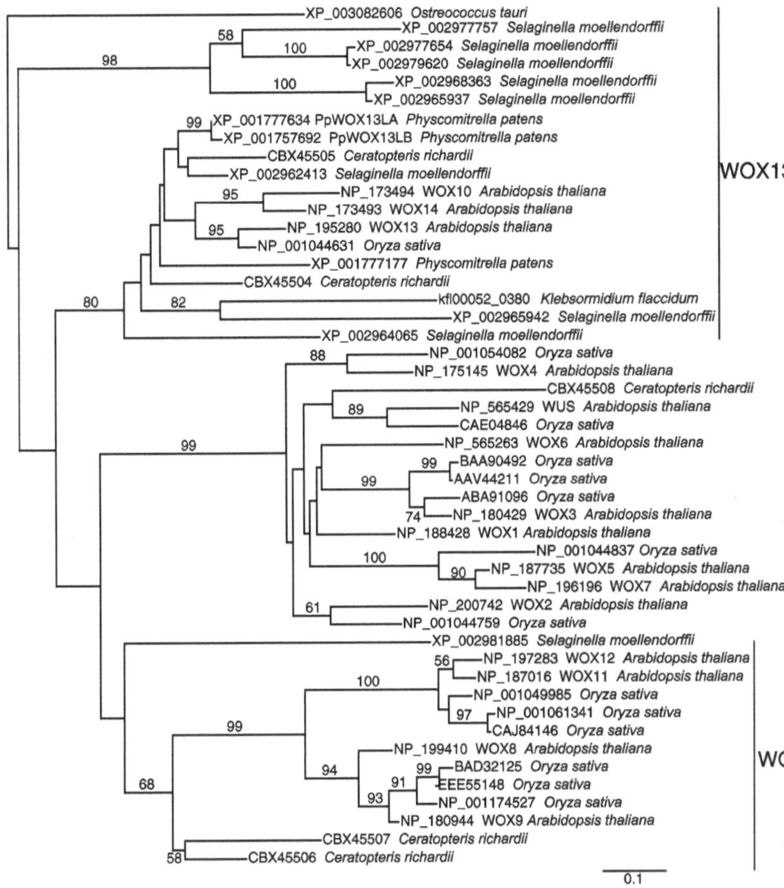

Figure 4 A maximum likelihood phylogenetic tree of WOX genes constructed using 62 conserved amino acid sequences in representative green plants. Bootstrap probabilities of more than 50% are shown on the branches. The horizontal branch length is proportional to the estimated evolutionary distance. The *bar* represents 0.1 substitutions per site. The *brackets* to the right indicate the subfamilies of the WOX gene subfamily.

meristems of the sporophyte body in flowering plants and specify stem cell identity in the respective meristem. Namely, *WUS* specifies stem cell identity in the shoot apical meristem (Laux et al., 1996), *WOX5* in the root meristem (Sarkar et al., 2007), *WOX3* in leaf marginal meristem (Nakata et al., 2012; Shimizu et al., 2009) and *WOX4* in the vascular meristem (Hirakawa, Kondo, & Fukuda, 2010; Ji, Shimizu, Sinha, & Scanlon, 2010). Two members of the WOX9-like subfamily, *WOX8* and *WOX9*, redundantly regulate zygote development and embryo axis formation (Breuninger, Rikirsch, Hermann, Ueda, & Laux, 2008; Ueda, Zhang,

& Laux, 2011) and embryo development, in combination with *WOX1* and *WOX2* in the WUS-like subfamily (Wu, Chory, & Weigel, 2007). Members of the WOX13-like subfamily function in replum development and dehiscence in *A. thaliana* fruits and influence flowering time, lateral root formation and fertility (Deveaux et al., 2008; Romera-Branchat, Ripoll, Yanofsky, & Pelaz, 2013).

WOX genes also occur in non-seed plants. WUS subfamily genes diverged from a single gene in the last common ancestor of ferns and seed plants, and the members of the WUS subfamily were dramatically more abundant in seed plants based on a phylogenetic analysis of WOX genes from a fern (*Ceratopteris richardii*), a lycophyte (*Selaginella moellendorffii*) and a moss (*P. patens*) (Hedman, Zhu, von Arnold, & Sohlberg, 2013; Nardmann & Werr, 2012). Several duplications of members of the WUS-like subfamily and their subfunctionalization in different stem cell niches might have contributed to the success of seed plants (Nardmann & Werr, 2012). To decipher the functional similarity among member of the WUS-like subfamily, we need to develop assays to probe the functions of *WUS* orthologs in ferns. Since transformation techniques have been established in ferns (Muthukumar, Joyce, Elless, & Stewart, 2013; Plackett et al., 2014), such assays should be available in the near future.

The function of WOX13-like members in nonflowering plants was investigated using the moss *P. patens*. In contrast to vascular plants, which depend on complex multicellular meristems, growth of *P. patens* depends on single apical stem cells that undergo asymmetric cell divisions in both the gametophyte and sporophyte generation (Kofuji & Hasebe, 2014). The *P. patens* genome contains three genes that belong to the ancestral WOX13-like subfamily. Two *P. patens* WOX13-like genes are expressed in most tissues in both the haploid and diploid generations and function in new stem cell formation in both of these generations, such as during zygote development and regeneration after leaf detachment or protoplast isolation (Sakakibara et al., 2014).

What was the origin of the WOX gene family? In the green algal lineage, the genome of the unicellular green alga *C. reinhardtii* does not have any WOX-like sequence, but the charophycean algae *K. flaccidum* has a single WOX13-like gene (Fig. 4) and the unicellular marine prasinophyceae green alga *Ostreococcus tauri* has a single WOX13-like gene that is constitutively expressed and not cell-cycle regulated (Deveaux et al., 2008). Mukherjee et al. (2009) speculated that WOX genes were already present in the last common ancestor of the green plant lineage and were successively lost in the *Chlamydomonas* lineage. The WOX gene family was expanded in land plants, extensively so in the seed plant lineage. Members of the WOX gene family presumably play important roles in establishing multicellularity in land plants via producing different meristems in the different parts of the body. The establishment of the WUS-like and WOX9-like subfamilies during vascular plant evolution and the diversification of these genes and their expression patterns by gene duplication and subsequent subfunctionalization might underlie diversification of meristems in the diploid generation (Fig. 4).

WUS regulates meristem size in the *A. thaliana* shoot apical meristem by communicating with a receptor kinase pathway composed of two leucine-rich repeat (LRR) receptor kinases, CLAVATA1 (CLV1) and CLAVATA2 (CLV2) and their ligand CLV3, a small secreted protein (Brand, Fletcher, Hobe, Meyerowitz, & Simon, 2000; Clark, Williams, & Meyerowitz, 1997; Schoof et al., 2000). CLV3 acts noncell autonomously to regulate *WUS* expression via interaction with CLV1 (Lenhard & Laux, 2003). This negative feedback mechanism restricts the size of meristems. *Arabidopsis thaliana* genome encodes several *CLV1* and *CLV3* (the CLV3 gene family is known as CLE (CLAVATA3/endosperm surrounding region)) paralogs. Different combination with signalling peptides, LRR receptor kinases and WOX transcription factors regulate cellular proliferation and differentiation in different meristems. Networks composed of CLE40-ACR4-WOX5 function in the identification of stem cell activity of the root meristem (Stahl, Wink, Ingram, & Simon, 2009) while TDIF-TDR-WOX4 determine the secondary growth of vascular tissues (Hirakawa et al., 2010). The duplication of genes encoding these signalling molecules and their cooption into different meristems might have contributed to increase the complexity of the flowering plant sporophyte body.

It is unclear at which time the gene regulatory network comprising CLV1, CLV3 and WOX was established. *CLV1* and *CLV2* orthologs apparently exist in *S. moellendorffii*, but these orthologs in *P. patens* are obscure, though the *P. patens* genome encodes LRR receptor kinases. The CLE sequence exists both in *S. moellendorffii* and *P. patens* (Banks et al., 2011; Floyd & Bowman, 2007). Functional analysis of these signalling factors in non-seed plants is expected to elucidate the origin of the CLV1-CLV3-WOX signalling network.

2.3 bHLH: The Evolution of the Rooting System

In Section 2.1, I introduced the hypothesis that the establishment of new gene regulatory networks contributed to the complexity of the sporophyte body. Here I describe another hypothesis, in which recruitment of preexisting genes that originally functioned in the gametophyte body promoted morphological diversification in the sporophyte body during land plant evolution.

The acquisition of tissues or organs that anchor the plant body in the ground was an important innovation for terrestrialization of plants. Various rooting structures were identified in the extinct and extant land plants (Kenrick & Crane, 1997a). The origin of roots is still unresolved (Friedman et al., 2004), though roots are thought to be evolved during Devonian Period, approximately 400 million years ago, based on fossil records (Kenrick, 2013). Roots are supposed to have evolved independently in the sporophytes of at least two distinct early vascular plant lineages, the lycophytes and the euphyllophytes, which include ferns, gymnosperms and flowering plants (Raven & Edwards, 2001), or to have originated in the common ancestor of vascular plants (Schneider, Pryer, Cranfill, Smith, & Wolf, 2002). Future comparative studies of the genes underlying development of roots in these two plant lineages may prove useful to the resolution of questions whether root organ(s) are homologous in vascular plants (Friedman et al., 2004).

Before the acquisition of roots, protonemata and rhizoids, which are the tip-growing filamentous structures that differentiate in the gametophyte body and serve as a rooting system. Rhizoids develop on the free-living gametophytes of vascular and nonvascular plants and existed on both the gametophytes and rootless sporophytes of extinct Rhyniophytes and Zosterophyllophyta (Hueber & Banks, 1979; Jones & Dolan, 2012). Vascular plants form root hairs, which are tip-growing filamentous cells that differentiate from the epidermal cells of roots in the sporophyte body, and these enhance nutrient acquisition and anchoring (Jones & Dolan, 2012). A recent comparative genetic analysis of *P. patens* and *A. thaliana* revealed conservation of gene regulatory networks that function in rhizoid differentiation in *P. patens* gametophytes and root hair differentiation in *A. thaliana* sporophytes (Menand et al., 2007; Tam, Catarino, & Dolan, 2015).

The bHLH proteins form a superfamily of transcription factors that bind as dimers to specific DNA target sites and share an ancient origin with other eukaryotes. Ancestral eukaryotes had a single bHLH gene, and the unicellular red alga *C. merolae* possesses a single bHLH gene, while the unicellular green algae *O. tauri* and *C. reinhardtii* have one and three bHLH genes, respectively (Table 1; Pires & Dolan, 2010). In land plants, the bHLH family expanded extensively to constitute 26 subfamily in organisms with more complex body plans and developed important roles in various developmental and physiological signalling processes (Pires & Dolan, 2010). Two classes of bHLH genes contributed to the evolution of tissues that functioned as a rooting system.

Two members of the bHLH VIIIc subfamily, *AtRHD6* (*root hair defective 6*) and *RHD six-like1* (*AtRSL1*), were derived from a relatively recent duplication event (Heim et al., 2003) and act together to positively regulate root hair development (Menand et al., 2007). bHLH XI transcription factors encoded by *Lotus japonicus roothairless1-Like* (*LRL*) genes are also key regulators of root hair development in diverse flowering plants, including *Lotus japonicus*, *A. thaliana* and *Oryza sativa* (Ding et al., 2009; Karas et al., 2009). Dolan's group showed that orthologs of these bHLH members isolated from the moss *P. patens* function in the development of tip-growing protonema and rhizoids in the gametophyte generation (Jang, Yi, Pires, Menand, & Dolan, 2011; Menand et al., 2007; Tam et al., 2015). These results suggest that these bHLH members functioned in the haploid generation (gametophyte) of early land plants, where they controlled the formation of cells with a rooting function. Then, during subsequent radiation of land plants, these genes were coopted for the development of the diploid generation (sporophyte) of vascular plants, where they controlled the development of rhizoids in the gametophyte and root hairs in the sporophyte. Based on their findings, the authors hypothesized that the increase in morphological complexity of the sporophyte body resulted at least in part from the recruitment of regulatory genes from the gametophyte to the sporophyte (Menand et al., 2007; Pires & Dolan, 2012; Pires et al., 2013). Considering the process of transition from the gametophyte to the sporophyte, there are suitable plants to examine this hypothesis. Extant lycophytes (clubmosses and quillworts) and monilophytes (ferns and horsetails) develop both free-living gametophytes and free-living sporophytes. These gametophytes and sporophytes grow in close contact with the soil and develop rhizoids and root hairs, respectively (Jones & Dolan, 2012). These taxa are good candidates to investigate whether the molecular mechanisms underlying rhizoid differentiation in the gametophyte generation and root hair differentiation in the sporophyte generation are common and to identify the regulatory genes in each generation.

Charophycean alga *Chara* (Charophytales) and *Spirogyra* (Zygnematales) produce rhizoids in the gametophytes (Lewis & McCourt, 2004). The charophycean alga *Klebsormidium* (Klebsomidiales), which consists of multicellular and non-branching filaments without differentiated or specialized cells such as rhizoids (Hori et al., 2014), does not contain the orthologs to bHLH VIIIc and bHLH XI members in the genome. Examination of the *Chara* and *Spirogyra* genomes will reveal whether the acquisition of bHLH VIIIc and bHLH XI members are involved in the origin of rhizoids.

2.4 bHLH: The Evolution of Stomata

The evolution of stomata represents another example of bHLH genes having conserved functions for specialized cell differentiation. When the ancestors of land plants colonized land, they developed a cuticle to adapt to the dry environment and differentiated stomata, a special apparatus for gas exchange. Stomata are the pore structures differentiated on the epidermis of land plant sporophytes used for the control of CO_2-O_2 exchange and water transpiration. A pair of highly specialized guard cells surrounds the pore and regulates the opening and closure of stomata. Stomata are produced in vascular plants, hornworts and mosses, but not in liverworts (Kenrick & Crane, 1997a). The cladistics analyses of basal land plants present a single origin of stomata and imply subsequent losses in various species, especially submerged aquatic plants, in mosses, hornworts and vascular plants (Kenrick & Crane, 1997a, p. 269). Phylogenomic results using transcriptomic data for 92 streptophyte taxa support a sister relationship between mosses and liverworts (Wickett et al., 2014). If the stomata of mosses and vascular plants are homologous, stomata likely originated in the common ancestor of all land plants and were subsequently lost in liverworts. A single origin of stomata at the base of the land plants and their subsequent loss in various lineages is thought to be the simplest scenario, but the other possible scenario of independent origins of stomata in the vascular plant, and moss and hornwort lineages was discussed (Garbary & Renzaglia, 1998). Several recent studies of the gene regulatory networks governing stomata differentiation support the single origin of stomata in all land plants.

Arabidopsis thaliana genetics studies have revealed that three closely related bHLH proteins, SPEECHLESS (SPCH), MUTE and FAMA, function as positive regulators that direct three consecutive cell fate decisions during stomatal development. Stomatal guard cells are differentiated from epidermal cells via consecutive asymmetric cell divisions. These bHLH genes promote cell fate transition, leading to mature stomatal guard cells; the transition from meristemoid mother cells to meristemoids is regulated by SPCH, meristemoids to guard mother cells by MUTE and guard mother cells to guard cells by FAMA (Fig. 6A; Pillitteri & Torii, 2007). The three bHLH genes *SPCH*, *MUTE* and *FAMA* belong to the bHLH Ia subfamily

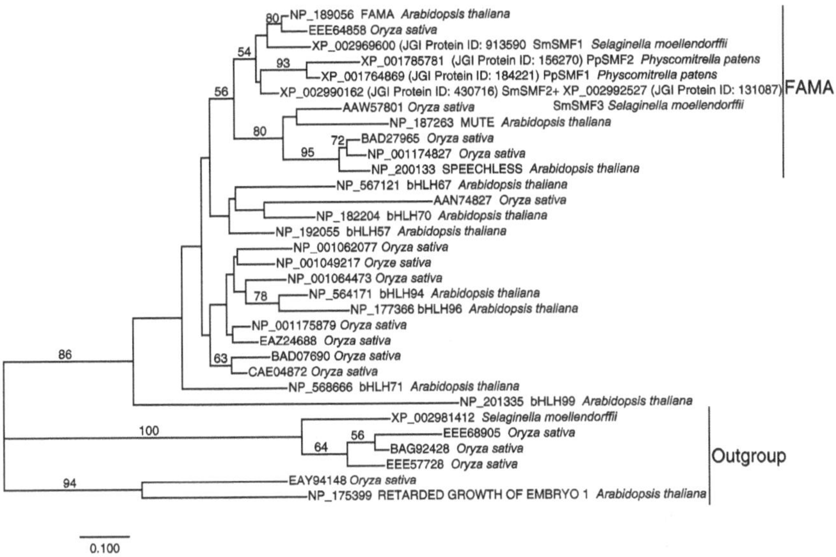

Figure 5 A maximum likelihood phylogenetic tree of *FAMA* genes constructed using 48 conserved amino acid sequences in representative green plants. Bootstrap probabilities of more than 50% are shown on the branches. The horizontal branch length is proportional to the estimated evolutionary distance. The *bar* represents 0.1 substitutions per site. The *bracket* to the right indicates the FAMA clade and outgroup.

and appear to have been derived from genomic duplication events (Pillitteri & Torii, 2007). Both the expression patterns and functional changes seem to have occurred between the stomatal bHLHs, given that their expression patterns do not significantly overlap and that they cannot functionally substitute for one another.

To decipher the evolutionary history and diversity of these groups of bHLHs, MacAlister and Bergmann (2011) conducted a comparative analysis of the SPCH, MUTE and FAMA sequences from nonflowering plants and identified two bHLH Ia members in *P. patens* and three in *S. moellendorffii*. They demonstrated that bHLH Ia members from the moss *P. patens* can partially complement the *mute* and *fama* mutant phenotypes and recapitulate gain-of-function phenotypes of bHLH Ia genes in multiple steps in the stomatal lineage in *A. thaliana* (Fig. 6B). Their data suggested that *P. patens* bHLH Ia members contained broad specificity to the targets to both FAMA and MUTE and the duplication followed by specialization provided the three (now nonoverlapping) functions of the flowering plant stomatal bHLH proteins (Fig. 5).

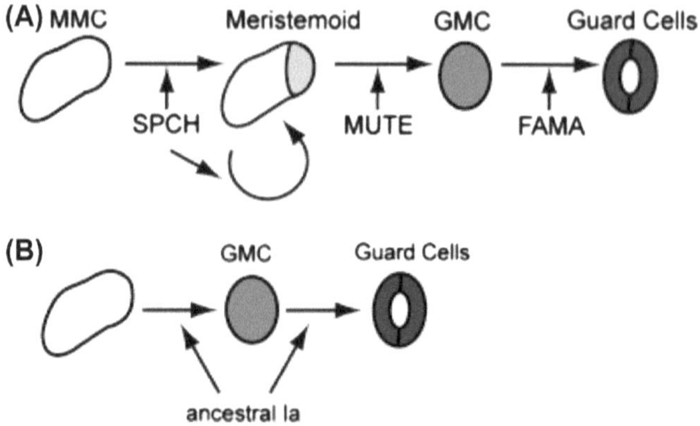

Figure 6 A model of genetic control of stomatal cell lineage cell division. (A) Stages of divisions during stomatal development. A protodermal cell is converted into a meristemoid mother cell (MMC) through an undefined mechanism. The MMC undergoes an asymmetric entry division to create a meristemoid (light grey). Meristemoids are self-renewing and can undergo one to three rounds of amplifying divisions before differentiating into a guard mother cell (GMC, dark grey). The stomatal lineage ground cells produced from amplifying divisions can initiate an entry division that is always oriented away from an existing GMC or stomata (spacing division). GMCs divide symmetrically to produce two guard cells (black), which eventually form the mature stomata. Site of stomatal bHLH protein action: the three bHLH proteins work in a three-step transcriptional cascade starting at the transition from MMC to meristemoid by SPCH, meristemoid to GMC by MUTE and GMC to guard cell by FAMA. *Arrows* indicate points of gene action. (B) A model of ancestral bHLH Ia function. A single, multifunctional Ia member drives both the specification of GMC identity and guard cell differentiation. *Modified from MacAlister, C.A., & Bergmann, D.C. (2011). Sequence and function of basic helix-loop-helix proteins required for stomatal development in Arabidopsis are deeply conserved in land plants.* Evolution and Development, 13*(2), 182–192.*

Comparative analysis of stomatal behaviour, ie, opening and closure, in *P. patens* and flowering plants was performed. Chatter et al. (2011) investigated stomatal abscisic acid (ABA) and CO_2 responses using *P. patens* and reported that the stomatal apertures of *P. patens* respond similarly to those reported for stomata on the sporophytes of vascular plants. Their results indicated that the regulatory mechanisms controlling stomatal behaviour were conserved between moss and flowering plants.

Based on these comparative analyses of bHLH Ia genes and physiological analysis, the moss *P. patens* and flowering plants likely share the molecular basis underlying stomatal differentiation and behaviour. However, it remains an open question whether the bHLH Ia members of early land plants functioned as stomatal regulators. Furthermore, it is not clear whether these bHLH Ia genes function only in the sporophytes or whether they have unknown functions in the gametophytes. *Physcomitrella patens* is a good candidate to examine the function of these regulators and to address these questions.

A search for bHLH Ia genes in the *K. flaccidum* genome revealed that it contains bHLH genes, but the FAMA ortholog gene could not be identified (Fig. 5). Presumably, genes underlying stomatal development originated in the land plant lineage. Liverwort sporophytes do not produce stomata, but it is still unclear whether the FAMA clade gene exists in the liverwort lineage. A liverwort *Marchantia polymorpha* genome information will be published soon (US DOE Joint Genome Institute, John L. Bowman and Takayuki Kohchi). Examination of the liverwort genome will reveal the question. The gametophytes of complex thallose liverworts, such as *M. polymorpha*, possess pores that open into air chambers and function as a gas exchange apparatus. However, the mechanism underlying opening and closure of these air pores differs from that in stomata (Ziegler, 1987). Gas exchange apparatuses that develop in the different generations, stomata and air pores, were acquired through convergent evolution. It would be of interest to compare the differentiation mechanisms of these apparatuses. The molecular mechanism underlying air pore differentiation has been examined using the model liverwort, *M. polymorpha* (Ishizaki et al., 2013). Further analysis of air pore development using *M. polymorpha* will elucidate how the developmental mechanism in stomata of flowering plant sporophytes differs from that in air pores of liverwort gametophytes. Functional analysis of bHLH Ia members in bryophytes will reveal how the gas exchange apparatus evolved.

2.5 *VNS* Genes: The Evolution of Conducting Tissue

The development of vascular tissue was one of the most important innovations in vascular plants, allowing for long-distance nutrient and water transport and providing rigid structural support. The evolution of vascular tissue enabled the sporophyte body of vascular plants to enlarge, and thus change the terrestrial landscape, in addition to supporting the adaptation of plants to dry land. Bryophytes are nonvascular land plants; however, some bryophytes produce specialized cells that conduct nutrients and water. Vascular plants differentiate vascular tissues in the sporophyte body, but, if present, specialized conducting tissues are restricted to the gametophyte generation in liverworts, while specialized conducting tissues occur in both generations in mosses. Conducting tissues have not been identified in the hornworts (Ligrone, Ducket, & Renzaglia, 2000).

In bryophytes, specialized cell types that mediate water transport have been characterized; hydroid cells conduct water internally and mature hydroids share several characteristics with xylem vessels from vascular plants, such as an elongated shape and programmed cytoplasmic lysis (Hébant, 1977; Ligrone, Vaughn, Renzaglia, Knox, & Duckett, 2002). However, unlike xylem vessels, hydroids lack distinct pits on their lateral walls and lignified secondary walls (Kenrick & Crane, 1997a). Stereid cells with thickened cell walls are hypothesized to function as structural support. Therefore, it remains unclear whether hydroids and stereids share an evolutionary lineage with xylem vessels and fibre cells, respectively.

Provascular cell specification in flowering plants involves the auxin signalling pathway and specific transcription factors (Ilegems et al., 2010). Both the auxin signalling pathway and HD-Zip III transcription factors are conserved among land plants, *Chara* and *Klebsormidium* (Floyd, Zalewski, & Bowman, 2006; Hori et al., 2014; Rensing et al., 2008). Investigating whether these genes are involved in the specification of conducting cells in nonvascular plants may provide insight into the molecular basis of the evolution of conducting tissues.

Xu et al. (2014) documented a striking discovery about the evolution of water-conducting tissues. A group of NAC transcription factors belonging to the VNS (VND (vascular-related NAC-domain), NST (NAC secondary wall thickening promoting factor), SMB (Sombrero)) clade regulated differentiation of water-conducting cells in vascular plants (Mitsuda et al., 2007; Yamaguchi et al., 2010, 2011; Zhong, Demura, & Ye, 2006). However, the path by which plants with these cells evolved from their nonvascular ancestors was unclear. Xu et al. (2014) investigated the function of *P. patens VNS* genes. Loss-of-function mutants of *P. patens VNS* genes formed abnormal water-conducting and supporting cells in the gametophytes, as well as malformed sporophyte cells, and overexpression of these genes induced ectopic differentiation of water-conducting cell-like cells. Their results indicate that the transcriptional regulation and cellular function of water-conducting cells were conserved between *P. patens* and *A. thaliana*. The conserved genetic basis suggests roles for NAC proteins in the adaptation of plants to land (Xu et al., 2014).

The *M. polymorpha* genome contains a single *VNS* gene (Nakano et al., 2015; Xu et al., 2014), although *M. polymorpha* lacks conducting tissues. The question now arises which function the *M. polymorpha VNS* ortholog conducts. It would also be interesting to examine whether the molecular basis underlying differentiation of a water-conducting system is conserved in leafy liverworts, which poses water-conducting tissues in the gametophyte body.

The *VNS* ortholog was not reported in the charophycean alga *K. flaccidum* or in other green algae (Table 1; Nakano et al., 2015). The acquisition of the *VNS* clade gene seems to be highly correlated with the establishment of water-conducting tissue.

The sieve element-like conducting parenchyma cells, the leptoids in Polytrichales, are recognized as the nutrient-conducting tissue in mosses, having differentiated both in the gametophyte and the sporophyte seta (Hébant, 1977; Ligrone et al., 2000). Other types of nutrient-conducting tissues, including mature protonemal and rhizoid cells, are postulated to function in the long-distance symplasmic transport of nutrients (Duckett, Schmid, & Ligrone, 1998). Although leptoid cells share many similarities with sieve elements (Esau, 1969; Ligrone et al., 2000; Scheirer, 1990), it was thought to be difficult to examine the homology between nutrient-conducting cells of bryophytes and sieve cells of vascular plants, because nutrient-conducting cells of bryophytes show highly distinctive characteristics (Ligrone et al., 2000). Orthologs of a gene involved in the specification of phloem in *A. thaliana*, *APL* (*altered phloem development*), which encodes an MYB transcription factor (Bonke, Thitamadee, Mähönen, Hauser, & Helariutta, 2003), are present in *P. patens* and *S. moellendorffii* (Banks et al., 2011). Molecular comparative functional analyses of genes promoting phloem differentiation in bryophytes with those in vascular plants, like *VNS* of water-conducting tissue, may provide insight into the evolution of nutrient-conducting tissues.

2.6 *Floricaula/Leafy*: Sporophyte Developmental Gene

Floricaula (*FLO*) and *Leafy* (*LFY*) are transcription factors that play a central role in flower development and were originally identified *Antirrhinum majus* (Coen et al., 1990) and in *A. thaliana* (Schultz & Haughn, 1991), respectively. FLO/LFY triggers the development of floral meristems and controls the patterning of floral organs by activating floral organ identity genes (Moyroud, Tichtinsky, & Parcy, 2009). FLO/LFY orthologs also function in other aspects of sporophyte development, including leaf shape and plant or inflorescence architecture (Moyroud et al., 2009). LFY/FLO orthologs are highly conserved in the streptophytes, including gymnosperms, ferns, lycophytes, bryophytes and charophytes (Sayou et al., 2014), but are not present in the green algae *C. reinhardtii* or *O. tauri*, or in the red alga *C. merolae*. FLO/LFY is thought to have been acquired in the common ancestor of the streptophyte lineage (Table 1; Fig. 7). The function of the FLO/LFY ortholog in charophycean alga is unknown.

FLO/LFY is not part of an extended gene family such as the homeobox genes or bHLH genes described in this chapter. *FLO/LFY* is mostly found as a single copy gene. Several species exhibit two or three *FLO/LFY*-like genes, but phylogenetic analysis suggests that these paralogs were duplicated recently, due to polyploidization or small-scale duplication events. It is unclear why *FLO/LFY* genes are mostly maintained as single copies in the genome (Moyroud et al., 2009).

To examine whether the functions of *FLO/LFY* genes are conserved, Maizel et al. (2005) conducted an investigation using nonflowering plant orthologs. They examined the functional conservation of FLO/LFY orthologs as floral meristem genes in *A. thaliana*. Expression of flowering plant FLO/LFY orthologs under the control of *Arabidopsis LFY* promoter fully complemented the *A. thaliana lfy* mutant, whereas gymnosperm genes provided only partial rescue. Some orthologs from the early diverged group, namely *CrLFY2* from the fern *C. richardii* and *AlLFY4* from the fern *Angiopteris lygodiifolia* had some ability to rescue the *lfy* mutation in *A. thaliana*, but to a lesser extent than the gymnosperm orthologs, whereas *P. patens LFY1* and *LFY2* failed to rescue the mutation. This gradient of complementation reflects the phylogenetic distance from flowering plants. Interestingly, the degree of complementation also correlates with the degree of activation ability of LFY target genes. *Arabidopsis thaliana* LFY activates 16 genes, while the expression of 15 and 13 of these genes is

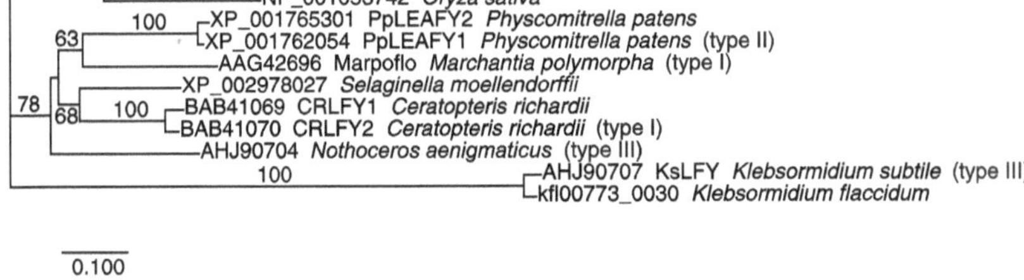

Figure 7 A maximum likelihood phylogenetic tree of *LFY* genes constructed using 236 conserved amino acid sequences in representative green plants. Bootstrap probabilities of more than 50% are shown on the branches. The horizontal branch length is proportional to the estimated evolutionary distance. The *bar* represents 0.1 substitutions per site. The types of DNA-binding specificity are shown at the right to gene name based on Sayou et al. (2014). (See colour plate)

significantly induced by the flowering plant orthologs UNI and ALF, respectively. The degree of activation is reduced according to phylogenetic distance from *A. thaliana*. WelNDLY, a gymnosperm ortholog, induces two targets, whereas CrLFY2, from ferns, induces only one. None of the LFY targets respond to *P. patens* LFY1 (Maizel et al., 2005). In contrast to flowering plant *FLO/LFY* genes, the *P. patens* LFY failed to activate floral organ identity genes, due to differences in two amino acids in the C domain (Maizel et al., 2005).

What is the function of nonflowering plant FLO/LFY orthologs? This question was studied using the two FLO/LFY paralogs encoded in the *P. patens* genome, ie, *P. patens LFY1* and *2*. These paralogs are expressed in both the gametophyte and sporophyte tissues, and double mutants lacking both of these genes produced normal gametophytes, but seemed to have a defect in the first step of sporophyte development. After fertilization, the zygote of the *P. patens lfy1 lfy2* double mutant was arrested at the single-cell stage. The moss FLO/LFY orthologs are thus thought to control the first sporophyte cell division and function throughout sporophyte development (Tanahashi, Sumikawa, Kato, & Hasebe, 2005). This finding shows that the function of FLO/LFY in basal land plant lineages is different from that of flowering plant orthologs, though both have important roles in sporophyte development.

FLO/LFY binds DNA through a highly conserved dimeric DNA-binding domain (Hames et al., 2008). Sayou et al. (2014) examined LFY/FLO orthologs from algal species and proposed that FLO/LFY proteins can be classified into three groups based on their DNA-binding specificity. Most LFY proteins from land plants, including flowering plants, gymnosperms, ferns and liverworts, bind to the same DNA motif, the type I motif. *Physcomitrella patens* LFY orthologs bind to a different motif, type II. Hornwort and algal FLO/LFY orthologs bind to the third motif, type III (Fig. 7). The specificity of these proteins is determined by amino acid sequence differences in the DNA-binding domain. FLO/LFY proteins are thought to have changed their target specificity at least twice during evolution. The question now emerges as to whether these transitions were accompanied by a complete change in target gene sets or whether some *cis* elements coevolved with DNA-binding specificity (Sayou et al., 2014). Functional analysis and screening of the genes that function downstream of the bryophyte and charophycean algal orthologs should be conducted to decipher the evolutionary history of *FLO/LFY*.

3. THE EVOLUTION OF HORMONE SIGNALLING

To integrate the development of a multicellular body in the sporophyte of flowering plants, cell-to-cell communication through signal molecules, such as plant hormones, is critical. Several plant hormones, such as auxin, ABA and cytokinin were already established in bryophytes (Rensing et al., 2008) and these hormones are important for gametophyte development in bryophytes (Bennett et al., 2014; Flores-Sandoval, Eklund, & Bowman, 2015; Kato et al., 2015; Khandelwal et al., 2010; von Schwartzenberg et al., 2007; Viaene et al., 2014). By contrast, components of the gibberellin signalling pathway are considered to have been acquired in steps during the evolutionary transitions from nonvascular to vascular plants (Hirano et al., 2007; Yasumura, Crumpton-Taylor, Fuentes, & Harberd, 2007). Flores-Sandoval et al. (2015) showed that the *M. polymorpha* genome encodes all the components of the auxin signalling pathway known in flowering plant systems, and that orthologs exist as a single copy in the genome, suggesting that the ancestor of land plants presumably contained a minimal auxin signalling unit. Subsequently, the copies of each component of this signalling unit diversified and different combinations of these components produced the wide range of auxin-binding affinities in the flowering plant lineages (Villalobos et al., 2012) to contribute to the complexity of the sporophyte body plan.

How did plant hormone signalling arise? Several plant hormones have been detected in both unicellular and multicellular algae (Le Bail et al., 2010; Tarakhovskaya, Maslov, & Shishova, 2007), but their functions in algae are largely unknown. Wang, Liu, Li, and Han (2015) conducted the comprehensive search of the signalling components of nine major plant hormones, ABA, auxin, brassinosteroid, cytokinin, ethylene, gibberellin, jasmonic acid, salicylic acid and strigolactone using genome or transcriptome information of 14 plant species including the red alga *C. merolae*, green algae, charophycean algae and land plants. They demonstrated that auxin, cytokinin and strigolactone signalling pathways originated in the common ancestor of streptophyte and that ABA, jasmonic acid and salicylic acid signalling pathways arose in the last common ancestor of land plants (Wang et al., 2015). They also supposed that the orthologs of these signalling components were more likely to retain similar functions in the course of evolution. Functional analysis of these hormone signalling components in charophycean alga will reveal the physiological function of these signalling components and plant hormones in these algae.

4. FUTURE PERSPECTIVES IN EVOLUTIONARY DEVELOPMENTAL BIOLOGY RESEARCH

In this chapter, I highlight several transcription factors that contributed to the morphological evolution of land plants and indicate differences and conserved aspects of the gene regulatory networks controlled by these transcription factors among the streptophytes. The emergence of next-generation sequencing technology has enabled genome-wide comparative analyses, which have valuable applications in evolutionary developmental research. Transcriptome analysis using different lineages facilitates identification of differentially expressed genes, and thus provides comprehensive information about gene regulatory networks, as comparative transcriptome analysis of different developmental stages revealed which developmental stages were the most conserved during vertebrate and *A. thaliana* embryogenesis (Irie & Kuratani, 2011; Quint et al., 2012). Such approaches comparing different generations and different meristematic cell or tissues have examined in different taxa of land plants (Frank et al., 2015; O'Donoghue et al., 2013; Ortiz-Ramírez et al., 2016; Szövényi, Rensing, Lang, Wray, & Shaw, 2011) and will provide a new insight into the evolution of gene regulation.

ChIP (chromatin immunoprecipitation) sequencing analysis enables the identification of the direct targets of each transcription factor (Kaufmann et al., 2010). Such new techniques have applications in plant evolutionary developmental biology research and will ultimately reveal how gene regulatory networks evolved in land plants (Muino et al., 2015).

Another major advance in genomic modification technology was recently reported. The CRISPR (clustered regularly interspaced short palindromic repeats)/Cas9 system, in which a short guide RNA targets a specific genome sequence that is edited by a Cas9 endonuclease (Cong et al., 2013), has been used for mutagenesis of basal land plants (Sugano et al., 2014). Such new technologies will reveal important developmental genes and enable testing of key hypotheses regarding the acquisition of morphological innovations. These new technologies will also stimulate the development of novel approaches to investigate the morphological evolution of land plants.

ACKNOWLEDGEMENT

I thank Dr. Mitsuyasu Hasebe, Dr. John L. Bowman and Dr. Tomoaki Nishiyama for critical reading of the manuscript draft and Dr. Misato Ohtani for discussion on the *VNS* genes.

REFERENCES

Abe, J., Hori, S., Tsuchikane, Y., Kitao, N., Kato, M., & Sekimoto, H. (2011). Stable nuclear transformation of the *Closterium peracerosum-strigosum-littorale* complex. *Plant and Cell Physiology, 52*(9), 1676–1685.

Banks, J. A., Nishiyama, T., Hasebe, M., Bowman, J. L., Gribskov, M., de Pamphilis, C. … Grigoriev, I. V. (2011). The Selaginella genome identifies genetic changes associated with the evolution of vascular plants. *Science, 332*(6032), 960–963.

Bennett, T. A., Liu, M. M., Aoyama, T., Bierfreund, N. M., Braun, M., Coudert, Y. … Harrison, C. J. (2014). Plasma membrane-targeted PIN proteins drive shoot development in a moss. *Current Biology, 24*(23), 2776–2785.

Birol, I., Raymond, A., Jackman, S. D., Pleasance, S., Coope, R., Taylor, G. A. … Jones, S. J. (2013). Assembling the 20 Gb white spruce (*Picea glauca*) genome from whole-genome shotgun sequencing data. *Bioinformatics, 29*(12), 1492–1497.

Bonke, M., Thitamadee, S., Mähönen, A. P., Hauser, M. T., & Helariutta, Y. (2003). APL regulates vascular tissue identity in *Arabidopsis*. *Nature, 426*, 181–186.

Bower, F. O. (1890). On antithetic as distinct from homologous alternation of generations in plants. *Annals of Botany, 4*(15), 347–370.

Brand, U., Fletcher, J. C., Hobe, M., Meyerowitz, E. M., & Simon, R. (2000). Dependence of stem cell fate in Arabidopsis on a feedback loop regulated by CLV3 activity. *Science, 289*(5479), 617–619.

Breuninger, H., Rikirsch, E., Hermann, M., Ueda, M., & Laux, T. (2008). Differential expression of *WOX* genes mediates apical-basal axis formation in the *Arabidopsis* embryo. *Developmental Cell, 14*, 867–876.

Byrne, M. E., Barley, R., Curtis, M., Arroyo, J. M., Dunham, M., Hudson, A., & Martienssen, R. A. (2000). *Asymmetric leaves1* mediates leaf patterning and stem cell function in *Arabidopsis*. *Nature, 408*, 967–971.

Chater, C., Kamisugi, Y., Movahedi, M., Fleming, A., Cuming, A. C., Gray, J. E., & Beerling, D. J. (2011). Regulatory mechanism controlling stomatal behavior conserved across 400 million years of land plant evolution. *Current Biology, 21*, 1–5.

Clark, S. E., Williams, R. W., & Meyerowitz, E. M. (1997). The *CLAVATA1* gene encodes a putative receptor kinase that controls shoot and floral meristem size in Arabidopsis. *Cell, 89*, 575–585.

Cock, J. M., Sterck, L., Rouze, P., Scornet, D., Allen, A. E., Amoutzias, G. … Wincker, P. (2010). The *Ectocarpus* genome and the independent evolution of multicellularity in brown algae. *Nature, 465*, 617–621.

Coen, E. S., Romero, J. M., Doyle, S., Elliott, R., Murphy, G., & Carpenter, E. (1990). *floricaula*: a homeotic gene required for flower development in *Antirrhinum majus*. *Cell, 63*(6), 1311–1322.

Collén, J., Porcel, B., Carré, W., Ball, S. G., Chaparro, C., Tonon, T. … Boyen, C. (2013). Genome structure and metabolic features in the red seaweed *Chondrus crispus* shed light on evolution of the Archaeplastida. *Proceedings of the National Academy of Sciences of the United States of America, 110*(13), 5247–5252.

Conant, G. C., & Wolfe, K. H. (2008). Turning a hobby into a job: how duplicated genes find new functions. *Nature Reviews Genetics, 9*, 938–950.

Cong, L., Ran, F. A., Cox, D., Lin, S., Barretto, R., Habib, N. … Zhang, F. (2013). Multiplex genome engineering using CRISPR/Cas systems. *Science, 339*(6121), 819–823.

Derelle, E., Ferraz, C., Rombauts, S., Rouzé, P., Worden, A. Z., Robbens, S. … Moreau, H. (2006). Genome analysis of the smallest free-living eukaryote *Ostreococcus tauri* unveils many unique features. *Proceedings of the National Academy of Sciences of the United States of America, 103*(31), 11647–11652.

Derelle, R., Lopez, P., Le Guyader, H., & Manuel, M. (2007). Homeodomain proteins belong to the ancestral molecular toolkit of eukaryotes. *Evolution and Development, 9*(3), 212–219.
Deveaux, Y., Toffano-Nioche, C., Claisse, G., Thareau, V., Morin, H., Laufs, P.… Lecharny, A. (2008). Genes of the most conserved WOX clade in plants affect root and flower development in Arabidopsis. *BMC Evolutionary Biology, 8*, 291.
Ding, W. N., Yu, Z. M., Tong, Y. L., Huang, W., Chen, H. M., & Wu, P. (2009). A transcription factor with a bHLH domain regulates root hair development in rice. *Cell Research, 19*(11), 1309–1311.
Duckett, J. G., Schmid, A., & Ligrone, R. (1998). Protonemal morphogenesis. In N. W. Ashton, J. G. Duckett, & J. W. Bates (Eds.), *Bryology for the twenty-first century* (pp. 223–246). Leeds: Maney Publishing and the British Bryological Society.
Esau, K. (1969). *The phloem*. Berlin: Borntraeger.
Flores-Sandoval, E., Eklund, D. M., & Bowman, J. L. (2015). A simple auxin transcriptional response system regulates multiple morphogenetic processes in the liverwort *Marchantia polymorpha*. *PLoS Genetics, 11*(5), e1005207.
Floyd, S. K., & Bowman, J. L. (2007). The ancestral developmental tool kit of land plants. *International Journal of Plant Sciences, 168*, 1–35.
Floyd, S. K., Zalewski, C. S., & Bowman, J. L. (2006). Evolution of class III homeodomain–leucine zipper genes in streptophytes. *Genetics, 173*, 373–388.
Frank, M. H., Edwards, M. B., Schultz, E. R., McKain, M. R., Fei, Z., Sørensen, I.… Scanlon, M. J. (2015). Dissecting the molecular signatures of apical cell-type shoot meristems from two ancient land plant lineages. *New Phytologist, 207*(3), 893–904.
Friedman, W. E. (2013). One genome, two ontogenies. *Science, 339*(6123), 1045–1046.
Friedman, W. E., Moore, R. C., & Purugganan, M. D. (2004). The evolution of plant development. *American Journal of Botany, 91*(10), 1726–1741.
Furumizu, C., Alvarez, J. P., Sakakibara, K., & Bowman, J. L. (2015). Antagonistic roles for KNOX1 and KNOX2 genes in patterning the land plant body plan following an ancient gene duplication. *PLoS Genetics, 11*(2), e1004980.
Garbary, D. J., & Renzaglia, K. S. (1998). Bryophyte phylogeny and the evolution of land plants: evidence from developmental and ultrastructure. In J. W. Bates, N. W. Ashton, & J. G. Duckett (Eds.), *Biology for twenty-first century* (pp. 45–63). Maney Publishing and the British Bryological Society.
van der Graaff, E., Laux, T., & Rensing, S. A. (2009). The WUS homeobox-containing (WOX) protein family. *Genome Biology, 10*(12), 248.
Graham, L. E. (1993). *Origin of land plants*. New York, USA: John Wiley and Sons, Inc.
Graham, L. E., Cook, M. E., & Busse, J. S. (2000). The origin of plants: body plan changes contributing to a major evolutionary radiation. *Proceedings of the National Academy of Sciences of the United States of America, 97*(9), 4535–4540.
Haecker, A., Gross-Hardt, R., Geiges, B., Sarkar, A., Breuninger, H., Herrmann, M., & Laux, T. (2004). Expression dynamics of *WOX* genes mark cell fate decisions during early embryonic patterning in *Arabidopsis thaliana*. *Development, 131*, 657–668.
Hake, S., Smith, H. M. S., Holtan, H., Magnani, E., Mele, G., & Ramirez, J. (2004). The role of KNOX genes in plant development. *Annual Review of Cell and Developmental Biology, 20*, 125–151.
Hames, C., Ptchelkine, D., Grimm, C., Thevenon, E., Moyroud, E., Gerard, F.… Müller, C. W. (2008). Structural basis for LEAFY floral switch function and similarity with helix-turn-helix proteins. *EMBO Journal, 27*, 2628–2637.
Hay, A., & Tsiantis, M. (2010). KNOX genes: versatile regulators of plant development and diversity. *Development, 137*, 3153–3165.
Hébant, C. (1977). *The conducting tissues of bryophytes*. J. Cramer.

Hedman, H., Zhu, T., von Arnold, S., & Sohlberg, J. J. (2013). Analysis of the *WUSCHEL-RELATED HOMEOBOX* gene family in the conifer *picea abies* reveals extensive conservation as well as dynamic patterns. *BMC Plant Biology, 13*, 89.

Heim, M. A., Jakoby, M., Werber, M., Martin, C., Weisshaar, B., & Bailey, P. C. (2003). The basic helix-loop-helix transcription factor family in plants: a genome-wide study of protein structure and functional diversity. *Molecular Biology and Evolution, 20*(5), 735–747.

Hirakawa, Y., Kondo, Y., & Fukuda, H. (2010). TDIF peptide signaling regulates vascular stem cell proliferation via the *WOX4* homeobox gene in *Arabidopsis*. *Plant Cell, 22*, 2618–2629.

Hirano, K., Nakajima, M., Asano, K., Nishiyama, T., Sakakibara, H., Kojima, M.… Matsuoka, M. (2007). The GID1-mediated gibberellin perception mechanism is conserved in the lycophyte *Selaginella moellendorffii* but not in the bryophyte *Physcomitrella patens*. *Plant Cell, 19*, 3058–3079.

Hori, K., Maruyama, F., Fujisawa, T., Togashi, T., Yamamoto, N., Seo, M.… Ohta, H. (2014). *Klebsormidium flaccidum* genome reveals primary factors for plant terrestrial adaptation. *Nature Communications, 5*, 3978.

Hueber, F. M., & Banks, H. P. (1979). *Serrulacaulis furcatus* gen. et sp. nov., a new zosterophyll from the lower Upper Devonian of New York State. *Review of Palaeobotany and Palynology, 28*, 169–189.

Ilegems, M., Douet, V., Meylan-Bettex, M., Uyttewaal, M., Brand, L., Bowman, J. L., & Stieger, P. A. (2010). Interplay of auxin, KANADI and Class III HD-ZIP transcription factors in vascular tissue formation. *Development, 137*, 975–984.

Irie, N., & Kuratani, S. (2011). Comparative transcriptome analysis reveals vertebrate phylotypic period during organogenesis. *Nature Communications, 2*, 248.

Ishizaki, K., Chiyoda, S., Yamato, K. T., & Kohchi, T. (2008). Agrobacterium-mediated transformation of the haploid liverwort *Marchantia polymorpha* L., an emerging model for plant biology. *Plant and Cell Physiology, 49*(7), 1084–1091.

Ishizaki, K., Mizutani, M., Shimamura, M., Masuda, A., Nishihama, R., & Kohchi, T. (2013). Essential role of the E3 ubiquitin ligase NOPPERABO1 in schizogenous intercellular space formation in the liverwort *Marchantia polymorpha*. *Plant Cell, 25*, 4075–4084.

Jang, G., Yi, K., Pires, N. D., Menand, B., & Dolan, L. (2011). RSL genes are sufficient for rhizoid system development in early diverging land plants. *Development, 138*, 2273–2281.

Ji, J., Shimizu, R., Sinha, N., & Scanlon, M. J. (2010). Analyses of *WOX4* transgenics provide further evidence for the evolution of the *WOX* gene family during the regulation of diverse stem cell functions. *Plant Signaling & Behavior, 5*(7), 916–920.

Jones, V. A. S., & Dolan, L. (2012). The evolution of root hairs and rhizoids. *Annals of Botany, 110*, 205–212.

Karas, B., Amyot, L., Johansen, C., Sato, S., Tabata, S., Kawaguchi, M., & Szczyglowski, K. (2009). Conservation of *Lotus* and *Arabidopsis* basic helix-loop-helix proteins reveals new players in root hair development. *Plant Physiology, 151*(3), 1175–1185.

Karol, K. G., McCourt, R. M., Cimino, M. T., & Delwiche, C. F. (2001). The closest living relatives of land plants. *Science, 294*(5550), 2351–2353.

Kato, H., Ishizaki, K., Kouno, M., Shirakawa, M., Bowman, J. L., Nishihama, R., & Kohchi, T. (2015). Auxin-mediated transcriptional system with a minimal set of components is critical for morphogenesis through the life cycle in *Marchantia polymorpha*. *PLoS Genetics, 11*(5), e1005084.

Kaufmann, K., Muiño, J. M., Østerås, M., Farinelli, L., Krajewski, P., & Angenent, G. C. (2010). Chromatin immunoprecipitation (ChIP) of plant transcription factors followed by sequencing (ChIP-SEQ) or hybridization to whole genome arrays (ChIP-CHIP). *Nature Protocols, 5*(3), 457–472.

Kenrick, P. (2013). The origin of roots. In A. Eshel, & T. Beeckman (Eds.), *Plant roots: The hidden half* (4th ed.). CRC Press.

Kenrick, P., & Crane, P. R. (1997a). *The origin and early diversification of land plants: A cladistic study*. Washington, DC, USA: Smithsonian Institution Press.

Kenrick, P., & Crane, P. R. (1997b). The origin and early evolution of plants on land. *Nature, 389*, 33–39.

Khandelwal, A., Cho, S. H., Marella, H., Sakata, Y., Perroud, P. F., Pan, A., & Quatrano, R. S. (2010). Role of ABA and ABI3 in desiccation tolerance. *Science, 327*(5965), 546.

Kofuji, R., & Hasebe, M. (2014). Eight types of stem cells in the life cycle of the moss *Physcomitrella patens*. *Current Opinion in Plant Biology, 17*, 13–21.

Kumaran, M. K., Bowman, J. L., & Sundaresan, V. (2002). YABBY polarity genes mediate the repression of KNOX homeobox genes in *Arabidopsis*. *Plant Cell, 14*, 2761–2770.

Lang, D., Weiche, B., Timmerhaus, G., Richardt, S., Riano-Pachon, D. M., Correak, L. G. G. … Rensing, S. A. (2010). Genome-wide phylogenetic comparative analysis of plant transcriptional regulation: a timeline of loss, gain, expansion, and correlation with complexity. *Genome Biology and Evolution, 2*, 488–503.

Laux, T., Mayer, K. F., Berger, J., & Jurgens, G. (1996). The WUSCHEL gene is required for shoot and floral meristem integrity in Arabidopsis. *Development, 122*(1), 87–96.

Le Bail, A., Billoud, B., Kowalczyk, N., Kowalczyk, M., Gicquel, M., Le Panse, S. … Charrier, B. (2010). Auxin metabolism and function in the multicellular brown alga *Ectocarpus siliculosus*. *Plant Physiology, 153*, 128–144.

Lee, J.-H., Lin, H., Joo, S., & Goodenough, U. (2008). Early sexual origins of homeoprotein heterodimerization and evolution of the plant KNOX/BELL family. *Cell, 133*, 829–840.

Lenhard, M., & Laux, T. (2003). Stem cell homeostasis in the *Arabidopsis* shoot meristem is regulated by intercellular movement of CLAVATA3 and its sequestration by CLAVATA1. *Development, 130*, 3163–3173.

Lewis, L. A., & McCourt, R. M. (2004). Green algae and the origin of land plants. *American Journal of Botany, 91*(10), 1535–1556.

Ligrone, R., Ducket, J. G., & Renzaglia, K. S. (2000). Conducting tissues and phyletic relationships of bryophytes. *Philosophical Transactions of the Royal Society of London Series B Biological Sciences, 355*, 795–813.

Ligrone, R., Vaughn, K. C., Renzaglia, K. S., Knox, J. P., & Duckett, J. G. (2002). Diversity in the distribution of polysaccharide and glycoprotein epitopes in the cell walls of bryophytes: new evidence for the multiple evolution of water-conducting cells. *New Phytologist, 156*, 491–508.

Long, J. A., Moan, E. I., Medford, J. I., & Barton, M. K. (1996). A member of the KNOTTED class of homeodomain proteins encoded by the *STM* gene of *Arabidopsis*. *Nature, 163*, 66–69.

MacAlister, C. A., & Bergmann, D. C. (2011). Sequence and function of basic helix-loop-helix proteins required for stomatal development in Arabidopsis are deeply conserved in land plants. *Evolution and Development, 13*(2), 182–192.

Maizel, A., Busch, M. A., Tanahashi, T., Perkovic, J., Kato, M., Hasebe, M., & Weigel, D. (2005). The floral regulator LEAFY evolves by substitutions in the DNA binding domain. *Science, 308*(5719), 260–263.

Matsuzaki, M., Misumi, O., Shin-i, T., Maruyama, S., Takahara, M., Miyagishima, S. … Kuroiwa, T. (2004). Genome sequence of the ultrasmall unicellular red alga *Cyanidioschyzon merolae* 10D. *Nature, 428*, 653–657.

Menand, B., Yi, K., Jouannic, S., Hoffmann, L., Ryan, E., Linstead, P. … Dolan, L. (2007). An ancient mechanism controls the development of cells with a rooting function in land plants. *Science, 316*(5830), 1477–1480.

Merchant, S. S., Prochnik, S. E., Vallon, O., Harris, E. H., Karpowicz, S. J., Witman, G. B. … Grossman, A. R. (2007). The *Chlamydomonas* genome reveals the evolution of key animal and plant functions. *Science, 318*(5848), 245–250.

Mishler, B. D., & Churchill, S. P. (1985). Transition to a land flora: phylogenetic relationships of the Green Algae and Bryophytes. *Cladistics, 1*, 305–328.

Mitsuda, N., Iwase, A., Yamamoto, H., Yoshida, M., Seki, M., Shinozaki, K., & Ohme-Takagi, M. (2007). NAC transcription factors, NST1 and NST3, are key regulators of the formation of secondary walls in woody tissues of *Arabidopsis*. *Plant Cell, 19*, 270–280.

Moyroud, E., Tichtinsky, G., & Parcy, F. (2009). The LEAFY floral regulators in angiosperms: conserved proteins with diverse roles. *Journal of Plant Biology, 52*, 177–185.

Muino, J. M., de Bruijn, S., Pajoro, A., Geuten, K., Vingron, M., Angenent, G. C., & Kaufmann, K. (2015). Evolution of DNA-binding sites of a floral master regulatory transcription factor. *Molecular Biology and Evolution, 32*(12), msv210.

Mukherjee, K., Brocchieri, L., & Bürglin, T. R. (2009). A comprehensive classification and evolutionary analysis of plant homeobox genes. *Molecular Biology and Evolution, 26*(12), 2775–2794.

Muthukumar, B., Joyce, B. L., Elless, M. P., & Stewart, C. N., Jr. (2013). Stable transformation of ferns using spores as targets: *Pteris vittata* and *Ceratopteris thalictroides*. *Plant Physiology, 163*, 648–658.

Nakamura, Y., Sasaki, N., Kobayashi, M., Ojima, N., Yasuike, M., Shigenobu, Y. … Ikeo, K. (2013). The first symbiont-free genome sequence of marine red alga, susabinori (*Pyropia yezoensis*). *PLoS One, 8*(3), e57122.

Nakano, Y., Yamaguchi, M., Endo, H., Rejab, N. A., & Ohtani, M. (2015). NAC-MYB-based transcriptional regulation of secondary cell wall biosynthesis in land plants. *Frontiers in Plant Science, 6*. http://dx.doi.org/10.3389/fpls.2015.00288.

Nakata, M., Matsumoto, N., Tsugeki, R., Rikirsch, E., Laux, T., & Okada, K. (2012). Roles of the middle domain–specific *WUSCHEL-RELATED HOMEOBOX* genes in early development of leaves in *Arabidopsis*. *Plant Cell, 24*, 519–535.

Nardmann, J., & Werr, W. (2012). The invention of WUS-like stem cell-promoting functions in plants predates leptosporangiate ferns. *Plant Molecular Biology, 78*, 123–134.

Nishiyama, T. (2007). Evolutionary developmental biology of nonflowering land plants. *International Journal of Plant Sciences, 168*(1), 37–47.

Nystedt, B., Street, N. R., Wetterbom, A., Zuccolo, A., Lin, Y.-C., Scofield, D. G. … Jansson, S. (2013). The Norway spruce genome sequence and conifer genome evolution. *Nature, 497*, 579–584.

O'Donoghue, M.-T., Chater, C., Wallace, S., Gray, J. E., Beerling, D. J., & Fleming, A. J. (2013). Genome-wide transcriptomic analysis of the sporophyte of the moss *Physcomitrella patens*. *Journal of Experimental Botany, 64*(12), 3567–3581.

Ortiz-Ramírez, C., Hernandez-Coronado, M., Thamm, A., Catarino, B., Wang, M., Dolan, L. … Becker, J. D. (2016). A transcriptome atlas of *Physcomitrella patens* provides insights into the evolution and development of land plants. *Molecular Plant, 9*(2), 205–220.

Pillitteri, L. J., & Torii, K. U. (2007). Breaking the silence: three bHLH proteins direct cell-fate decisions during stomatal development. *BioEssays, 29*, 861–870.

Pires, N., & Dolan, L. (2010). Origin and diversification of basic-Helix-Loop-Helix proteins in plants. *Molecular Biology and Evolution, 27*(4), 862–874.

Pires, N. D., & Dolan, L. (2012). Morphological evolution in land plants: new designs with old genes. *Philosophical Transactions of the Royal Society of London Series B Biological Sciences, 367*, 508–518.

Pires, N. D., Yi, K., Breuninger, H., Catarino, B., Menand, B., & Dolan, L. (2013). Recruitment and remodeling of an ancient gene regulatory network during land plant evolution. *Proceedings of the National Academy of Sciences of the United States of America, 110*(23), 9571–9576.

Plackett, A. R. G., Huang, L., Sanders, H. L., & Langdale, J. A. (2014). High-efficiency stable transformation of the model fern species *Ceratopteris richardii* via microparticle bombardment. *Plant Physiology, 165*, 3–14.

Price, D. C., Chan, C. X., Yoon, H. S., Yang, E. C., Qiu, H., Weber, A. P. M.… Bhattacharya, D. (2012). *Cyanophora paradoxa* genome elucidates origin of photosynthesis in algae and plants. *Science, 335*, 843–847.

Pringsheim, N. (1878). Uber Sprossung der Moosfruchte und den Generationswechsel der Thallophyten. *Jahrbücher für wissenschaftliche Botanik, 11*, 1–46.

Prochnik, S. E., Umen, J., Nedelcu, A. M., Hallmann, A., Miller, S. M., Nishii, I.… Rokhsar, D. S. (2010). Genomic analysis of organismal complexity in the multicellular green alga *Volvox carteri*. *Science, 329*(5988), 223–226.

Quint, M., Drost, H.-G., Gabel, A., Ullrich, K. K., Bönn, M., & Grosse, I. (2012). A transcriptomic hourglass in plant embryogenesis. *Nature, 490*, 98–101.

Raven, J. A., & Edwards, D. (2001). Roots: evolutionary origins and biogeochemical significance. *Journal of Experimental Botany, 52*, 381–401.

Rayko, E., Maumus, F., Maheswari, U., Jabbari, K., & Bowler, C. (2010). Transcription factor families inferred from genomesequences of photosynthetic stramenopiles. *New Phytologist, 188*, 52–66.

Rensing, S. A. (2014). Gene duplication as a driver of plant morphogenetic evolution. *Current Opinion in Plant Biology, 17*, 43–48.

Rensing, S. A., Lang, D., Zimmer, A. D., Terry, A., Salamov, A., Shapiro, H.… Boore, J. L. (2008). The Physcomitrella genome reveals evolutionary insights into the conquest of land by plants. *Science, 319*(5859), 64–69.

Romera-Branchat, M., Ripoll, J. J., Yanofsky, M. F., & Pelaz, S. (2013). The *WOX13* homeobox gene promotes replum formation in the *Arabidopsis thaliana* fruit. *Plant Journal, 73*, 37–49.

Sakakibara, K., Ando, S., Yip, H. K., Tamada, Y., Hiwatashi, Y., Murata, T.… Bowman, J. L. (2013). KNOX2 transcription factors regulate the haploid to diploid morphological transition in land plants. *Science, 339*(6123), 1067–1070.

Sakakibara, K., Nishiyama, T., Deguchi, H., & Hasebe, M. (2008). Class 1 KNOX genes are not involved in shoot development in the moss *Physcomitrella patens* but do function in sporophyte development. *Evolution and Development, 10*, 555–566.

Sakakibara, K., Reisewitz, P., Aoyama, T., Friedrich, T., Ando, S., Sato, Y.… Laux, T. (2014). *WOX13-like* genes are required for reprogramming of leaf and protoplast cells into stem cells in the moss *Physcomitrella patens*. *Development, 141*, 1660–1670.

Sarkar, A. K., Luijten, M., Miyashima, S., Lenhard, M., Hashimoto, T., Nakajima, K.… Laux, T. (2007). Conserved factors regulate signalling in *Arabidopsis thaliana* shoot and root stem cell organizers. *Nature, 446*, 811–814.

Sayou, C., Monniaux, M., Nanao, M. H., Moyroud, E., Brockington, S. F., Thévenon, E.… Dumas, R. (2014). A promiscuous intermediate underlies the evolution of LEAFY DNA binding specificity. *Science, 343*(6171), 645–648.

Schaefer, D. G., & Zrÿd, J. P. (1997). Efficient gene targeting in the moss *Physcomitrella patens*. *Plant Journal, 11*(6), 1195–1206.

Scheirer, D. C. (1990). Mosses. In H. D. Behnke, & R. D. Sjolund (Eds.), *Sieve elements: Comparative structure, induction and development* (pp. 19–31). Berlin Heidelberg: Springer-Verlag.

Schneider, H., Pryer, K. M., Cranfill, R., Smith, A. R., & Wolf, P. G. (2002). Evolution of vascular plant body plans: a phylogenetic perspective. In R. M. Bateman, Q. C. B. Cronk, & J. A. Hawkins (Eds.), *Developmental genetics and plant evolution* (pp. 330–364). New York, New York, USA: Taylor & Francis.

Schönknecht, G., Chen, W.-H., Ternes, C. M., Barbier, G. G., Shrestha, R. P., Stanke, M.… Weber, A. P. (2013). Gene transfer from bacteria and archaea facilitated evolution of an extremophilic eukaryote. *Science, 339*, 1207–1210.

Schoof, H., Lenhard, M., Haecker, A., Mayer, K. F., Jurgens, G., & Laux, T. (2000). The stem cell population of Arabidopsis shoot meristems in maintained by a regulatory loop between the *CLAVATA* and *WUSCHEL* genes. *Cell, 100*, 635–644.

Schultz, E. A., & Haughn, G. W. (1991). *LEAFY*, a homeotic gene that regulates inflorescence development in Arabidopsis. *Plant Cell, 3*(8), 771–781.

von Schwartzenberg, K., Núñez, M. F., Blaschke, H., Dobrev, P. I., Novák, O., Motyka, V., & Strnad, M. (2007). Cytokinins in the bryophyte *Physcomitrella patens*: analyses of activity, distribution, and cytokinin oxidase/dehydrogenase overexpression reveal the role of extracellular cytokinins. *Plant Physiology, 145*(3), 786–800.

Shimizu, R., Ji, J., Kelsey, E., Ohtsu, K., Schnable, P. S., & Scanlon, M. J. (2009). Tissue specificity and evolution of meristematic WOX3 function. *Plant Physiology, 149*, 841–850.

Sørensen, I., Fei, Z., Andreas, A., Willats, W. G. T., Domozych, D. S., & Rose, J. K. C. (2014). Stable transformation and reverse genetic analysis of *Penium margaritaceum*: a platform for studies of charophyte green algae, the immediate ancestors of land plants. *Plant Journal, 77*, 339–351.

Stahl, Y., Wink, R. H., Ingram, G. C., & Simon, R. (2009). A signaling module controlling the stem cell niche in *Arabidopsis* root meristems. *Current Biology, 19*, 909–914.

Sugano, S. S., Shirakawa, M., Takagi, J., Matsuda, Y., Shimada, T., Hara-Nishimura, I., & Kohchi, T. (2014). CRISPR/Cas9-mediated targeted mutagenesis in the liverwort *Marchantia polymorpha* L. *Plant and Cell Physiology, 55*(3), 475–481.

Szövényi, P., Rensing, S. A., Lang, D., Wray, G. A., & Shaw, A. J. (2011). Generation-biased gene expression in a bryophyte model system. *Molecular Biology and Evolution, 28*(1), 803–812.

Tam, T. H. Y., Catarino, B., & Dolan, L. (2015). Conserved regulatory mechanism controls the development of cells with rooting functions in land plants. *Proceedings of the National Academy of Sciences of the United States of America, 112*(29), E3959–E3968.

Tanahashi, T., Sumikawa, N., Kato, M., & Hasebe, M. (2005). Diversification of gene function: homologs of the floral regulator FLO/LFY control the first zygotic cell division in the moss *Physcomitrella patens*. *Development, 132*, 1727–1736.

Tarakhovskaya, E. R., Maslov, Y. I., & Shishova, M. F. (2007). Phytohormones in algae. *Journal of Plant Physiology, 54*, 163–170.

Ueda, M., Zhang, Z., & Laux, T. (2011). Transcriptional activation of *Arabidopsis* axis patterning genes WOX8/9 links zygote polarity to embryo development. *Developmental Cell, 20*, 264–270.

Viaene, T., Landberg, K., Thelander, M., Medvecka, E., Pederson, E., Feraru, E. … Friml, J. (2014). Directional auxin transport mechanisms in early diverging land plants. *Current Biology, 24*(23), 2786–2791.

Villalobos, L. I. A. C., Lee, S., De Oliveira, C., Ivetac, A., Brandt, W., Armitage, L. … Estelle, M. (2012). A combinatorial TIR1/AFB–Aux/IAA co-receptor system for differential sensing of auxin. *Nature Chemical Biology, 8*, 477–485.

Vollbrecht, E., Veit, B., Sinha, N., & Hake, S. (1991). The developmental gene *Knotted-1* is a member of a maize homeobox gene family. *Nature, 350*, 241–243.

Wang, C., Liu, Y., Li, S.-S., & Han, G.-Z. (2015). Insights into the origin and evolution of the plant hormone signaling machinery. *Plant Physiology, 167*, 872–886.

Wickett, N. J., Mirarab, S., Nguyen, N., Warnow, T., Carpenter, E., Matasci, N. … Leebens-Mack, J. (2014). Phylotranscriptomic analysis of the origin and early diversification of land plants. *Proceedings of the National Academy of Sciences of the United States of America*, E4859–E4868.

Wu, X., Chory, J., & Weigel, D. (2007). Combinations of WOX activities regulate tissue proliferation during Arabidopsis embryonic development. *Developmental Biology, 309*, 306–316.

Xu, B., Ohtani, M., Yamaguchi, M., Toyooka, K., Wakazaki, M., Sato, M. ... Demura, T. (2014). Contribution of NAC transcription factors to plant adaptation to land. *Science, 343*(6178), 1505–1508.

Yamaguchi, M., Goué, N., Igarashi, H., Ohtani, M., Nakano, Y., Mortimer, J. C. ... Demura, T. (2010). VASCULAR-RELATED NAC-DOMAIN6 and VASCULAR-RELATED NAC-DOMAIN7 effectively induce transdifferentiation into xylem vessel elements under control of an induction system. *Plant Physiology, 153*, 906–914.

Yamaguchi, M., Mitsuda, N., Ohtani, M., Ohme-Takagi, M., Kato, M., & Demura, T. (2011). VASCULAR-RELATED NAC-DOMAIN7 directly regulates the expression of a broad range of genes for xylem vessel formation. *Plant Journal, 66*, 579–590.

Yasumura, Y., Crumpton-Taylor, M., Fuentes, S., & Harberd, N. P. (2007). Step-by-step acquisition of the gibberellin-DELLA growth-regulatory mechanism during land-plant evolution. *Current Biology, 17*(14), 1225–1230.

Zhong, R., Demura, T., & Ye, Z.-H. (2006). SND1, a NAC domain transcription factor, is a key regulator of secondary wall synthesis in fibers of *Arabidopsis*. *Plant Cell, 18*, 3158–3170.

Ziegler, H. (1987). The evolution of stomata. In E. Zeiger, G. D. Farquhar, & I. R. Cowan (Eds.), *Stomatal function* (pp. 29–57). Stanford University Press.

CHAPTER TWO

Coevolution of Organelle RNA Editing and Nuclear Specificity Factors in Early Land Plants

M. Schallenberg-Rüdinger*,§, V. Knoop*,1
*Universität Bonn, Bonn, Germany
§University of Marburg, Marburg, Germany
[1]Corresponding author: E-mail: volker.knoop@uni-bonn.de

Contents

Advances in Botanical Research, Volume 78
ISSN 0065-2296
http://dx.doi.org/10.1016/bs.abr.2016.01.002

Abstract

RNA editing alters genomic information by site-specific conversion of cytidines into uridines in land plant chloroplast and mitochondrial transcripts. Key to this process are nuclear-encoded RNA-binding pentatricopeptide repeat (PPR) proteins imported into the two endosymbiotic organelles. The model moss *Physcomitrella patens* has been proven useful in identifying those specificity factors and has now become the first organism with a full mutual assignment of specific PPR proteins and organelle RNA editing sites. With only 11 mitochondrial and two chloroplast RNA editing events, however, *Physcomitrella* is at the lower end of the tremendously variable editing frequencies among land plants. At the other end of the spectrum, thousands of editing sites change genetic information in both organelles of club moss (*Selaginella*) species. Moreover, reverse U-to-C editing operates in parallel to C-to-U in other lycophyte genera, in hornworts and in ferns. In contrast to dozens of editing factors meantime characterized for C-to-U editing in model taxa like *Arabidopsis*, *Oryza* or *Physcomitrella*, no factors are yet identified for U-to-C editing. Gains and losses of RNA editing sites and adaptations of their cognate nuclear specificity factors are an outstanding opportunity to study 500 million years of coevolution between the three genetic systems in plant cells. A particular interest in this subject also results from necessary improvements of a recently proposed PPR–RNA binding code. We here summarize the state of research, highlighting the particular value of nonseed plants in studies of RNA editing and give an outlook on the likely future developments of the field.

1. RNA EDITING IN PLANT ORGANELLES: INTRODUCTION AND STATE OF THE ART

More than 60 years after the discovery of the double helix structure of DNA by Watson and Crick in 1953 it is evident that many twists and turns contribute to protein biogenesis. Different forms of nucleotide sequence alterations on genomic and transcript level are common in many lineages of life. One of the detours on transcript level is called RNA editing. RNA editing describes changes of genetic information on RNA level, which could be directly encoded in the DNA.

1.1 RNA Editing: A Vast Field

The first case of RNA editing was recognized 1986 in the kinetoplastid protozoon *Trypanosoma brucei*. Rob Benne and colleagues noticed that something was wrong with the mitochondrial *cox2* gene, which showed reading frame shifts. On transcript level, however, they were gone. Insertions of four uridines were identified to correct the defect (Benne et al., 1986). Many other mitochondrial mRNAs in *Trypanosoma* and other kinetoplastid protozoa were likewise identified as targets for U insertions or deletions, all being necessary to regain the accurate genetic information (Feagin, Abraham, & Stuart, 1988; Simpson & Shaw, 1989). Antisense guide RNAs (gRNAs) transcribed from maxicircle DNAs were found to direct the editing process as templates (Blum, Bakalara, & Simpson, 1990; Sturm & Simpson, 1990).

Shortly thereafter, another form of RNA editing, cytidine-to-uridine conversion, was discovered (Powell et al., 1987; Hospattankar, Higuchi, Law, Meglin, & Brewer, 1987). A C-to-U exchange was first identified in the human nuclear transcript of *apoB* and drew particular attention. The cytidine substitution generated an early stop codon in the *apoB* transcript. Unedited and edited versions were found to be expressed in different organs (liver and intestine). Due to the fact that this editing site was also present in mammalian model organisms like mice and rats, the editing machinery leading to the precise C-to-U exchange was quickly identified and characterized within the next years (Cho, Blanc, & Davidson, 2007). A zinc-dependent cytidine deaminase named APOBEC-1 as the key player catalyzes the C-to-U deamination process and sequence recognition is provided via an interplay of an 11-nt target sequence in the vicinity of the editing site and the RNA-binding "APOBEC-1 complementing factor" (reviewed in Smith, Gott, & Hanson, 1997). Only a few other C-to-U editing sites were identified in nuclear transcripts of mammals, however, making the *apoB* transcript case more of an exception than the tip of an iceberg (Legendre, Forstera, Juttner, & Meier, 2009; Meier et al., 2005; Navaratnam et al., 1991). The many different, independently evolved RNA-editing machineries in different organisms have been reviewed elsewhere (Chateigner-Boutin & Small, 2011; Knoop, 2011). We will here focus on plant-type organelle RNA editing.

1.2 Plant Organelle RNA Editing: What, Where, When

By the end of 1989, three independent research groups in Canada, France and Germany, working on the flowering plant models wheat (*Triticum sativum*) and evening primrose (*Oenothera berteriana*), respectively, discovered the first editing sites in land plant mitochondria (Covello & Gray, 1989; Gualberto, Lamattina, Bonnard, Weil, & Grienenberger, 1989; Hiesel, Wissinger, Schuster, & Brennicke, 1989). As in the case of *apoB*, cytidines were found to be edited into uridines. Like in the kinetoplastids, mitochondrial transcripts were targeted and not nuclear ones. Shortly after this discovery, C-to-U editing sites were also identified in transcripts of the second endosymbiotic organelle of plants, the plastid (Hoch, Maier, Appel, Igloi, & Kössel, 1991). In contrast to nuclear C-to-U RNA editing in mammals, cytidines to be edited were found in nearly all mitochondrial and chloroplast mRNAs (Chateigner-Boutin & Small, 2010; Finster, Legen, Qu, & Schmitz-Linneweber, 2012; Shikanai, 2015; Takenaka, Zehrmann, Verbitskiy, Härtel, & Brennicke, 2013). After the complete sequence of the mitochondrial DNA of *Arabidopsis* became available (Unseld, Marienfeld, Brandt, & Brennicke, 1997), 441 C-to-U editing sites were determined in its mitochondrial transcriptome (Giegé & Brennicke, 1999). In the chloroplast, in contrast, only 34 sites of RNA editing were shown to affect transcripts (Chateigner-Boutin & Small, 2007; Tillich et al., 2005).

Editing was found to be fast-evolving with variation in the editing pattern to exist even between different *Arabidopsis* ecotypes (Tillich et al., 2005; Zehrmann, van der Merwe, Verbitskiy, Brennicke, & Takenaka, 2008). Numbers of approximately 30–50 editing sites in plastid and some 300–500 editing sites in mitochondrial transcripts were identified in diverse other flowering plants including monocots like *Zea mays* (maize, Tillich, Schmitz-Linneweber, Herrmann, & Maier, 2001) *Oryza sativa* (rice, Notsu et al., 2002) or the orchid *Phalaenopsis aphrodite* (Zeng, Liao, & Chang, 2007) and dicots like *Pisum sativum* (pea, Inada, Sasaki, Yukawa, Tsudzuki, & Sugiura, 2004), *Nicotiana tabacum* (tobacco, Sasaki, Yukawa, Miyamoto, Obokata, & Sugiura, 2003), *Brassica napus* (rapeseed, Handa, 2003) or *Beta vulgaris* (sugar beet, Mower & Palmer, 2006).

With phylogenetically wider taxon sampling, organellar RNA editing was early found to be present in all major land plant clades including liverworts, mosses, hornworts, lycophytes, ferns and gymnosperms (Freyer, Kiefer-Meyer, & Kössel, 1997; Malek et al., 1996; Sper-Whitis, Moody, & Vaughn, 1996; Steinhauser, Beckert, Capesius, Malek, & Knoop, 1999). Only in one subclade of liverworts, the "complex-thalloid" Marchantiidae, RNA editing was found to be completely lost (Groth-Malonek, Wahrmund, Polsakiewicz, & Knoop, 2007; Rüdinger, Volkmar, Lenz, Groth-Malonek, & Knoop, 2012). In contrast, RNA editing in hornworts, ferns and in lycophytes exceeds the numbers in flowering plants and reaches extreme levels with thousands of sites edited in the organelles of the lycophytes *Selaginella* or *Isoetes* (Grewe et al., 2011; Hecht, Grewe, & Knoop, 2011; Oldenkott, Yamaguchi, Tsuji-Tsukinoki, Knie, & Knoop, 2014).

In hornworts, ferns and *Isoetes*, the near-omnipresent C-to-U editing is found to be accompanied by 'reverse' editing, i.e. uridine-to-cytidine conversions (Duff, 2006; Duff & Moore, 2005; Grewe et al., 2011; Steinhauser et al., 1999; Vangerow, Teerkorn, & Knoop, 1999; Wolf, Rowe, & Hasebe, 2004; Yoshinaga, Iinuma, Masuzawa, & Uedal, 1996).

With even more studies on plant organelle RNA editing and vast numbers of editing sites in at least some species, the need for a simple and clear editing site nomenclature became obvious. Such a nomenclature (described in Table 1) was introduced together with the characterization of the complete mitochondrial editome of the moss *Physcomitrella patens*. This model plant and its relative *Funaria hygrometrica* are at the very low end of editing frequencies with only 11 and 8 C-to-U sites in mitochondrial coding regions, respectively (Rüdinger, Funk, Rensing, Maier, & Knoop, 2009; Rüdinger, Szövényi, Rensing, & Knoop, 2011) and two editing sites in the chloroplasts (Miyata & Sugita, 2004; Miyata, Sugiura, Kobayashi, Hagiwara, & Sugita, 2002).

In general, most RNA editing sites are located in coding regions of messenger RNAs and affect the first or the second position of a codon. These conversions always result in changes of amino acid identity except for the silent first codon position editing of YUR leucine codons. Editing may generate start (ACG > AUG) or stop (CAA > UAA, CAG > UAG, CGA > UGA) codons (Fig. 1). As an example for the editing site nomenclature we consider the *P. patens cox1* gene that is edited at position 755 in its transcript, counting from the first nucleotide of its AUG start codon. This C-to-U editing event (labelled eU for *e*diting to *U*) results in a serine (S) to leucine (L) codon identity change. The editing site is accordingly labelled cox1eU755SL.

In species with more overall editing, intron sequences were also found to be edited. Such sites in group II introns in mitochondria of the evening primrose *Oenothera* affect highly structured domains V and VI essential for splicing functionality (Binder, Marchfelder, Brennicke, & Wissinger, 1992; Börner, Mörl, Wissinger, Brennicke, & Schmelzer, 1995). The other type of organellar introns, group I introns, were likewise found to be affected by editing (Bégu, Castandet, & Araya, 2011). Similarly, transfer RNAs can be affected by RNA editing to allow for proper folding and processing (Binder, Marchfelder, & Brennicke, 1994; Kunzmann, Brennicke, & Marchfelder, 1998; Marchfelder, Brennicke, & Binder, 1996; Maréchal-Drouard, Ramamonjisoa, Cosset, Weil, & Dietrich, 1993). The number of editing sites outside of coding regions generally increases with the

Table 1 Assignment of DYW-type PPR Proteins and Editing Events in *Physcomitrella patens*

Localization	PPR/Phypa	No of PPRs	No of Introns	Function/Editing Site(s)	Editing Site Conservation		Phenotype of KO	Study
					Funaria	Mosses		
Plastid								
	45/156474	25	1	rps14cpeU2TMp	Yes	?	†	Ichinose et al. (2014)
				rps14cpeU-1p5	Yes	?		
Mitochondrial	56/169414	15	3	nad3eU230SLp	No	No	+	Ohtani et al. (2010)
				nad4eU272SL	No	No		
	65/175641	15	1 + (1)	ccmFCeU103PS	Yes	Yes	+++	Schallenberg-Rüdinger, Kindgren, et al. (2013), Ichinose et al. (2013)
	71/181369	17	0	ccmFCeU122SF	Yes	Yes/No	++	Tasaki et al. (2010)
	77/191892	26	1	cox2eU370RW	Yes	Yes	++	Ohtani et al. (2010)
				cox3eU733RW	Yes	Yes		
	78/192620	21	4	cox1eU755SL	Yes	Yes/No	0	Rüdinger, Szövényi, et al. (2011), Uchida et al. (2011)
				rps14eU137SLp	No	Yes		
	79/192787	17	1	nad5eU598RC	Yes	Yes	+	Uchida et al. (2011)
	91/211228	19	1	nad5eU730RW	Yes	Yes	+++	Ohtani et al. (2010)
	98/224041	21	1 + (1)	atp9eU92SL	Yes	Yes	†	Schallenberg-Rüdinger, Kindgren, et al. (2013), Ichinose et al. (2013)
	43/154890	25	1	cox1 intron splicing	Yes	–	++	Ichinose et al. (2012)

Introns are located in the region encoding the amino-terminal targeting signal except of one in PPR_65 and PPR_98, respectively (in parentheses). Editing site nomenclature is composed of the gene name, followed by an 'e' for editing, the nucleotide identity resulting from editing (here: 'U'), the position of the edited nucleotide starting with the adenine of the start codon and the resulting amino acid codon exchange or its position relative to the CDS if in UTRs and optionally the percentage of editing behind a 'p'. A "yes" indicates conservation of a cytidine in at least 10 of 13 available moss mtDNAs (Liu et al., 2014), "yes/no" indicates multiple C/T (gain/loss) scenarios during moss evolution. Phenotype characteristics of KO plants are indicated as '0' for no '+' for a modest, '++' for a strong, '+++' for a very strong macroscopic phenotype and '†' for lethal.

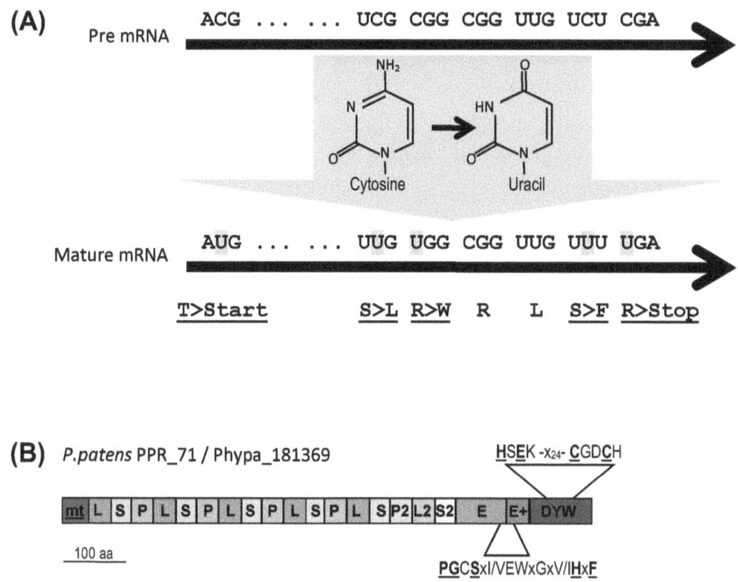

Figure 1 (A) Numerous cytidines are specifically converted into uridines in plant chloroplast and mitochondrial transcripts. The C-to-U editing events mainly serve to change codon identities in the 1st and 2nd codon position to restore evolutionarily conserved amino acid identities but may also occur in silent codon positions and in introns, UTRs, rRNAs and tRNAs. (B) Pentatricopeptide Repeat (PPR) proteins are key to the specific recognition of the organelle RNA editing sites. Exemplarily shown is the structure of PPR_71 a typical "DYW-type" PPR protein crucial for editing event ccmFCeU122SF in mitochondria of the moss *Physcomitrella patens*. The DYW domain is characterized by a conserved cytidine deaminase motif. A conserved motif at the border of the E and E+ domains linking to the upstream PPR array is shown to be crucial for editing in angiosperms. The PPR arrays of editing factors are of the PLS-type featuring alternations of long (L, 35/36 aa) and short (S, 31 aa) variants of the classic P-type PPRs of 35 aa. The terminal PLS2 triplet is characterized by a deviating repeat consensus. (See colour plate)

number in mRNA exons. The quillwort *Isoetes engelmannii* with its altogether more than 1700 sites of mitochondrial RNA editing features 10 positions of RNA editing in the tRNA for proline alone (Grewe, Viehoever, Weisshaar, & Knoop, 2009). Ribosomal RNA editing is rare but can also be identified in species with overall high editing frequencies. In the spike moss, *Selaginella moellendorffii*, with more than 2000 editing sites in its mitochondrial transcriptome, 13 RNA editing sites were found in the two mitochondrial encoded rRNAs (Hecht et al., 2011). Like the overwhelming majority of mRNA editing events, the editing sites in introns, tRNAs or rRNAs reconstitute evolutionarily conserved positions and in correcting the genetic information contribute to the proper function of the affected molecule.

Plant organelle RNA editing appears to be an efficient posttranscriptional process. Sequencing of cDNA with very high coverage is needed to identify remaining unedited Cs in transcript pools at most sites. However, RNA editing efficiencies may also be much lower at other sites. This is frequently observed for editing of third codon positions, called silent editing. High-quality RNAseq data are needed to reliably determine RNA editing efficiencies at individual positions (Bentolila, Oh, Hanson, & Bukowski, 2013; Ruwe, Castandet, Schmitz-Linneweber, & Stern, 2013). Pitfalls of the RNA-seq technology in its current state should be kept in mind though.

Those sites which are edited to low degrees in the transcript pool may be functionally neutral and reflect a lack of specificity of the RNA editing machinery. For example, in the chloroplast of *P. patens*, an editing site is located two nucleotides upstream of the essential start codon editing rps14cpeU2TM of *rps14*. That upstream site is only found to be edited in ca. 5% of the transcripts (Miyata et al., 2002). According to the suggested nomenclature (Lenz et al., 2010; Rüdinger et al., 2009), this information may be summarized as rps14cpeU-1p5 (Table 1). The extraordinary frequency of RNA editing in the *Selaginella uncinata* chloroplast (3400 events) allowed evaluating sites of silent editing in more detail. Here, they were noticed to dominantly occur in the immediate vicinity of nonsilent editing sites (Oldenkott et al., 2014).

The degree of editing at some sites has occasionally been reported to change across tissues, development or environmental conditions. For example, some chloroplast RNA-editing sites are fully edited in *Arabidopsis thaliana* leaves, but only partially edited in nonphotosynthetic tissues (Bock, Hagemann, Kössel, & Kudla, 1993; Chateigner-Boutin & Hanson, 2003; Peeters & Hanson, 2002). Reduced editing efficiency at particular sites can also be observed when chloroplast development is impaired or under mild stress conditions like elevated temperatures (Halter, Peeters, & Hanson, 2004; Kakizaki, Yazu, Nakayama, Ito-Inaba, & Inaba, 2012; Karcher & Bock, 2002; Nakajima & Mulligan, 2001; Tseng et al., 2010). Similarly, editing efficiency may vary between different isolates or ecotypes (Bentolila, Chateigner-Boutin, & Hanson, 2005; Tillich et al., 2005; Zehrmann et al., 2008). Especially, a tissue-dependent effectiveness of editing generating a start or stop codon suggested a regulatory function of editing in plant organelles. Differences in the editing efficiencies of rps14cpeU2TM of *P. patens* and ndhDeU2TM of tobacco, respectively, were observed under different developmental and environmental conditions (Hirose & Sugiura, 1997; Miyata & Sugita, 2004). Whether the different degree of editing has indeed a regulatory function or is only a result of altered plastid metabolism remains an open question. Notably, it has been shown that an unedited ACG threonine codon may also be used as an alternative start codon (Zandueta-Criado & Bock, 2004). Evidence is lacking that partial editing of internal nonsilent sites plays an important functional role. Protein versions of unedited and edited transcripts of *rps12* in maize and petunia mitochondria were identified immunologically (Lu, Wilson, Phreaner, Mulligan, & Hanson, 1996; Phreaner, Williams, & Mulligan, 1996), but no functional consequence is evident.

Editing sites in mRNAs of plant mitochondria and chloroplasts can be reliably predicted benefiting from the high degree of conservation in organellar coding sequences. The mitochondrial genomes of species lacking organelle RNA editing, e.g. the marchantiid liverwort *Marchantia polymorpha* (Oda et al., 1992) can be used as references to identify T-to-C exchanges which could be rescued via RNA editing in query sequences from taxa showing editing. With more completely sequenced organellar genomes and corresponding editomes, the set of reference genomes for homology-based search for editing sites was extended. The first publicly available prediction tool for plants PREP (predictive RNA editors for plants) focused on C-to-U editing and included available data for plant mitochondria (Mower, 2005) and chloroplasts (Mower, 2009), at that time mainly angiosperm sequences. The later developed tool PREPACT (Plant RNA Editing Prediction and Analysis Computer Tool) extended prediction to U-to-C editing (see Section 4), incorporated reference genomes and editomes of diverse plants along land plant phylogeny and finally allowed to use entire new organelle genomes as queries to simultaneously identify protein-coding genes and predict candidate editing sites (Lenz & Knoop, 2013; Lenz et al., 2010).

1.3 Elucidating the Plant Organelle RNA-Editing Machinery

Soon after the discovery of RNA editing in land plant organelles, several studies identified the cytidine-to-uridine conversion to be based on cytidine deamination (Blanc, Litvak, & Araya, 1995; Rajasekhar & Mulligan, 1993; Yu & Schuster, 1995). Cytidines with labelled α-phosphates were observed to be converted into uridines, excluding the alternative scenario of excisions of cytidine and insertion of uridine residues. C-to-U deaminases involved in uridine and cytidine biosynthesis suggested themselves as potential enzymes. In mammals, the identified deamination enzyme APOBEC-1 indeed is a zinc-dependent cytidine deaminase (Smith et al., 1997). Although nine putative cytidine deaminases are encoded in the *A. thaliana* genome, none of them could be linked to the C-to-U RNA-editing process in organelles (Faivre-Nitschke, Grienenberger, & Gualberto, 1999). Intriguingly, one of them was later shown to be an adenosine deaminase responsible for A-to-I editing in the ACG anticodon of the plastid tRNA for arginine, the second type of RNA editing in land plant organelles discovered later (Delannoy et al., 2009; Karcher & Bock, 2009). This enzyme, as well as all C-to-U deaminases in general, requires zinc ions as cofactors. Zinc dependency and supply of other cofactors for C-to-U editing was tested in in vitro systems of mitochondria and plastids established in different laboratories. Whereas a study on plastidial extract of *A. thaliana* clearly identified zinc to be necessary for RNA editing (Hegeman, Hayes, & Hanson, 2005), another in vitro study on pea mitochondria showed that zinc chelators do not reduce RNA editing (Takenaka & Brennicke, 2003). Other supplies were found to enhance the in vitro RNA-editing reaction, such as ATP, NTPs or dNTPs. The positive effect was finally explained with energy supply of helicases unwinding the RNA targets rather than direct participation of these factors in the editing process (Takenaka & Brennicke, 2003). Since stable transformation of plant mitochondria is not possible to this day, the in vitro system represented an excellent alternative to study RNA editing in mitochondrial lysates. Nevertheless, results had to be interpreted carefully as competence of mitochondrial lysates to edit an inserted artificial RNA molecule is limited, e.g. only some but not all editing sites are reliably edited and editing never reaches full efficiency (Takenaka, Zehrmann, Verbitskiy, et al., 2013). Most often, only 10–20% of transcripts are edited (Takenaka & Brennicke, 2003). Still, the successful editing of in vitro transcripts showed the general ability of editing independent of the transcription machinery.

In parallel to the mitochondrial in vitro systems, *in organello* assays based on mitochondrial electroporation were established to introduce modified genes or transcripts into flowering plant mitochondria (Farré & Araya, 2001; Staudinger & Kempken, 2003). In parallel to the in vitro systems with plastid lysates established for several model plant systems like tobacco (Hirose & Sugiura, 2001), pea (Miyamoto, Obokata, & Sugiura, 2002; Nakajima & Mulligan, 2005), maize (Hayes & Hanson, 2007a) and *Arabidopsis* (Hegeman et al., 2005), transplastomic approaches in tobacco and *Arabidopsis* were used to study RNA editing in this organelle. The transplastomic approaches showed that approximately 20 nucleotides upstream of the editing site are sufficient for editing (Bock, Hermann, & Kössel, 1996; Chaudhuri, Carrer, & Maliga, 1995; Chaudhuri & Maliga, 1996). Independent studies confirmed that 15–12 nucleotides upstream of the mitochondrial and plastid editing sites are most important (Choury, Farré, Jordana, & Araya, 2004; Farré, Leon, Jordana, & Araya, 2001; Neuwirt, Takenaka, van der Merwe, & Brennicke, 2005; Reed, Peeters, & Hanson, 2001; Takenaka, Neuwirt, & Brennicke, 2004) but the wider sequence environment was also shown to influence editing of at least some sites in both organelles (Hayes & Hanson, 2007b; Sasaki, Yukawa, Wakasugi, Yamada, & Sugiura, 2006; Staudinger, Bolle, & Kempken, 2005). In cauliflower mitochondria, for example, 20 nucleotides upstream of the investigated *atp9* editing site were crucial for editing, but higher RNA editing efficiencies were only reached with further upstream sequence extensions to 40–70 nucleotides (Neuwirt et al., 2005; Van der Merwe, Takenaka, Neuwirt, Verbitskiy, & Brennicke, 2006). Higher efficiency of editing with an extended region upstream of the editing sites was also observed for sites in the plastid transcript *rpoB* of tobacco (Hayes, Reed, Hegeman, & Hanson, 2006). Influence of distal sequence elements might be linked to better accessibility of the RNA editing site due to structural changes of the complete target transcript, potentially via binding of other proteins.

General features of *cis*-elements upstream of different plastid or mitochondrial editing sites could not be identified, suggesting individual *trans*-factors to recognize the different target sequences. Similarity of some *cis*-elements, however, was determined assuming the assignment of single *trans*-factors to more than one target (Chateigner-Boutin & Hanson, 2003). First biochemical evidence of *trans*-factor proteins acting on *cis*-elements came from UV cross-linking experiments in the in vitro system of tobacco plastids (Hirose & Sugiura, 2001).

No in vitro, in vivo or *in organello* system had been successfully established for nonangiosperm plants in the early years of RNA editing research. In 2004, plastid transformation in the moss, *P. patens*, was described for the first time (Sugiura & Sugita, 2004). After complete sequencing of the *P. patens* chloroplast (Sugiura, Kobayashi, Aoki, Sugita, & Sugita, 2003), mitochondrial (Terasawa et al., 2006) and nuclear genomes (Rensing et al., 2008), studies on the RNA-editing machinery of the moss were initiated (see Section 3).

1.4 A Key Role for Pentatricopeptide Proteins in Plant Editing

The complete sequence of the nuclear genome of the model flowering plant *A. thaliana* (Brassicaceae) brought forth the discovery of one of the largest protein-encoding gene families in land plants, the pentatricopeptide repeat (PPR) protein family (Aubourg, Boudet, Kreis, & Lecharny, 2000; Small

& Peeters, 2000). The expansion of that protein family in land plants with more than 400 members in angiosperms, the prediction of most members to be targeted to mitochondria and plastids and their modular protein structure (Lurin et al., 2004; Manna, 2015; Shikanai, 2006) favoured these proteins to act as *trans*-factors in organelle RNA maturation processes in general and in the RNA-editing process in particular. PPR proteins are characterized by tandem arrays (2–26) of the degenerate 35-amino acid PPR motif-suggested to specifically bind to RNA sequence targets (Small & Peeters, 2000). In the following years, PPR proteins were indeed demonstrated to participate in diverse RNA processing steps like splicing (De Longevialle et al., 2007; Schmitz-Linneweber et al., 2006), intergenic cleavage (Hashimoto, Endo, Peltier, Tasaka, & Shikanai, 2003), 5′-end processing (Jonietz, Forner, Hölzle, Thuss, & Binder, 2010), RNA stabilization (Pfalz, Bayraktar, Prikryl, & Barkan, 2009) and translation (Manavski, Guyon, Meurer, Wienand, & Brettschneider, 2012; Schmitz-Linneweber, Williams-Carrier, & Barkan, 2005).

In 2005, the first PPR protein was genetically identified as an editing factor in chloroplasts (Kotera, Tasaka, & Shikanai, 2005). CRR4 (named for chlororespiratory reduction phenotype) was identified via screening ethane methyl sulfonate mutants for chlorophyll fluorescence defects. The defect was caused by impaired CRR4, characterized as being responsible for the RNA editing event reintroducing the start codon in the *ndhD* mRNA in *A. thaliana* plastids. The *ndhD* locus encodes subunit D of the NADH dehydrogenase complex. As of writing this manuscript, more than 60 PPR editing factors have already been identified affecting individual or up to 10 editing sites simultaneously in plastids or mitochondria. While the majority of editing factors has been characterized in the key model flowering plant *A. thaliana*, the moss *P. patens* has become the first genetic model system with a full mutual assignment of editing factors to all its organelle editing sites (see Section 3). Other than *Arabidopsis* (Barkan & Small, 2014; Finster et al., 2012; Shikanai, 2015) and *Physcomitrella* (Sugita, Ichinose, Ide, & Sugita, 2013), a few editing factors have also been genetically characterized in maize (Sosso et al., 2012), tabacco (Okuda, Habata, Kobayashi, & Shikanai, 2008) and rice (Kim et al., 2009).

All site-specific plant organelle RNA editing factors hitherto characterized are members of a specific PPR protein subfamily (Lurin et al., 2004). Besides the classic 35-amino acid long P-type PPR motifs, these PLS-type proteins feature PPR variants of the L- (for long) and S-type (for short), which differ in sequence conservation and lengths (Fig. 1). Very dominantly, these PPR variants occur in PLS_n arrangements, giving the subfamily the name PLS. Characteristically, the ultimate C-terminal PLS triplet varies in sequence conservation and is labelled PLS2 (Fig. 1). In *A. thaliana* 200 of the altogether 450 PPR proteins are PLS-type proteins. In *P. patens*, only 16 of the 105 PPR proteins are of the PLS-type, correlating well with the low number of editing sites identified in that species (O'Toole et al., 2008; Sugita et al., 2013). The majority of the plant PLS-type PPR proteins feature additional conserved domains at their carboxy-termini: the 'E' domain (E for extension, ~90 amino acids), the 'E+' domain (~30 amino acids) and the 'DYW' domain (~100 amino acids) as optional, successive extensions, always in this order. The DYW domain is named after the three conserved amino acids aspartate (D), tyrosine (Y) and tryptophan (W) frequently occurring at its very end.

A very recent publication has re-evaluated the PLS protein makeup and suggested new domains E1 and E2 to replace the former E domain, to extend the DYW domain upstream to essentially incorporate the former E+ domain and identified a new type of PPR labelled "SS"-type (Cheng et al., 2016).

The DYW domain attracted particular attention regarding its structural and sequence similarity to cytidine deaminases (Iyer, Zhang, Rogozin, & Aravind, 2011; Salone et al., 2007). Notably, the zinc-binding motif $HxEx_nCxxC$ with its histidines and cysteines coordinating the Zn^{2+} ion is highly conserved in DYW domains (Fig. 1). The capacity of the DYW domain to bind zinc ions was only recently demonstrated (Boussardon et al., 2014; Hayes, Giang, Berhane, & Mulligan, 2013). Importantly, the evolutionary distribution of organelle RNA editing and nuclear-encoded DYW domains correlates well along land plant phylogeny (Rüdinger, Polsakiewicz, & Knoop, 2008; Salone et al., 2007), with no evidence for DYW domains in the marchantiid liverworts or green algae, which lack RNA editing in their organelles (see Section 4). Conversely, numbers of DYW-domain containing proteins seem to increase with organelle RNA editing frequencies (Rüdinger et al., 2012). Land plants with high amounts of RNA editing like the lycophyte *Selaginella moellendorffii* (Hecht et al., 2011) feature high numbers of DYW-type PPR proteins in their genome (Banks et al., 2011). In *Physcomitrella*, all PLS-type proteins with canonical DYW domains have been shown to be site-specific RNA editing factors. The only member of the group with a mutated DYW domain, PPR_43, has been demonstrated to be a splicing factor instead (see Section 3, Ichinose, Tasaki, Sugita, & Sugita, 2012).

Despite all the evidence, however, a cytidine deaminase activity has not yet been biochemically confirmed for the DYW domain. Instead, DYW domains involved in RNA editing *in planta*, did not show deaminase activity in vitro (Nakamura & Sugita, 2008; Okuda et al., 2009) but were rather shown to cleave RNA. This appears to be congruent with the function of DYW-type PPR protein CRR2 in *A. thaliana* plastids, which was shown to be involved in intergenic cleavage (Hashimoto et al., 2003). Ironically, CRR2 was the first DYW-type PPR protein characterized, but to this date also remained the only example of a DYW protein involved in RNA cleavage. Similar to the above example of the *Physcomitrella* splicing factor PPR_43, CRR2 may be a degenerated DYW-type protein. Moreover, RNA degradation might be merely a side effect of the in vitro experiment explainable by the absence of proteins protecting the RNA.

1.5 Complicating the Plant Editing Machinery: The DYW Domain Issue

Not all site-specific RNA editing factors identified are DYW-type PPR proteins. For example, CRR4, the first editing factor to be cloned, is an E+ protein lacking a DYW domain (Kotera et al., 2005). Moreover, some editing sites assigned to specific DYW-type PPR proteins in *A. thaliana* are still efficiently edited when the DYW domain of the appropriate protein is mutated or truncated in complementation experiments of the respective mutants, e.g. CRR28, CRR22, OTP82, ELI1, MEF11 (Hayes et al., 2013; Okuda et al., 2009, 2010; Zehrmann et al., 2011). Other DYW-type PPR proteins like the first identified mitochondrial editing factor MEF1 (mitochondrial editing factor) characterized as recognition factor for three editing sites rps4eU956SL, nad7eU963FF and nad2eU1160SL (Zehrmann, Verbitskiy, Härtel, Brennicke, & Takenaka, 2010; Zehrmann, Verbitskiy, van der Merwe, Brennicke, & Takenaka, 2009) were shown to require the DYW domain for complete editing. Partial restoration of editing, however, was also possible with the truncated protein version. In yet other cases, a mutated DYW domain led to abolished editing of some but not all assigned sites (Hayes, Dang, Diaz, & Mulligan, 2015; Verbitskiy, Zehrmann, van der Merwe, Brennicke, & Takenaka, 2010; Wagoner, Sun, Lin, & Hanson, 2015). For some proteins like RARE1 or QED1 (Wagoner et al., 2015), the complementation of the corresponding KO plant with a truncated version of the DYW domain is not feasible. These examples already indicate different relevance of the DYW domain in diverse DYW-type proteins in *A. thaliana*.

Particularly intriguing is the function of plastid DYW1, an isolated DYW domain in *Arabidopsis*. This small protein of 239 amino acids consists of a conserved DYW domain preceded by a short C terminal part of the E+

domain but carries neither an E domain nor evident PPR motifs. The hypothesis of DYW1 being the universal cytidine deaminase enzyme did not hold true, but DYW1 was instead found to be exclusively required for editing of the *ndhD* start codon in combination with CRR4 (Boussardon et al., 2012). An artificial protein fusion of both proteins reconstituted editing in the loss-of-function background. These experiments demonstrated that the DYW1 domain can operate both as a translational fusion or, as in the natural situation, in *trans*. Mutagenesis of highly conserved amino acid residues in the putative zinc-binding motif of DYW1 caused impairment of editing demonstrating their importance (Boussardon et al., 2014). The DYW1 example strongly supports the idea that E and E+, and also at least some DYW-type editing factors can recruit other DYW domains in *trans* for catalyzing the C-to-U conversion (Chateigner-Boutin & Small, 2010; Knoop, 2011). Such a scenario (Fig. 2) could also explain findings of two different DYW-type PPR proteins being essential for the same editing site in *A. thaliana* plastids (Hammani et al., 2009; Robbins, Heller, & Hanson, 2009; Tseng et al., 2010; Yu, Jiang, Chong, & Yang, 2009). Interestingly, in both cases as yet discovered, one of the two participating proteins is AtECB2/VAC1. Its DYW domain terminates two amino acids early with D, suggesting possibly impaired function, as it was also shown in experiments with DYW1 in which the DYW amino acid triplet was truncated (Boussardon et al., 2014). Hence, recruitment of another functional DYW domain in *trans* may be necessary in such cases (Hammani et al., 2009). However, as the PPR array of AtECB2/VACI does not suggest particular affinity to the sequences upstream of the specific editing sites accDeU794SL and ndhFeU290SL (for recognition code see Section 2), it is likely that this protein has a more general effect like stabilizing the affected transcripts or inducing particular folding rather than recognition of the particular editing sites. Targeting prognoses indeed suggest binding of AtECB2/VACI to multiple regions elsewhere in the *accD* and *ndhF* transcripts and rather match the other participating proteins RARE1 and OTP84 with accDeU794SL or ndhFeU290SL, respectively (unpublished observations). If true, such an RNA-binding or stabilizing function of ECB2/VAC1 may reflect a further type of functional shift in proteins with a degenerated DYW domain (Fig. 2).

The two short DYW-type PPR proteins MEF8 and MEF8S in *Arabidopsis* were the first mitochondrial PPR proteins, for which redundant functionality could be demonstrated (Verbitskiy, Zehrmann, Härtel, Brennicke, & Takenaka, 2012). The one or the other can be mutated without any

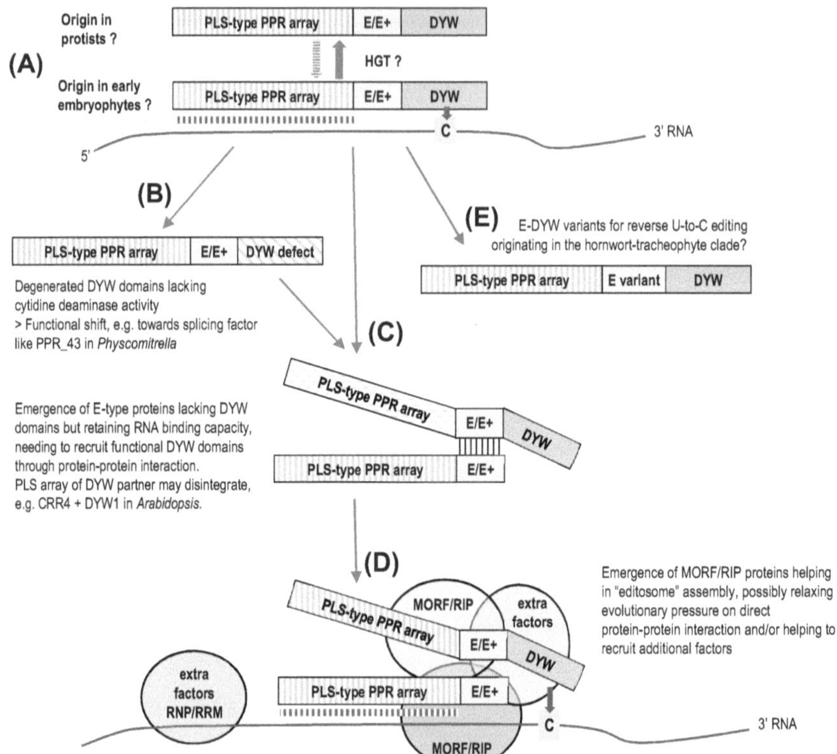

Figure 2 Hypothetical scenarios for the evolution of DYW-type PPR proteins acting as editing factors as discussed in the text. (A) DYW-type PPR proteins may have originated in land plants or in protists and horizontal gene transfer (HGT) in the one or the other direction may explain their current distribution in the living world. (B) Degeneration of the cytidine deaminase functionality in the DYW domain may cause functional shifts towards intron splicing, endonucleolytic cleavage or RNA stabilization. (C) Loss of DYW domains in E-type proteins characterized as editing factors may be explained by their ability to recruit DYW domains *in trans*. (D) Other than E− or DYW-type PPR proteins, MORF/RIP proteins play a key role in angiosperm RNA editing, likely by promoting the interactions of more than one E− or DYW-type protein on their RNA target. Further protein factors (like e.g. cpRNPs, ORRMs or OZs) may contribute either by separately binding to the target transcript and making the editing site accessible or by directly interacting in a complex editosome. (E) An alteration in the terminal E/E+/DYW domains may explain the later origin of reverse U-to-C RNA editing in some land plant clades. (See colour plate)

significant phenotype except reduced editing at two sites in *nad5* and *nad6*, whereas the double mutant plant is embryo lethal. This observation in combination with the particularly short PPR arrays of MEF8 and MEF8Ss with only five PPRs may suggest that they act to supply a DYW domain *in trans*, comparable to DYW1 in chloroplasts (Boussardon et al., 2012;

Chateigner-Boutin et al., 2013). However, the mechanisms of protein–protein interaction between E-type PPR proteins and DYW domains are unclear. The boundary of E and E+ domain features a highly conserved 15 amino acid long motif (PGCSxI/VEWxGxV/IHxF, Fig. 1). With that motif truncated or deleted, E- or DYW-type PPR proteins lose their ability to edit their corresponding sites (Chateigner-Boutin et al., 2013; Okuda, Myouga, Motohashi, Shinozaki, & Shikanai, 2007). The E/E+ domains have similarity with two tandem tetratricopeptide repeat (TPR) motifs (unpublished observations). TPRs are related to PPRs but mediate protein–protein instead of protein–RNA interaction (D'Andrea & Regan, 2003; Das, Cohen, & Barford, 1998; Karpenahalli, Lupas, & Soding, 2007; Small & Peeters, 2000). Hence, they may provide a surface for protein–protein interactions to attract DYW domains *in trans*. A recent study has shown that the E domains of plastid and mitochondrial PPR proteins are not functionally replaceable whereas the E domains of PPR proteins within one organelle are (Chateigner-Boutin et al., 2013). Hence, the editing apparatus may be differently adapted for protein–protein interactions in chloroplasts and mitochondria. Interestingly though, there is the recent example of the dual-targeted E-type protein AEF1 (atpF editing factor 1) involved in editing of both a plastid (atpFeU92PL) and a mitochondrial (nad5eU1580SL) editing site (Yap et al., 2015). It will be interesting to see whether the E domain of AEF1 will be shown to recruit different DYW domains *in trans* in the two organelles.

The recently suggested PPR-RNA recognition code (Barkan et al., 2012, Section 2), provides an additional argument supporting the idea that editing at some sites requires the interaction of different factors in *trans*. The PPR-RNA recognition code allows nucleotide sequence target prognoses based on two crucial amino acid residues (1′ and 6 positions, see Section 2 and Fig. 3) in each PPR. In some cases, however, the PPR proteins assigned to cognate editing sites do not match the upstream recognition sequences (Glass, Härtel, Zehrmann, Verbitskiy, & Takenaka, 2015; Hammani et al., 2009; Kim et al., 2009). These observations could be explained with the assigned PPR proteins serving as the DYW donors and not as the sequence recognition factors.

1.6 Yet Further Complicating Plant RNA Editing: Complex 'Editosomes'

Yet other findings make plant RNA editing even more complex, since factors other than PLS-type PPR proteins were likewise shown to influence editing,

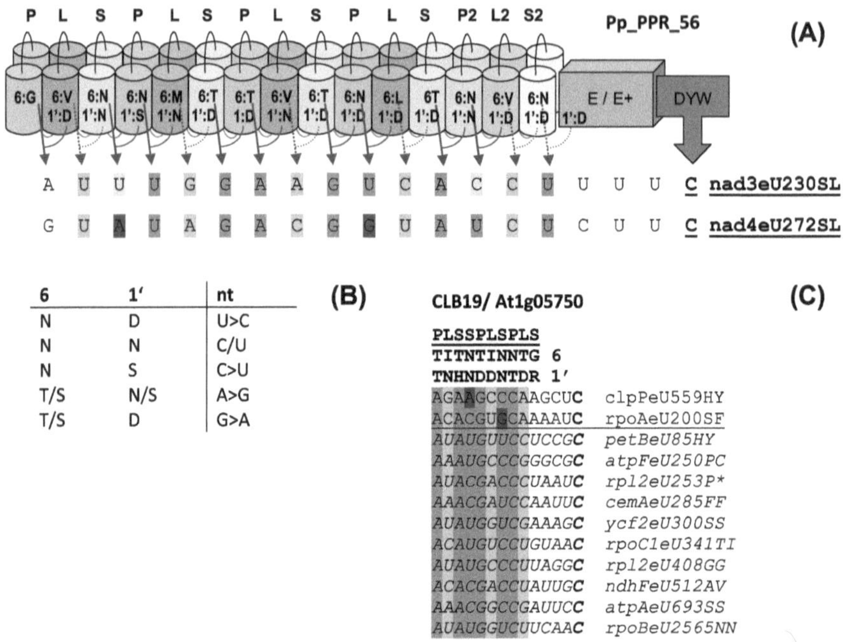

6	1′	nt
N	D	U>C
N	N	C/U
N	S	C>U
T/S	N/S	A>G
T/S	D	G>A

Figure 3 (A) Key amino acid positions 6 and 1′ of DYW-type PPR protein PPR_56 of *Physcomitrella patens* are aligned to the upstream target sequences of their assigned RNA editing sites nad3eU230SL and nad4eU272SL with the last S repeat opposite the 4th nucleotide upstream of the editing sites (C̲). Green shading indicates matches of amino acids in positions 6/1′ with nucleotides according to the rules described in (B) Mismatches are shaded in red. Pyrimidines (C, U) are less well distinguished than purines (A, G). Hence, pyrimidine transition mismatches are shaded in yellow. Nucleotides at positions assigned to L motifs are shaded in grey, as their role in recognition is minor regarding the rules proposed in Barkan et al. (2012). (B) Combinations of amino acids at positions 6 and 1′ in PPR motifs that specify binding to specific bases proposed in Barkan et al. (2012). (C) Alignment of E-type PPR protein CLB19 of *Arabidopsis thaliana* with the upstream target sequences of its assigned RNA editing sites clpPeU559HY and rpoAeU200SF and 10 putative target sequences in plastid coding sequences. Shading rules are listed in (A). (See colour plate)

at least in angiosperms. Chloroplast ribonucleoproteins (cpRNPs) are RNA-binding proteins containing RNA recognition motifs (RRMs) and are highly abundant in plastids. The nucleo-cytosolic RRM proteins were shown to dissolve secondary structures of particular RNAs (Dreyfuss, Kim, & Kataoka, 2002), the precise functioning of cpRNPs is largely unknown. At least one cpRNP, cp31, with a domain rich in acidic amino acid residues was shown in tobacco to be crucial for two editing sites in *ndhB* and *psbL*. Immunodepletion of the factor in in vitro experiments resulted in lack of editing (Hirose &

Sugiura, 2001). The disruption of cp31A, an ortholog of the tobacco protein in *A. thaliana*, reduced efficiency of editing at multiple sites but was not found to be crucial for individual editing sites (Tillich et al., 2009). These slightly different findings in the two angiosperms suggest that cp31 might influence editing more or less indirectly via improving accessibility of the key PPR editing factors to certain editing sites, maybe by dissolving secondary structures like their counterparts in the cytosol.

Other than the RNP/RRM proteins a new small protein gene family was discovered to influence editing both in plastids and mitochondria on a large scale. Two research groups named this protein family 'MORFs' (Multiple Organellar RNA editing Factors, Takenaka et al., 2012) or 'RIPs' (RNA editing factor Interacting Proteins, Bentolila et al., 2012), respectively. The *Arabidopsis* MORF/RIP gene family consists of 10 members targeted to mitochondria (5—6), plastids (2) or both (2). Disruption of individual MORF/RIP proteins leads to loss of editing at some sites and reduction of editing at others. MORF/RIP proteins form specific homo- and heteromeric protein—protein interactions (Zehrmann et al., 2015) and interact with PLS proteins identified as editing factors in a specific manner (Glass et al., 2015; Härtel et al., 2013). Their common feature is a conserved central domain of 100 amino acids, called the 'MORF box' or the 'RIP domain', respectively. This domain does not show any sequence similarity to other known protein domains, but might be essential for the interaction with PPR proteins. It seems likely that MORF/RIP proteins function as connector between E proteins and DYW type proteins, but this needs to be further evaluated. MORF/RIP proteins appear to be an evolutionary novelty arising in the seed plant (spermatophyte) lineage since unequivocal homologues can be identified only in available angiosperm and gymnosperm genome or transcriptome data.

Another protein, ORRM1, was likewise found to play an important role in editing of maize and *A. thaliana* plastids. The protein contains two truncated MORF/RIP boxes in combination with an RRM domain (Sun et al., 2013). Loss of that protein abolishes editing of diverse sites almost completely. Other RRM-domain containing proteins in mitochondria were likewise shown to be important for efficient editing and interaction of those with specific MORF/RIP proteins further supported their function in the mitochondrial editosome in *A. thaliana*. However, in contrast to ORRM1 they do not seem to interact with PPR proteins assigned to the same editing sites (Shi, Hanson, & Bentolila, 2015).

With a zinc finger containing protein (OZ1, organelle zinc finger 1) identified to influence the editing of most sites in chloroplasts, a member of a fourth protein family was recently described as being involved in the editing machinery of angiosperms (Sun et al., 2015). The organelle zinc finger editing factor family (OZ) contains four members targeted to organelles. OZ1 interacts with PPR proteins and ORRM1, but not with MORFs. The interaction of OZ1 with other editing factors suggests a direct participation in the editosome of chloroplasts. However, OZ1 as well as ORRMs might also influence editing indirectly like suggested for CP31.

Among all editing helper proteins, the OZ family is the only one with clear orthologs in the genomes of early branching land plants like the moss *P. patens* or the lycophyte *S. moellendorffii*. In contrast, MORF/RIPs are exclusively found in seed plants and ORRM1 orthologs even appear to be restricted to angiosperms (unpublished observations). This absence of those proteins outside of spermatophytes is particularly surprising, regarding the high number of editing sites in organelles of several early land plant lineages like hornworts (Duff, 2006; Duff & Moore, 2005), ferns (Vangerow et al., 1999; Wolf et al., 2004) or lycophytes (Grewe et al., 2011; Hecht et al., 2011; Oldenkott et al., 2014), which need to be targeted specifically (see Section 4). The *Selaginella* genome (Banks et al., 2011), to give one concrete example, encodes several hundred PLS proteins, but these would need to address the approximate 10-fold amount of RNA editing sites (c. 6000 altogether) in the two endosymbiotic organelles of *Selaginella* species.

2. PENTATRICOPEPTIDE REPEAT PROTEINS FOR RNA RECOGNITION: THE BINDING CODE

The seminal study reporting on PPR proteins already included a prediction of their structure based on the sequence similarity to TPR proteins (Small & Peeters, 2000). A first X-ray crystallographic structure analysis of TPR proteins was already available at that time (Das et al., 1998) and protein folding of PPR proteins was predicted to be similar to TPR protein folding. Individual repeats form pairs of antiparallel α-helices and the arrays of consecutive repeats form a superhelix enclosing a groove or tunnel. Accordingly, TPR and PPR proteins are defined as α-solenoid proteins (Kobe & Kajava, 2000). The main differences between the two repeat types are the amino acid residues in the central groove, which specifically bind to their interaction partners. In TPR proteins, which interact with other proteins, these residues vary depending on its target. In PPR proteins the groove was predicted to be hydrophilic with a positively charged bottom.

2.1 Pentatricopeptide Repeat Motifs Bind to RNA Specifically

In the beginning it was merely speculative that these characteristic differences would allow PPR proteins to specifically bind to RNA rather than to protein interaction partners (Small & Peeters, 2000). For several PPR proteins, however, such as the *Drosophila* bicoid stability factor, Radish p67 and Hcf152 of maize (Lahmy et al., 2000; Mancebo, Zhou, Shillinglaw, Henzel, & Macdonald, 2001; Meierhoff, Felder, Nakamura, Bechtold, & Schuster, 2003; Nakamura, Meierhoff, Westhoff, & Schuster, 2003) their binding to specific RNAs could be demonstrated quickly. All of these are 'P-type' PPR proteins consisting entirely of the canonical 35 aa PPRs. In the following years, evidence for direct interaction of RNAs and PPR proteins was mainly found using two different experimental setups: co-immunoprecipitation (IP) experiments with organellar extracts or classic in vitro binding assays with recombinant proteins (Barkan & Small, 2014). In contrast to IP experiments with the canonical P-type PPR proteins, those with PLS proteins, which are the key players in organelle editing, could not be performed successfully (Chateigner-Boutin et al., 2008). This may be due to different binding properties of the two PPR protein subtypes. The PLS-type PPR proteins functioning in RNA editing may bind only very transiently to their targets since a stable binding would hinder subsequent mRNA translation at the ribosome. P-type PPR proteins, in contrast, may bind to their targets more stably and with higher affinity. Their main tasks appear to be preventing transcripts from cleavage or secondary structure folding and most of them bind in noncoding regions (Barkan & Small, 2014). The detection of RNA footprints of P-type but not PLS-type PPR proteins in *A. thaliana* (Ruwe & Schmitz-Linneweber, 2012), barley and maize chloroplasts (Zhelyazkova et al., 2012) fully confirmed this assumption. CRR2 was the only PLS-type protein, for which a small RNA covering its putative target region, the *rps7-ndhB* spacer region, was detected (Ruwe & Schmitz-Linneweber, 2012). However, CRR2 is also unique as a PLS-type PPR protein with a DYW domain functioning in a process other than RNA editing (Hashimoto et al., 2003; Shikanai, 2015).

In vitro RNA-binding assays were successfully performed using recombinant PLS proteins fused to tags like maltose-binding protein, thioredoxin or glutathione S-transferase (Kindgren, Yap, Bond, & Small, 2015; Okuda, Nakamura, Sugita, Shimizu, & Shikanai, 2006; Tasaki, Hattori, & Sugita, 2010). For CRR4, the first editing factor identified (Kotera et al., 2005), binding to its target RNA 25 nucleotides upstream and 5 nucleotides downstream of ndhDeU2TM was shown by EMSA (electromobility shift assays) studies (Okuda et al., 2006). RNA binding of PLS-type DYW protein PPR_71 of *P. patens* to its target ccmFCeU122SF was similarly confirmed (Tasaki et al., 2010).

2.2 The RNA-Pentatricopeptide Repeat Recognition Code

The principle of RNA-PPR protein binding is similar to the sequence recognition code of two other α-solenoid protein groups, the TALE (transcription activator-like effector) and PUF (Pumilio and FBF homology protein) proteins (Filipovska & Rackham, 2012). TALEs are proteins secreted by *Xanthomonas* bacteria, which infect a broad range of flowering plants. The characteristic TAL repeats of their avirulence proteins bind to specific sequences of host DNA molecules (Boch et al., 2009; Moscou & Bogdanove, 2009). PUF proteins are involved in regulation of development via activation or repression of translation. The PUF domain consisting of eight 36-amino acid repeats specifically binds to nuclear RNA targets in an antiparallel manner. For both protein classes the nucleic acid recognition code was elucidated when researchers still brooded over the RNA recognition of PPR proteins (Boch et al., 2009; Cheong & Hall, 2006; Filipovska, Razif, Nygård, & Rackham, 2011; Moscou & Bogdanove, 2009). Individual amino acid residues at specific positions in each repeat were identified to be crucial for binding to individual nucleotides. This one-repeat-one-nucleotide binding feature turned out to be the key property for PPR proteins, too.

A coevolution analysis of a subset of PPR proteins and their RNA targets led to the identification of key amino acid positions in PPR motifs as likely critical for RNA binding (Fujii, Bond, & Small, 2011). A subset of P-type PPR proteins linked to cytoplasmic male sterility suppression in flowering plants is encoded by the so-called restorer of fertility (Rf) genes. Analysis of 212 genes from 13 different species suggested their rapid evolution under diversifying selection. Amino acid residues 1, 3 and 6 of their PPR motifs revealed highest probabilities of diversifying selection (i.e. highest rates of nonsynonymous over synonymous substitutions), suggesting their importance for RNA target recognition (Fig. 3). Their predicted position in the central groove of the protein likely facing the RNA ligand further supported their sequence-specific binding to their transcript targets.

Editing factors in *A. thaliana* chloroplasts affecting more than one target allowed insights on *cis*-element recognition (Hammani et al., 2009). Simple nucleotide identities of *cis* elements were insufficient to explain protein–RNA binding specificities. The inclusion of additional characteristics of RNA bases (double vs triple hydrogen bond-forming bases and purines vs pyrimidines), however, increased the specificity significantly. Consideration of the latter was confirmed by the recognition code elucidated 3 years later.

The RNA-binding surface of a P-type PPR protein HCF152 in maize chloroplasts was characterized with a series of experiments using mini-PPR proteins with two successive PPR motifs each for in vitro binding studies (Kobayashi et al., 2012). Mutagenesis experiments and structural modelling led to the identification of five amino acid positions involved in RNA binding. Besides positions 1, 3, 6 at that time already previously proposed to be crucial for nucleotide recognition (Fujii et al., 2011), positions 10 and 14 were also shown to interact with the RNA. Recognition of an individual nucleotide is provided by positions 3, 6, 10, 14 of one PPR motif and position 1 of the following PPR motif (hence called 1′) using the classification of PPR motifs based on PROSITE (Lurin et al., 2004). Note that the PPR motif based on Pfam modelling is shifted by two amino acids towards the C terminus. Accordingly, all amino acids relevant for binding to a specific nucleotide are included in one Pfam-modelled PPR motif unit (Kobayashi et al., 2012). However, this needs adaptation for the length-variable motifs of PLS-type PPRs since the position equivalent to 1′ would need to be labelled as position 30 rather than 34 in S-type motifs. To suggest a common label for the different PPR types, that position was alternatively labelled 'ii', as the position located two residues before the first amino acid of the next PPR motif (Kobayashi et al., 2012). Both models have since been used in parallel. Here we continue to use the PROSITE model numbering (Fig. 3).

Ultimately a code for PPR-RNA recognition in which positions 6 and 1′ play the key role was proposed (Barkan et al., 2012). This study was based on a combination of computational and experimental data on P-type PPR protein PPR10 of maize (Pfalz et al., 2009). The chloroplast-localized PPR10 protein contains 19 PPR motifs and binds at two different transcript targets with significant sequence similarity, *psaJ* and *atpH* (Prikryl, Rojas, Schuster, & Barkan, 2011). Correlating the 24 nt RNA footprint and the 17 core PPR motifs of PPR10, only positions 6 and 1′ showed strong co-occurrence of RNA bases and amino acid residues when the PPRs were aligned with the RNA sequences in colinear orientation. The recognition code of combined PPR positions 6 + 1′ could be defined to follow these rules: T(S) + N: A > G, T + D: G > A, N + S: C > U, N + D: U > C, N + N: U/C (Fig. 3B). Accordingly, as a first approximation a near-binary code resulted with T versus N in position 6 distinguishing purines (A/G) versus pyrimidines (C/U) and D versus N in position 1′ distinguishing keto bases (G/U) versus the amino bases (A/C). An experimental validation was performed by in vitro binding studies with modified PPR motifs and RNA targets. Gaps in the RNA target as well as in the PPR motif stretch that do not contribute to the above interactions are tolerated or even required to relax structural constraints. The simple two-amino-acid-positions-for-one-nucleotide code largely appeared to be valid for other P-type PPR proteins and their RNA targets, although ca. one-third of amino acid combinations at position 6/1′ could not be explained by that code concept. Moreover, the code concept could be applied to known PLS-type PPR proteins acting as editing factors recognizing different target sites when the ultimate carboxy-terminal PPR ('S2') motif is aligned with the fourth nucleotide upstream of the edited cytidine residue (position −4, Fig. 3A). However, the basic code was valid only for P- and S-type but neither for the L-type nor for the S2-type PPRs in the PPR arrays (Fig. 3A). It is possible that L-type PPRs function as spacers to reduce structural constraints or binding affinity of PLS-type PPR proteins rather than being involved in sequence-specific target recognition (Barkan et al., 2012; Barkan & Small, 2014).

Two further studies addressed the RNA recognition of PLS-type PPR proteins specifically. Yagi, Hayashi, Kobayashi, Hirayama, and Nakamura (2013) studied 24 proteins with altogether 327 PLS-type PPRs of *A. thaliana*, which were aligned with their target RNA in various configurations. They likewise identified PPR amino acid residues at position 6 and 1′ as highly essential, but also at position 3. Accordingly, a 3-amino-acid recognition code was proposed with position 3 increasing target assignment accuracy. Adding position 3 improved the distinct assignment of nucleotides for specific amino acid combinations (3/6/1′ − FTN: A, VTD: G, YND: U). In case of two asparagines (NN) at positions 6 and 1′ the amino acid at position 3 can even differentiate binding to a purine or a pyrimidine (FNN: G/A, INN: U/C). Furthermore, nucleotide recognition for at least a subset of the L-type PPRs was suggested, with e.g. proline at position 6 in L-type repeats preferentially recognizing U (FPD: U). Takenaka, Zehrmann, Brennicke, and Graichen (2013) suggested that not only L-type repeats, but also the L2- and S2-PPRs at the end of the PLS-type PPR arrays contribute to specific RNA recognition.

2.3 PPR Crystal Structure X-ray Analyses Confirm the RNA-Binding Code

The protein-only RNase P1 PRORP1 (Howard, Lim, Fierke, & Koutmos, 2012) of *A. thaliana* processing the 5′-end of precursor tRNAs in organelles and the mitochondrial RNA polymerase of humans mtRNAP (Ringel et al., 2011) were the first two PPR proteins for which a crystal structure could be obtained. PRORP1 contains 5.5 consecutive PPRs in combination with a central and a metallonuclease domain, whereas mtRNAP holds a unique helical domain with two PPR motifs. Both could neither be classified as P-type nor as PLS-type PPR proteins, rather representing two exceptional cases of proteins containing PPRs. Nevertheless, the crystal structure of their PPR stretches confirmed the prediction of structural folding of PPR proteins proposed previously (Delannoy, Stanley, Bond, & Small, 2007; Fujii et al., 2011; Small & Peeters, 2000). Further details about the recognition of RNA targets, however, were not disclosed in the two studies. The crystal structures of two further PPR proteins, both of which contain canonical P-type PPRs, were published recently. Whereas the five PPRs in the thylakoid assembly 8 like (THA8L) protein of *A. thaliana* were crystallized without its target RNA (Ban et al., 2013), the crystal structure of PPR-10 of maize was reported both in its RNA-free and RNA-bound state (Yin et al., 2013). The PPR-10 protein bound to an 18 nt long *psaJ* transcript fragment revealed conformational changes in contrast to the unbound formation. The basic recognition code proposed in Barkan et al. (2012) was confirmed in this study, although fewer sequence-specific protein–RNA contacts as proposed were indicated. The amino acid in position 3, most often large and hydrophobic, was shown to stack between adjacent bases, comparable to amino acid residues in PUF-RNA complexes (Barkan & Small, 2014). The PPR-10-RNA complex was present as an antiparallel homodimer. Similarly a dimerization of PPR protein THA8 of *Brachypodium distachyon* forming an asymmetric protein-RNA complex with the RNA molecule at the dimer interface was observed in crystal structure analysis (Ke et al., 2013). It was doubted, that the proteins effectively function as dimers under physiological conditions. N-terminal truncation, a point mutation beneficial for crystallogenesis and micromolar concentration of protein in solution were assumed to promote the dimerization of PPR-10 (Li et al., 2014). The interaction of PPR-10 with its second target *atpH*, however, was determined to be in a monomeric manner (Gully, Cowieson, et al., 2015; Yin et al., 2013). Gully and colleagues further presented a crystal structure of PPR-10 with a broad

RNA–protein interface confirming the PPR10-RNA recognition in Barkan et al. (2012).

Experiments to solve the crystal structure of a PLS-type protein, ideally with a carboxy-terminal DYW domain, are still pending. Its results would certainly, among other open points, also contribute to answer the still open questions of the role of the L motifs, the PLS2 triplet and the C-terminal domains.

2.4 Open Questions of Recognition and Functioning

Since publication of the PPR-RNA code, the assignment of editing factors to their target sequences largely confirmed the code predictions in several studies (Bentolila, Babina, Germain, & Hanson, 2013; Chateigner-Boutin et al., 2013; Schallenberg-Rüdinger, Kindgren, Zehrmann, Small, & Knoop, 2013; Wagoner et al., 2015; Yap et al., 2015). However, in numerous cases, better matches to the determined targets could be expected. CLB19 of *A. thaliana*, for example, is assigned to chloroplast editing events clpPeU559HY and rpoAeU200SF (Chateigner-Boutin et al., 2008) but the corresponding alignments show only five matches (excluding the L motifs and S2) and one mismatch for each target, respectively (Kindgren et al., 2015, Fig. 3C). The CLB19 protein was the main subject in one of two recent studies intensively testing the proposed codes experimentally (Kindgren et al., 2015; Okuda et al., 2014). Both studies confirmed the code proposed by Barkan et al. (2012), including the assumption that residues at positions 6 and 1′ of L, L2 and S2 motifs do not affect the RNA binding affinity, at least in in vitro binding assays. As already shown for P-type PPR proteins (Barkan et al., 2012; Fujii, Sato, & Shikanai, 2013), only some PPRs in a full array may be essential for binding to a specific target. This goes well with the presence of editing factors recognizing editing sites with diverse target sequences (eg, Glass et al., 2015; Hammani et al., 2009; Härtel et al., 2013; Okuda & Shikanai, 2012). At least for editing factor CRR22, the alternative explanation that the editing factor functions as DYW domain donor rather than a specificity factor, can be ruled out, as CRR22 specifically binds to its diverse target sequences in vitro (Okuda & Shikanai, 2012). Conclusions that a few motifs are sufficient to bind to RNA targets and mismatches are tolerated to some degree (Kindgren et al., 2015) seem to be in stark contrast to accurate editing of specific cytidines in plant organelles. In *P. patens* mitochondria, for example, a putative sequence with a C at a position to be edited (cobeU930FF) would match PPR_77 even better than the assigned editing sites cox2eU370RW and cox3eU733RW considering the rules of the proposed code, but the cytidine is not edited *in planta* (Schallenberg-Rüdinger, Kindgren, et al., 2013). Putative binding sequences for CLB19 could likewise be identified in the *A. thaliana* chloroplast coding sequences (unpublished observations), but the cytidines downstream of the matching targets are not edited either (Bentolila, Oh, et al., 2013, Fig. 3C). The question remains, how specificity is provided and contribution of other amino acids, e.g. at position 3, 10 or 14 or even yet unrecognized positions might be part of the answer. In particular, the relevance of the L-type PPRs still remains puzzling. Their strong conservation in selected PLS protein orthologues along angiosperm evolution, eg, in CRR28 and RARE1 (Hein, Polsakiewicz, & Knoop, 2016) and also in the related mosses *Funaria* and *Physcomitrella* (Rüdinger, Szövényi, et al., 2011) again supports their as yet unknown, but important functionality.

Another as yet open question is the function of nucleotides surrounding the editing site. Experiments with in vitro editing systems already revealed that the immediate upstream nucleotides (−5 to −1) are not essential for binding to *trans*-factors (Miyamoto, Obokata, & Sugiura, 2002, 2004) but have an impact on the C-to-U nucleotide conversion. Most of the cytidines to be edited have a bias towards a pyrimidine upstream and a purine downstream (Giegé & Brennicke, 1999; Hirose, Kusumegi, Tsudzuki, & Sugiura, 1999; Tillich, Lehwark, Morton, & Maier, 2006). A study on the mitochondrial *cox*2 mRNAs in wheat showed that the +1/−1 dependency of RNA editing depends on the editing site itself (Castandet & Araya, 2011). Some sites are not edited when the −1 nucleotide is substituted (Bock, Hermann, & Fuchs, 1997; Bock et al., 1996; Miyamoto et al., 2004). A binding of the DYW domain itself up to position −3 upstream of the editing site was recently shown by in vitro binding assays (Okuda et al., 2014). It should indeed be expected that editing factors have a stronger preference for transcript targets in the unedited rather than the edited state. Such a preference has been seen for *P. patens* DYW-type PPR_71 showing higher affinity to the unedited than to the edited target (Tasaki et al., 2010).

The correct positioning of the putative catalytic DYW domain might be facilitated by the E domain, whose function is still unclear. An early study already revealed that the distance of editing site and 5′ upstream region is critical, as inclusion of additional cytidines led to editing of the incorrect position (Hermann & Bock, 1999). The E domain, not binding to the RNA but crucial for the editing functionality, might function as a spacer. In case of E proteins lacking the DYW domain, the E domain with its TPR-like structure might recruit heterologous DYW domains *in trans* either directly or via helper proteins, for example MORFs (Fig. 2). In those cases, the −1/+1 region might be not as relevant as for editing sites addressed exclusively by complete DYW-type PPR proteins.

The function of the carboxy-terminal PLS2 triplet of editing-type PPR proteins, which significantly differs in its sequence conservation from the upstream PLS motifs, is still enigmatic. On the one hand, it could contribute to nucleotide recognition as recently proposed (Takenaka, Zehrmann, Brennicke, et al., 2013). On the other hand, its unique conservation, its proximity to the C terminal domains and its dispensability for high-affinity binding (Okuda et al., 2014) rather suggests its contribution to the correct folding and positioning of the catalytic unit.

3. *PHYSCOMITRELLA PATENS*: A PARTICULARLY INTERESTING MODEL TO STUDY RNA EDITING

The advantageous features of the moss *P. patens* making it a key model organism are addressed in other chapters of this volume (e.g. the one by D. Lang and colleagues) and will not be detailed here. We will here focus on its role as a model organism to study RNA editing.

3.1 RNA Editing and Bryophytes

From an evolutionary perspective, research on bryophytes (liverworts, mosses and hornworts), representing the earliest emerging land plant lineages, are of particular interest to reconstruct the ancient evolutionary changes allowing the transition from water to land in the green plant lineage (Qiu et al., 2006; Wickett et al., 2014). With the colonization of land, the streptophyte algal ancestors of land plants had to adapt to extreme environmental changes, such as reduced water and nutrient availability, variations in temperature and increased exposure to radiation. Surprisingly, the origin of plant-type C-to-U RNA editing also appears to coincide with the water-to-land transition some 500 million years ago (Fujii & Small, 2011). It is intriguing to look for a functional role of the organelle deamination type of RNA editing in this transition, e.g. a possible repair mechanism for mutations or a response for increased organelle GC content (Malek et al., 1996; Smith, 2009). However, convincing evidence is lacking as yet. Equally well it may turn out that a novel deamination activity resulting from a demand for metabolic changes accidently operating on polyribonucleotides got locked in evolution once T-to-C mutations became fixed in the organelle genomes.

3.2 Funariaceae Feature Low Number of Editing Sites and Editing Factors

Physcomitrella patens drew attention as a model organism to study RNA editing given its low number of RNA editing sites in the first mitochondrial transcripts investigated, *nad5* and *nad2*, encoding for subunits of the NADH dehydrogenase (Beckert, Muhle, Pruchner, & Knoop, 2001; Steinhauser et al., 1999).

With the early availability of the *Physcomitrella* plastid genome (Sugiura et al., 2003), two editing sites were detected in plastid transcripts. The first site rps14cpeU2TM generates the start codon of *rps14* and was shown to be partially edited (Miyata et al., 2002). Additionally, rps14cpeU-1p5, a site one nucleotide upstream of the ACG start codon of *rps14* is only edited in approximately 5% of the transcripts. No further editing sites were identified in the plastid genome in later studies.

After sequencing of the mitochondrial genome (Terasawa et al., 2006) a similar picture of only rare RNA editing events emerged for the mitochondrial transcripts. In only 7 of 42 protein encoding genes a total of 11 editing sites were detected (Table 1), all of them in coding regions (Rüdinger et al., 2009). The low number of organellar editing sites in *Physcomitrella* promised a simple system of recognition by cofactors, easier to survey and to elucidate. Indeed, with the complete nuclear genome sequence of *Physcomitrella* available (Rensing et al., 2008) the number of only 16 PLS proteins identified turned out to be much lower than in other land plants. The PLS-type subfamily of PPR protein genes in *P. patens* consists of 10 complete DYW-type PPR proteins and 6 'pure' PLS-type proteins with P, L and S motifs only (O'Toole et al., 2008). Whereas in angiosperms the low number of introns in PPR protein genes was striking, in *P. patens* two-thirds of PPR genes turned out to contain introns. In the PLS subfamily, 9 of the 10 DYW-type PPR proteins contain 1–4 introns in the upstream region encoding the amino-terminal organelle targeting peptide. Two of the genes, PPR_98 (Phypa_224041) and PPR_65 (Phypa_175641) also have introns in their PPR arrays. The *Physcomitrella* PPR proteins are named after their gene number in the *P. patens* genome version v1.2 (www.cosmoss.de) and alternatively simply based on serial numbering of the 105 PPR proteins encoded in the *P. patens* genome (Table 1, O'Toole et al., 2008). Another feature setting the *P. patens* PLS proteins apart from their angiosperm counterparts is the length of their PPR arrays. Whereas in flowering plants like rice or *A. thaliana* the average number of PPRs per protein is 13–14, in

P. patens the editing factors contain 20 PPRs on average (Table 1, Takenaka et al., 2014).

3.3 The First Complete Assignment of DYW-Type PPR Proteins and Editing Sites in a Model Organism

The first RNA editing factor characterized in a non-angiosperm was PPR_71 (Phypa_181369) shown to be involved in the editing of site ccmFCeU122SF in *ccmFC*, encoding for a component of the cytochrome c biogenesis complex (Tasaki et al., 2010; Tasaki & Sugita, 2010). The loss-of-function mutant displayed poor protonema growth, a phenotype explainable with impaired cytochrome biogenesis. The binding of PPR_71 to the editing site environment (40 bp upstream and 6 bp downstream) of the assigned editing site could be verified by electromobility shift assays (Tasaki et al., 2010; Tasaki & Sugita, 2010). This was of main interest with respect to a second editing site in close proximity (19 nt upstream) of the assigned one. EMSA studies revealed a slightly better binding to the target sequence, when the upstream editing site is processed from C to U (Tasaki et al., 2010). This finding was in line with the later assignment of the editing factor PPR_65 (Phypa_175641) to the upstream editing site in *ccmFC* (ccmFCeU103PS). The PPR_65 KO plant lines showed abolished editing at both sites, although EMSA studies and PPR-RNA recognition code assignment clearly demonstrated that PPR_65 recognizes the upstream site exclusively (Ichinose, Sugita, Yagi, Nakamura, & Sugita, 2013; Schallenberg-Rüdinger, Kindgren, et al., 2013). Although a weak protein–protein interaction of PPR_65 and PPR_71 was detected in yeast two-hybrid experiments, changes in RNA secondary structure depending on the upstream editing site and increasing accessibility of the downstream site were more likely to contribute to the observation. The phenotype of the PPR_65 KO *Physcomitrella* plants was dramatic with development strongly retarded (Schallenberg-Rüdinger, Kindgren, et al., 2013). KO plants with disrupted DYW type PPR proteins PPR_77 (Phypa_191892) and PPR_91 (Phypa_211228), respectively, were likewise impaired in growth and shown to be involved in RNA editing of sites cox2eU370RW and cox3eU733RW (*cox* genes encode subunits of the cytochrome oxidase) and site nad5eU730RW, respectively (Ohtani et al., 2010). Editing of a second editing site in *nad5*, nad5eU598RC, does not seem to be equally important for the functionality of NADH dehydrogenase in mitochondria. Under normal growth conditions, the corresponding KO plant with editing factor PPR_79 (Phypa_192787) disrupted does not show a strong macroscopic phenotype (Uchida, Ohtani, Ichinose, Sugita,

& Sugita, 2011). A PPR_56 (Phypa_169414) knockout likewise causes no strong phenotype although editing of two sites, nad3eU230SL and nad4eU272SL, is defective in those plants (Table 1, Ohtani et al., 2010).

PPR_78 (Phypa_192620), was identfied as the editing factor for two editing sites, cox1eU755SL and rps14eU137SL in the *rps14* transcript encoding protein 14 of the small ribosomal subunit. The corresponding PPR_78 KO mutant did not show growth retardation under laboratory growth conditions (Table 1, Rüdinger, Szövényi, et al., 2011). This case is particularly surprising given the high conservation of amino acid residues affected by editing along evolution. Mild or even no macroscopic phenotypes were likewise observed for diverse editing mutants of *A. thaliana* affecting important mitochondrial and chloroplast genes (Chateigner-Boutin et al., 2013; Okuda et al., 2010; Takenaka, 2010; Takenaka, Verbitskiy, Zehrmann, & Brennicke, 2010). Whether these mutants show retardations under certain environmental or developmental conditions is an open question to be answered.

For PPR_98 (Phypa_224041) and the plastid-targeted PPR_45 (Phypa_156474) no knockout lines could be generated suggesting that these proteins are indispensable. A knockdown approach and EMSA-binding studies (Ichinose et al., 2013; Schallenberg-Rüdinger, Kindgren, et al., 2013) showed that PPR_98 addresses the remaining mitochondrial editing site atp9eU92SL and that PPR_45 targets the chloroplast editing site rps14cpeU2TM accompanied by upstream editing rps14cpeU-1p5 (Ichinose, Uchida, & Sugita, 2014), respectively. With these two DYW-type PPR proteins linked to their cognate editing sites, *P. patens* eventually became the first land plant with a complete set of editing factors and editing sites mutually assigned (Table 1). The meanwhile proposed RNA-PPR recognition code (Barkan et al., 2012; Yagi et al., 2013) confirms the assignments of mitochondrial editing factors and editing sites. However, PPR_77 matches the upstream recognition sequence only when shifted one nucleotide upstream (Schallenberg-Rüdinger, Kindgren, et al., 2013).

3.4 Further *Physcomitrella* PLS-Type Pentatricopeptide Repeats: Beyond Editing

One DYW type PPR protein in *Physcomitrella*, PPR_43 (Phypa_154890), could not be assigned to an RNA editing site. The mitochondrially localized PPR_43 protein was conspicuous in carrying a less conserved C terminal domain with indels in the DYW domain (Rüdinger, Szövényi, et al., 2011). PPR_43 KO analysis eventually characterized this factor as being required for splicing of the group II intron cox1i732g2 of cox1 pre-mRNA instead of RNA editing (Ichinose et al., 2012). Complementation studies with truncated versions of the PPR_43 gene showed that the E and the DYW domain are not essential to reconstitute splicing functionality. This result may suggest that PPR_43 may originally have been an editing factor, which has adapted to a new function as a splicing factor once its cognate editing site was lost through conversion to a thymidine (Fig. 2C). We currently explore this idea investigating a candidate editing site conserved in another moss that could be target of an ancestral PPR_43 version acting as an editing factor for that site.

Other than *Physcomitrella* PPR_43 only few PLS-type PPR proteins were characterized as functioning in other processes than RNA editing: as described, CRR2, identified as an RNA cleavage factor in *A. thaliana* chloroplasts (Hashimoto et al., 2003), OTP70, an E protein likewise promoting RNA splicing in plastid *rpoC1* transcripts in *A. thaliana* (Chateigner-Boutin et al., 2011) and another E-type protein, Rf1, linked to CMS (cytoplasmatic male sterility) fertility restoration in *Sorghum bicolor* (Klein et al., 2005).

In *P. patens* the function of the six 'pure' PLS proteins remains to be investigated. As the assignment of editing sites to editing factors is already completed, they supposedly function in other processes than editing. One of the six PLS proteins was recently found to be derived from a former DYW-type PPR protein. PPR_105 contains a truncated E domain disconnected from a DYW domain in the 3′ region (K. Ullrich, M. Schallenberg-Rüdinger, unpublished). PPR_105 is predicted to be targeted to chloroplasts (Sugita et al., 2013) and phylogenetic analysis places it closest related to plastid-targeted PPR_45 (unpublished observations). Microarray data reveal that the protein is moderately expressed under diverse growth conditions (Hiss et al., 2014).

3.5 RNA Editing and Its Cofactors: From *Physcomitrella* to Other Taxa

A comparative study with the related moss *F. hygrometrica* identified 9 clear orthologs of the 10 DYW-type PPR proteins of *P. patens* (Table 1). *Physcomitrella* PPR_56, however, could neither be identified in the high throughput RNA-seq data (Rüdinger, Szövényi, et al., 2011) of *Funaria* nor in the genomic data that became available later (P. Soevenij, M. Schallenberg-Rüdinger, unpublished). The absence of PPR_56 excellently matches the lack of its assigned editing sites nad3eU230SL and nad4eU272SL in *Funaria*, where thymidines are present in the mitochondrial genome at the corresponding positions (Table 1). A survey including 12 recently sequenced moss mitochondrial genomes of phylogenetically wide sampling (Liu, Medina, & Goffinet, 2014) suggests these two editing sites to be only recently emerged. They are conserved in some *Physcomitrium–Phycomitrella* species including *P. patens* subspecies *patens* and in the close relative *Aphanorrhegma serratum* but neither in the funariid moss *F. hygrometrica* nor in other moss families investigated (Beike et al., 2014). Gain and loss scenarios of the two editing sites within the *Physcomitrium–Phycomitrella* species complex support independent speciation of *Physcomitrella* and hybridization of *Physcomitrium* (Beike et al., 2014). Phylogenetic analysis displaying the relationships of *Funaria* and *Physcomitrella* DYW protein orthologs suggested PPR_56 to be possibly originating from PPR_45 (Rüdinger, Szövényi, et al., 2011).

Editing site rps14eU137SL is likewise lost in *F. hygrometrica*, but a clear ortholog of PPR_78 is encoded and expressed. This is no surprise considering that editing site cox1eU755SL, also assigned to PPR_78, is retained in *Funaria* (Table 1). Editing site rps14eU137SL appears to be an ancient editing site found in nearly all mosses except *Buxbaumia* and only recently lost in *Funaria* (unpublished observations). This might also be the reason, why the *Funaria* ortholog can efficiently complement the *Physcomitrella* PPR_78 KO plant with a regain of RNA editing at both sites (M. Schallenberg-Rüdinger, M. Hiss, B. Oldenkott, P. LeTrinh, V. Knoop, S. Rensing, unpublished). In angiosperms, only one pair of functional orthologs is characterized so far: DYW type PPR protein MEF29 in *A. thaliana* and PPR2263 in maize (Sosso et al., 2012). Both proteins edit the same sites, nad5eU1550TI and cobeU908PL. Complementation of mutant plants with the ortholog of the respective other species was not yet tested. Complementation assays of *A. thaliana* editing mutants with putative orthologs of other species failed to reconstitute editing in the case of MEF3 from *Vitis vinifera* (Verbitskiy, van der Merwe, Zehrmann, Härtel, & Takenaka, 2012) or resulted in less efficient editing, e.g. in the case of *Nicotiana* CRR4 orthologs (Okuda et al., 2009).

Physcomitrella patens editing sites are very differently conserved along moss phylogeny. In contrast to nad3eU230SL and nad4eU272SL which are only recently evolved and rps14eU137SL as an ancient and highly conserved editing site, cox1eU755SL seems to be lost and regained several times in moss evolution (unpublished observations, Table 1). The high flexibility in the latter case may be due to specificity factor PPR_78 simultaneously serving rps14eU137SL and therefore relaxing pyrimidine transitions but experiencing higher evolutionary pressure as long as one site needs editing.

4. EVOLUTION OF RNA EDITING: GAIN, LOSS AND ADAPTATION SCENARIOS IN LAND PLANT PHYLOGENY

4.1 C-to-U Editing: Gained Once, Lost Once

No evidence for plant-type organelle C-to-U RNA editing has as yet been found in algae, including green streptophyte algae supposedly related to the land plant lineage. Accordingly, the origin of organelle C-to-U RNA editing appears to be intimately connected to the origin of land plants (Fig. 4). Although it remains intriguing to speculate on the emergence of RNA editing in a functional relation to the alga—embryophyte transition at the origin of land plants, no convincing point has yet been made in support for such a scenario. The absence of RNA editing in the organelles of the model liverwort *M. polymorpha* (Oda et al., 1992; Ohyama et al., 1986) has initially been confusing, particularly after documentation of chloroplast and mitochondrial RNA editing in other bryophytes including liverworts (Freyer et al., 1997; Malek et al., 1996). All available evidence now supports the absence of C-to-U RNA editing in the entire clade of complex-thalloid, 'marchantiid' liverworts (Groth-Malonek et al., 2007; Rüdinger et al., 2012), likely a unique complete loss of C-to-U editing among land plants (Fig. 4). Notably, the earliest branching extant lineages both within the mosses (genus *Takakia*) and within the liverworts (genus *Haplomitrium*)

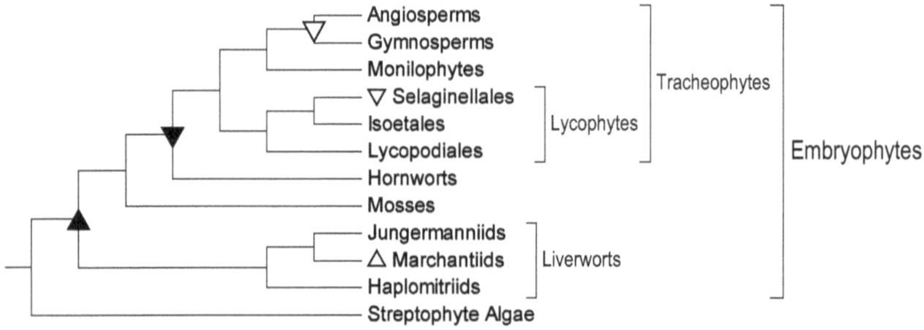

Figure 4 C-to-U organelle RNA editing is gained (*solid upward triangle*) with the origin of land plants (embryophytes) and retained in all plant clades except the marchantiid, complex-thalloid liverworts (*empty upward triangle*). Reverse U-to-C editing (*solid downward triangle*) appears to be gained as a molecular synapomorphy in the hornwort-tracheophyte clade and lost in the seed plant lineage and in Selaginellales (*open downward triangles*, phylogeny based on Qiu et al., 2006).

show particularly abundant RNA editing (Groth-Malonek et al., 2007; Sugita et al., 2006). Very similar observations have been made for flowering plants with significantly more RNA editing in early branching than in later emerging lineages (Fujii & Small, 2011; Hein et al., 2016; Mower, 2008; Rice et al., 2013; Richardson, Rice, Young, Alverson, & Palmer, 2013). In contrast, RNA editing varies widely among monilophytes (i.e. ferns *sensu lato*) and among lycophytes and rather increases than decreases with evolutionary diversification in these two clades (Guo, Grewe, & Mower, 2015). Hence, there is neither a gradual gain nor a gradual overall loss of editing sites along the course of plant evolution, but evolutionary trends depend on the specific plant clades under consideration.

Similarly, no general rules can be deduced for mitochondrial versus chloroplast RNA editing frequencies. As a rule of thumb, mitochondrial often exceeds chloroplast RNA editing by a factor of about 10. However, this does not hold true anymore once extremely high RNA editing frequencies are reached. For example, chloroplast RNA editing may even exceed mitochondrial RNA editing at ca. 3500 versus 2500 sites of C-to-U editing in the lycophyte genus *Selaginella* (Hecht et al., 2011; Oldenkott et al., 2014).

The numbers of DYW-type PPR proteins as the key factors of RNA editing site recognition increase with the numbers of editing sites in the two organelles (Rüdinger et al., 2012; Salone et al., 2007). Screening the currently available genome and transcriptome data at the NCBI or the OneKP project we generally find this trend well confirmed (unpublished observations). A one-to-one correspondence of editing cofactors and editing sites is not supposed to exist. Even in the simple, and now fully explored, RNA editing setup of *P. patens*, nine DYW-type proteins address altogether 13 organellar RNA editing sites given that four of them affect two editing events each (see Section 3). Several factors likely contribute to the fact that numbers of editing sites grow faster than the numbers of the nuclear-encoded editing factors. First, the number of editing sites affected by a single factor may be as high as seven as demonstrated for SLO2 (Zhu et al., 2012) and OGR1 (Kim et al., 2009) or even eight as shown for MEF13 (Glass et al., 2015), and possibly yet more for others. Second, many more nuclear-encoded factors may be dual targeted to chloroplasts and mitochondria as recently demonstrated for AEF1 (Yap et al., 2015). Third, different combinatorial effects based on the interactions of individual site-specific factors may play a role, particularly in seed plants where MORF/RIP proteins contribute to protein–protein interactions (Fig. 2D).

4.2 C-to-U Editing and DYW Proteins Also Outside of Land Plants

The intriguing correlation of DYW-type PPR proteins and organelle RNA editing turned out to be valid outside of the plant kingdom, too. The initial discovery of 10 DYW-type PPR proteins in the heterolobosean protist *Naegleria gruberi* (Knoop & Rüdinger, 2010) was followed by the identification of two mitochondrial RNA editing events at predicted positions (Rüdinger, Fritz-Laylin, Polsakiewicz, & Knoop, 2011). Hence, in this case, the number of DYW proteins exceeds the number of organelle editing sites by a factor of five, which leaves the function of most of them an open question. Genes for 16 DYW-type PPR proteins are similarly present in the slime mould (Amoebozoa) *Physarum polycephalum* (Schaap et al., 2015; Schallenberg-Rüdinger, Lenz, Polsakiewicz, Gott, & Knoop, 2013). The slime mould is notorious for its numerous types of RNA editing affecting mitochondrial transcripts (Bundschuh, Altmüller, Becker, Nürnberg, & Gott, 2011; Mahendran, Spottswood, & Miller, 1991). Only four events of C-to-U editing have been identified, however, while the C insertion type of RNA editing is vastly dominating in *Physarum* (Bundschuh et al., 2011). Like in *Naegleria*, it remains unclear if and how the DYW proteins outnumbering the C-to-U editing sites have adapted novel functionalities. Finally, the discovery of DYW domains in the enigmatic and isolated protist *Malawimonas jakobiformis* (Schallenberg-Rüdinger, Lenz, et al., 2013) suggested closer inspection of its mitochondrial transcriptome. RNA editing at expected sites could indeed be confirmed (V. Knoop, B.F. Lang, unpublished). Land plants, *Naegleria, Physarum* and *Malawimonas* occupy very distant branches in the evolution of eukaryotes and speculations on horizontal gene transfer of DYW-type PPR protein genes are obvious (Fig. 2A). However, a close physical interaction in nature is evident only for *Physarum* growing on decaying plant substrates and no evidence is found for evolutionary recent HGT between the four taxa. Consequently, it remains fully unclear at present to which degree horizontal gene transfers have contributed to the current distribution of DYW-type PPR protein genes and organelle RNA editing in nature. If any, HGT from living or decaying plant material to and among protists seems to be a more likely explanation than in the opposite direction (Fig. 2A).

So far, only a single DYW domain signature has been found in the streptophyte algal genus *Nitella* (Schallenberg-Rüdinger, Lenz, et al., 2013). This finding may certainly raise the possibility of an early evolutionary emergence of the DYW domain and/or DYW-type PPR proteins already in the algal ancestors of land plants. Further genome and transcriptome data will be highly valuable to further check upon this possibility.

4.3 Functional Shifts: Degenerating and Lost DYW Domains in PLS-Type PPR Proteins

Although no biochemical proof for a cytidine deaminase activity acting on polyribonucleotides is available for the DYW domain, its strong correlation with C-to-U editing inside and outside of plants and its zinc-binding cytidine deaminase features (Boussardon et al., 2014; Hayes et al., 2015, 2013; Hegeman et al., 2005; Salone et al., 2007) make it the prime candidate for the cytidine deaminating activity in RNAs. Evolutionarily puzzling, however, is its (near-) exclusive presence only in connection to PLS-type PPR repeats linked via the E/E+ domain. No evidence is found for originally free-standing DYW domains or DYW domains linked to other protein domains, including the canonical P-type PPR proteins that have an evolutionary wider distribution than PLS-type PPR proteins (Schallenberg-Rüdinger, Lenz, et al., 2013). As such, the origin of DYW domain and PLS-type PPR repeats appears to be intimately linked in evolution (Fig. 2A). PLS-type PPR proteins lacking a DYW domain ('E-type') or both the DYW and the preceding E/E+ domains appear to be evolutionary reductions in the course of plant evolution rather than evolutionarily ancestral (Fig. 2C and D). This is well supported by the small gene families in the protists only featuring DYW-type PLS proteins. Further support for the DYW domain as the cytidine deamination activity comes from examples of proteins with degenerated DYW domains having functionalities other than C-to-U editing (Fig. 2B, Section 3.4).

Apparently contradictory, however, are the cases of E-type proteins lacking a DYW domain, for which the first editing factor identified, CRR4, is a case in point. However, the editing activity of CRR4 was shown to rely on the orphan DYW1 protein in *Arabidopsis* lacking upstream PPR repeats (Boussardon et al., 2012). Artificial CRR4-DYW1 fusions were shown to reconstitute RNA editing in the respective *Arabidopsis* mutants. Similarly it may be likely that all other E-type proteins lacking a DYW domain and identified as specific RNA editing factors may merely serve in site-specific RNA recognition but recruit a DYW-type PPR protein *in trans* (Fig. 2C and D). The necessary protein—protein interaction may either be mediated directly between E-type and DYW-type proteins (possibly via their E domains) or with the help of MORF/RIP proteins (Fig. 2D). The former possibility may be dominating in editing-rich-taxa lacking MORF/RIP domains outside of seed plants such as lycophytes, hornworts and ferns, the latter may be dominating in spermatophytes.

4.4 Editing in Reverse: U-to-C in Three Nonspermatophyte Clades of Plants

The near-ubiquitous presence of C-to-U RNA editing in plant organelles is accompanied by the reverse process of U-to-C RNA editing in hornworts (Duff, 2006; Kugita, Yamamoto, Fujikawa, Matsumoto, & Yoshinaga, 2003; Steinhauser et al., 1999), in ferns (Vangerow et al., 1999; Wolf et al., 2004) and at least some lycophytes (Grewe et al., 2011). The presence of U-to-C editing has remained vague in the other land plant groups. No evidence of U-to-C editing has been reported for liverworts and mosses (Rüdinger et al., 2012) or in gymnosperms. Reinvestigating previous claims of U-to-C editing in angiosperms we now suggest that U-to-C RNA editing only originated in the common ancestor of hornworts and tracheophytes and was lost altogether in the seed plant (spermatophyte) lineage (Fig. 4; P. Gerke, N. Knie, A. Hein, M. Schallenberg-Rüdinger, V. Knoop, unpublished). This scenario would suggest reverse U-to-C editing to be a molecular synapomorphy in support of a joint hornwort-tracheophyte (HT) clade (Groth-Malonek, Pruchner, Grewe, & Knoop, 2005; Qiu et al., 2006). Necessarily, this evolutionary gain of a uridine amination or transamination activity should be accompanied by gain of corresponding co-factors specifically addressing sites of U-to-C editing. Given its later appearance subsequent to the origin of C-to-U editing, we speculate that reverse U-to-C editing may essentially be a modification of the former, biochemically inverting the direction of pyrimidine exchange, but using similar mechanisms of transcript site recognition. Indeed, we now hypothesize that alteration of a highly conserved motif within the E domain of certain DYW-type protein variants exclusively present in reverse editing taxa may be at the origin of reverse U-to-C editing (Fig. 2B; P. Gerke, N. Knie, A. Hein, M. Schallenberg-Rüdinger, V. Knoop, unpublished).

In this context, it is noteworthy to find that C-to-U editing is more widespread in nature than the reverse U-to-C editing. Liberating the amino group of cytidine for conversion is biochemically simpler than the reverse reaction which requires an appropriate amino group donor. This may explain why reverse editing occurs only in a subset of taxa also showing C-to-U editing and it will be interesting to see whether reverse editing may be introduced into taxa exclusively showing C-to-U editing by recombinant proteins featuring the divergent E domains present in hornworts and ferns.

4.5 RNA Editing – Why Dominating in mRNAs?

In all cases of "plant-type" organelle RNA editing, the numbers of C-to-U (and U-to-C) conversions in the exons of mRNAs strongly dominate over the more exceptional cases of RNA editing in introns, UTRs, rRNAs or tRNAs. Exceptionally higher editing in the latter RNA types is found where overall RNA editing frequencies are high, as for example documented by frequent mitochondrial tRNA editing in *I. engelmannii* (Grewe et al., 2011, 2009) or mitochondrial rRNA editing in *S. moellendorffii* (Hecht et al., 2011). We now also find exceptionally high RNA editing in chloroplast group II introns of *S. uncinata* (B. Oldenkott, N. Knie, V. Knoop, unpublished) which goes along with its record amount of mRNA exon editing (Oldenkott et al., 2014). The dominance of organelle RNA editing in mRNA exon regions over introns, tRNAs and rRNAs is likely due to the stable secondary structures mostly formed by the latter. Newly evolving editing factors or pre-existing ones changing their specificity more likely find new targets in unfolded RNAs with fully accessible nucleotides rather than in structured RNAs, which could only be accessed early upon transcription before base-pairs are formed or upon melting of secondary structures by helper proteins.

4.6 The Coevolution of RNA Editing Sites and Their Cofactors

Evidently, any gain of a novel, essential RNA editing site must be proceeded or at least be immediately accompanied by gain of an appropriate editing activity. This activity may either arise through functional shift or extended target specificity of an existing editing factor or by de novo creation of a new editing factor paralogue through gene duplication. Conversely, the loss of an editing site through a genomic pyrimidine transition can be followed by subsequent loss of the previously essential editing factor or its reassignment to other targets. Intriguingly, chloroplast and mitochondrial but not the nuclear transcripts of plants are subject to RNA editing and this may have its causes in the multiplicity of organelle DNA copies in each plant cell and their possible heteroplasmy. Any C-to-T or reverse transition in the organelle DNA will first exist in low copy numbers and functional consequences can be buffered by the hundreds of copies of organelle genomes not yet carrying this mutation before homoplasmy is achieved over a couple of generations. It will be interesting to see whether such heteroplasmy will exist for a certain time span in evolution for an RNA editing site vanishing or newly emerging.

An early example for gain of new editing sites and a new co-factor is the case of PPR_56 affecting mitochondrial RNA editing sites nad3eU230SL and nad4eU272SL in *P. patens* (Ohtani et al., 2010). An ortholog of PPR_56 is missing in *F. hygrometrica* and both editing sites are genomically encoded as T in this sister taxon within the Funariaceae (Rüdinger, Szövényi, et al., 2011). Similarly, both positions are genomically "pre-edited" in all other moss mtDNAs currently available (unpublished observations). Hence, a gain of PPR_56 and the two editing sites in *Physcomitrella* is a much more likely evolutionary scenario than a loss in *Funaria* and this has been investigated in more detail among the Funariaceae (Beike et al., 2014). While this may suggest a rather recent emergence also of PPR_56, the protein sequence is not particularly closely related to any other editing factor paralogue in *Physcomitrella* although it may originate from the more ancient PPR_45 paralogue (see Section 3.2).

Taxa representing evolutionary ancient lineages of angiosperms like *Liriodendron tulipifera* (Richardson et al., 2013) or *Amborella trichopoda* (Rice et al., 2013) feature significantly more mitochondrial RNA editing than recently diversified flowering plants. This is fully congruent with early suggestions of a gradual loss of mitochondrial RNA editing sites during angiosperm diversification (Fujii & Small, 2011; Mower, 2008). We recently found that the same holds true for chloroplasts with much more RNA editing in taxa representing early-emerging lineages, including *Amborella* (Hein et al., 2016). Moreover, we could find evidence for multiple independent losses of editing sites, always strictly coinciding with concomitant losses of the corresponding editing cofactor orthologues. Interestingly, once an editing factor addresses two sites, it is retained as long as one of the editing sites remains but vanishes only once when both are lost simultaneously (Hein et al., 2016).

5. LAND PLANTS AND RNA EDITING: OPEN ISSUES AND FUTURE PERSPECTIVES

5.1 RNA Editing Cofactors: Demands for New Bioinformatic Approaches

Tremendous progress has been made in the field of plant organelle RNA editing over the recent decade after identification a first PPR protein (CRR4) as a chloroplast RNA editing factor (Kotera et al., 2005). More than 60 PLS-type PPR proteins have meantime been characterized as site-specific RNA editing factors in *Arabidopsis*, rice, maize or *Physcomitrella*. New additions to this collection come regularly, making even the most recent compilations of site-specific editing factors quickly outdated (Barkan & Small, 2014; Finster et al., 2012).

We currently develop a new database module "EdiFacts" with the intention to summarize such editing factor information to be accessible for querying in a web interface to be integrated into the forthcoming version three of PREPACT (H. Lenz, A. Hein, V. Knoop, unpublished).

The integration of information on plant organelle editomes and the specificity factors recognized for individual editing sites will be mandatory to better understand RNA recognition mechanisms. Any amendments and suggestion for an improved PPR-RNA binding code (Barkan et al., 2012; Takenaka, Zehrmann, Brennicke, et al., 2013; Yagi et al., 2013) must explain why similar transcript sequences in the same organelle are not equally targeted by a given editing factor (Fig. 3C). This requires position-weighted querying of known editome references to score for true editing targets vs. alternative candidates. To this end, a new module "TargetScan" will also be integrated in the forthcoming version of PREPACT, allowing to query any combination of the editome references.

The extraction of relevant amino acid positions from PLS-type PPR proteins is currently cumbersome since the identification of PPRs in programs like TPRpred (Karpenahalli et al., 2007) is designed for classical P-type PPR repeats. We aim for the automatic extraction of relevant amino acid positions in PLS-type PPR arrays to be translated in position weighted target queries for automatic input into TargetScan (K. Ullrich, M. Schallenberg-Rüdinger, S. Rensing and V. Knoop).

5.2 Testing Models on the Evolution of RNA Editing and Its Cofactors

The introduction of recombinant RNA editing factors into wild-type plants and editing factor mutants of model species such as *Arabidopsis*, tobacco, rice, maize or *Physcomitrella* will give further insights into the mechanisms of RNA target recognition and base conversion. *Physcomitrella patens* with its now completed mutual assignment of organelle editing sites and nuclear editing factors is a particularly interesting model for several reasons. Firstly, it appears to represent an evolutionary ancient state of RNA editing machinery with only bona fide DYW-type PPR proteins acting on the RNA targets (Fig. 2). It is reasonable to assume that each editing factor contains the deaminase activity needed for C-to-U conversion at the specific targeted site(s). Knock-out and knock-down of the nine DYW-type editing factors in *Physcomitrella* has not revealed spurious side effects on the non-target editing sites. In particular, those editing deficient *Physcomitrella* mutants with no obvious phenotypes of developmental retardation are ideal candidates for complementation studies allowing to focus on the known target sites. Variation of the PLS-type PPR arrays with the intention to change target specificity can be done to check for conversion of specific novel cytidine targets or more broadly using transcriptome wide RNA-seq studies.

The liverwort *M. polymorpha* is currently emerging as a new alternative molecular genetic model system among bryophytes (Ishizaki, Nishihama, Yamato, & Kohchi, 2015) and this is of particular interest given the absence of RNA editing, and apparently also of corresponding co-factors, in the marchantiid liverwort clade (Fig. 4). The introduction of adequately designed RNA editing factors targeting cytidines in its organelle transcripts will reveal whether RNA editing can simply be engineered de novo in an organism. If so, this would support the idea that a simple horizontal gene transfer of DYW-type proteins may indeed allow for organelle RNA editing in the new host environment.

The flowering plant model systems will further help to reveal how different protein cofactors interact on individual editing sites. It will be particularly important to investigate how exactly MORF/RIP proteins interact with the sequence-specific E-type and DYW-type PPR editing factors (Fig. 2D). Whether direct interactions of DYW-type and truncated E-type proteins in the absence of the MORF/RIP helper proteins are an evolutionary intermediate (Fig. 2C), possibly represented by the lycophytes and ferns, is another evolutionarily interesting issue. Unfortunately, no amenable molecular genetic model taxon is currently available for these clades. A complicating issue in the monilophytes are the generally large and polyploid genomes. In contrast, the lycophyte genus *Selaginella* is characterized by consistently small genome sizes of 100–200 MBp (Little, Moran, Brenner, & Stevenson, 2007), a particularly mysterious fact given the abundance of cofactors that is supposedly needed to address the possibly around 6000 editing sites present in at least some *Selaginella* species. Particularly intriguing is a vast variability of organelle RNA editing patterns among *Selaginella* taxa (B. Oldenkott, N. Knie, V. Knoop, unpublished). Hence, even without Selaginellales as model taxa for reverse genetics they will be interesting candidates for massive bioinformatic studies to identify PPR proteins and their candidate target assignments. Aside from its vast amount, another peculiar feature of C-to-U editing in *Selaginella* may be noteworthy and possibly an indication for a slight shift in cytidine deamination functionality. Silent RNA editing sites in seed plants have mostly been found to be edited at low frequency and interpreted as a side effect of unspecific binding of editing factors mainly targeting relevant, non-silent sites elsewhere. In *Selaginella*, however, silent sites were identified in the immediate neighbourhood of non-silent editing sites rather indicating a sterical flexibility of the deaminase activity (Oldenkott et al., 2014).

Finally, the cofactors for reverse U-to-C editing occurring at least in hornworts, some lycophytes and in monilophytes (Fig. 4) remain to be identified. Again, the lack of appropriate model systems in these three plant groups is an issue. However, any candidate DYW-proteins with E-domain variants (Fig. 2E) could be tested in the above-mentioned established model systems. Lack of appropriate co-substrates acting as amino group donors may, however, preclude the straightforward establishment of U-to-C editing in the C-to-U editing models or in *Marchantia*. While reverse editing is frequently considered a diffuse side-effect in C-to-U editing taxa, we now observe U-to-C editing to even dominate over C-to-U RNA editing in some fern groups (N. Knie, S. Fischer, V. Knoop, unpublished).

REFERENCES

Aubourg, S., Boudet, N., Kreis, M., & Lecharny, A. (2000). *Arabidopsis thaliana*, 1% of the genome codes for a novel protein family unique to plants. *Plant Molecular Biology, 42*, 603–613.

Ban, T., Ke, J., Chen, R., Gu, X., Tan, M. H., Zhou, X. E. … Xu, H. E. (2013). Structure of a PLS-class pentatricopeptide repeat protein provides insights into mechanism of RNA recognition. *Journal of Biological Chemistry, 288*, 31540–31548.

Banks, J. A., Nishiyama, T., Hasebe, M., Bowman, J. L., Gribskov, M., dePamphilis, C. … Grugoriev, I. V. (2011). The *Selaginella* genome identifies genetic changes associated with the evolution of vascular plants. *Science, 332*, 960–963.

Barkan, A., Rojas, M., Fujii, S., Yap, A., Chong, Y. S., Bond, C. S., & Small, I. (2012). A combinatorial amino acid code for RNA recognition by pentatricopeptide repeat proteins. *PLoS Genetics, 8*, e1002910.

Barkan, A., & Small, I. (2014). Pentatricopeptide repeat proteins in plants. *Annual Review of Plant Biology, 65*, 415–442.

Beckert, S., Muhle, H., Pruchner, D., & Knoop, V. (2001). The mitochondrial nad2 gene as a novel marker locus for phylogenetic analysis of early land plants: a comparative analysis in mosses. *Molecular Phylogenetics and Evolution, 18*, 117–126.

Bégu, D., Castandet, B., & Araya, A. (2011). RNA editing restores critical domains of a group I intron in fern mitochondria. *Current Genetics, 57*, 317–325.

Beike, A. K., von Stackelberg, M., Schallenberg-Rüdinger, M., Hanke, S. T., Follo, M., Quandt, D. … Rensing, S. A. (2014). Molecular evidence for convergent evolution and allopolyploid speciation within the *Physcomitrium-Physcomitrella* species complex. *BMC Evolutionary Biology, 14*, 158.

Benne, R., Van Den Burg, J., Brakenhoff, J. P., Sloof, P., Van Boom, J. H., & Tromp, M. C. (1986). Major transcript of the frameshifted coxII gene from trypanosome mitochondria contains four nucleotides that are not encoded in the DNA. *Cell, 46*, 819–826.

Bentolila, S., Babina, A. M., Germain, A., & Hanson, M. R. (2013). Quantitative trait locus mapping identifies REME2, a PPR-DYW protein required for editing of specific C targets in *Arabidopsis* mitochondria. *RNA Biology, 10*, 1520–1525.

Bentolila, S., Chateigner-Boutin, A. L., & Hanson, M. R. (2005). Ecotype allelic variation in C-to-U editing extent of a mitochondrial transcript identifies RNA-editing quantitative trait loci in *Arabidopsis*. *Plant Physiology, 139*, 2006–2016.

Bentolila, S., Heller, W. P., Sun, T., Babina, A. M., Friso, G., van Wijk, K. J., & Hanson, M. R. (2012). RIP1, a member of an *Arabidopsis* protein family, interacts with the protein RARE1 and broadly affects RNA editing. *Proceedings of the National Academy of Sciences of the United States of America, 109*, E1453–E1461.

Bentolila, S., Oh, J., Hanson, M. R., & Bukowski, R. (2013). Comprehensive high-resolution analysis of the role of an *Arabidopsis* gene family in RNA editing. *PLoS Genetics, 9*, e1003584.

Binder, S., Marchfelder, A., & Brennicke, A. (1994). RNA editing of $tRNA^{Phe}$ and $tRNA^{Cys}$ in mitochondria of *Oenothera berteriana* is initiated in precursor molecules. *Molecular and General Genetics, 244*, 67–74.

Binder, S., Marchfelder, A., Brennicke, A., & Wissinger, B. (1992). RNA editing in trans-splicing intron sequences of nad2 mRNAs in *Oenothera* mitochondria. *Journal of Biological Chemistry, 267*, 7615–7623.

Blanc, V., Litvak, S., & Araya, A. (1995). RNA editing in wheat mitochondria proceeds by a deamination mechanism. *FEBS Letters, 373*, 56–60.

Blum, B., Bakalara, N., & Simpson, L. (1990). A model for RNA editing in kinetoplastid mitochondria: "guide" RNA molecules transcribed from maxicircle DNA provide the edited information. *Cell, 60*, 189–198.

Boch, J., Scholze, H., Schornack, S., Landgraf, A., Hahn, S., Kay, S. … Bonas, U. (2009). Breaking the code of DNA binding specificity of TAL-type III effectors. *Science, 326*, 1509–1512.

Bock, R., Hagemann, R., Kössel, H., & Kudla, J. (1993). Tissue- and stage-specific modulation of RNA editing of the psbF and psbL transcript from spinach plastids – a new regulatory mechanism? *Molecular & General Genetics, 240*, 238–244.

Bock, R., Hermann, M., & Fuchs, M. (1997). Identification of critical nucleotide positions for plastid RNA editing site recognition. *RNA, 3*, 1194–1200.

Bock, R., Hermann, M., & Kössel, H. (1996). In vivo dissection of *cis*-acting determinants for plastid RNA editing. *EMBO Journal, 15*, 5052–5059.

Börner, G. V., Mörl, M., Wissinger, B., Brennicke, A., & Schmelzer, C. (1995). RNA editing of a group II intron in *Oenothera* as a prerequisite for splicing. *Molecular & General Genetics, 246*, 739–744.

Boussardon, C., Avon, A., Kindgren, P., Bond, C. S., Challenor, M., Lurin, C., & Small, I. (2014). The cytidine deaminase signature HxE(x)nCxxC of DYW1 binds zinc and is necessary for RNA editing of ndhD-1. *New Phytologist, 203*, 1090–1095.

Boussardon, C., Salone, V., Avon, A., Berthome, R., Hammani, K., Okuda, K. … Lurin, C. (2012). Two interacting proteins are necessary for the editing of the NdhD-1 site in *Arabidopsis* plastids. *Plant Cell, 24*, 3684–3694.

Bundschuh, R., Altmüller, J., Becker, C., Nürnberg, P., & Gott, J. M. (2011). Complete characterization of the edited transcriptome of the mitochondrion of *Physarum polycephalum* using deep sequencing of RNA. *Nucleic Acids Research, 39*, 6044–6055.

Castandet, B., & Araya, A. (2011). The RNA editing pattern of *cox2* mRNA is affected by point mutations in plant mitochondria. *PLoS One, 6*.

Chateigner-Boutin, A.-L., Colas des Francs-Small, C., Fujii, S., Okuda, K., Tanz, S. K., & Small, I. (2013). The E domains of pentatricopeptide repeat proteins from different organelles are not functionally equivalent for RNA editing. *Plant Journal for Cell and Molecular Biology, 74*, 935–945.

Chateigner-Boutin, A. L., des Francs-Small, C. C., Delannoy, E., Kahlau, S., Tanz, S. K., de Longevialle, A. F. ... Small, I. (2011). OTP70 is a pentatricopeptide repeat protein of the E subgroup involved in splicing of the plastid transcript rpoC1. *Plant Journal for Cell and Molecular Biology, 65*, 532–542.

Chateigner-Boutin, A. L., & Hanson, M. R. (2003). Developmental co-variation of RNA editing extent of plastid editing sites exhibiting similar *cis*-elements. *Nucleic Acids Research, 31*, 2586–2594.

Chateigner-Boutin, A. L., Ramos-Vega, M., Guevara-García, A., Andrés, C., de la Luz Gutiérrez-Nava, M., Cantero, A. ... Leon, P. (2008). CLB19, a pentatricopeptide repeat protein required for editing of rpoA and clpP chloroplast transcripts. *Plant Journal, 56*, 590–602.

Chateigner-Boutin, A. L., & Small, I. (2007). A rapid high-throughput method for the detection and quantification of RNA editing based on high-resolution melting of amplicons. *Nucleic Acids Research, 35*, e114.

Chateigner-Boutin, A. L., & Small, I. (2010). Plant RNA editing. *RNA Biology, 7*, 213–219.

Chateigner-Boutin, A. L., & Small, I. (2011). Organellar RNA editing. *Wiley Interdisciplinary Reviews-RNA, 2*, 493–506.

Chaudhuri, S., Carrer, H., & Maliga, P. (1995). Site-specific factor involved in the editing of the psbL mRNA in tobacco plastids. *EMBO Journal, 14*, 2951–2957.

Chaudhuri, S., & Maliga, P. (1996). Sequences directing C to U editing of the plastid psbL mRNA are located within a 22 nucleotide segment spanning the editing site. *EMBO Journal, 15*, 5958–5964.

Cheng et al., 2016 (http://www.ncbi.nlm.nih.gov/pubmed/26764122)

Cheong, C.-G., & Hall, T. M. (2006). Engineering RNA sequence specificity of pumilio repeats. *Proceedings of the National Academy of Sciences of the United States of America, 103*, 13635–13639.

Cho, S. J., Blanc, V., & Davidson, N. O. (2007). Mouse models as tools to explore cytidine-to-uridine RNA editing. *Methods in Enzymology, 424*, 417–435.

Choury, D., Farré, J. C., Jordana, X., & Araya, A. (2004). Different patterns in the recognition of editing sites in plant mitochondria. *Nucleic Acids Research, 32*, 6397–6406.

Covello, P. S., & Gray, M. W. (1989). RNA editing in plant mitochondria. *Nature, 341*, 662–666.

D'Andrea, L. D., & Regan, L. (2003). TPR proteins: the versatile helix. *Trends in Biochemical Sciences, 28*, 655–662.

Das, A. K., Cohen, P. T. W., & Barford, D. (1998). The structure of the tetratricopeptide repeats of protein phosphatase 5: Implications for TPR-mediated protein-protein interactions. *EMBO Journal, 17*, 1192–1199.

De Longevialle, A. F., Meyer, E. H., Andres, C., Taylor, N. L., Lurin, C., Millar, A. H. ... Andrés, C. (2007). The pentatricopeptide repeat gene OTP43 is required for trans-splicing of the mitochondrial nad1 Intron 1 in *Arabidopsis thaliana*. *Plant Cell, 19*, 3256–3265.

Delannoy, E., Le Ret, M., Faivre-Nitschke, E., Estavillo, G. M., Bergdoll, M., Taylor, N. L. ... Gualberto, J. M. (2009). *Arabidopsis* tRNA adenosine deaminase arginine edits the wobble nucleotide of chloroplast tRNAArg(ACG) and is essential for efficient chloroplast translation. *Plant Cell, 21*, 2058–2071.

Delannoy, E., Stanley, W. A., Bond, C. S., & Small, I. D. (2007). Pentatricopeptide repeat (PPR) proteins as sequence-specificity factors in post-transcriptional processes in organelles. *Biochemical Society Transactions, 35*, 1643–1647.

Dreyfuss, G., Kim, V. N., & Kataoka, N. (2002). Messenger-RNA-binding proteins and the messages they carry. *Nature Reviews Molecular Cell Biology, 3*, 195–205.

Duff, R. J. (2006). Divergent RNA editing frequencies in hornwort mitochondrial *nad5* sequences. *Gene, 366*, 285–291.

Duff, R. J., & Moore, F. B. (2005). Pervasive RNA editing among hornwort *rbcL* transcripts except *Leiosporoceros*. *Journal of Molecular Evolution, 61*, 571–578.

Faivre-Nitschke, S. E., Grienenberger, J. M., & Gualberto, J. M. (1999). A prokaryotic-type cytidine deaminase from *Arabidopsis thaliana*. Gene expression and functional characterization. *European Journal of Biochemistry, 263*, 896–903.

Farré, J. C., & Araya, A. (2001). Gene expression in isolated plant mitochondria: high fidelity of transcription, splicing and editing of a transgene product in electroporated organelles. *Nucleic Acids Research, 29*, 2484–2491.

Farré, J. C., Leon, G., Jordana, X., & Araya, A. (2001). *cis* recognition elements in plant mitochondrion RNA editing. *Molecular and Cellular Biology, 21*, 6731–6737.

Feagin, J. E., Abraham, J. M., & Stuart, K. (1988). Extensive editing of the cytochrome c oxidase III transcript in *Trypanosoma brucei*. *Cell, 53*, 413–422.

Filipovska, A., & Rackham, O. (2012). Modular recognition of nucleic acids by PUF, TALE and PPR proteins. *Molecular BioSystems, 8*, 699.

Filipovska, A., Razif, M. F., Nygård, K. K., & Rackham, O. (2011). A universal code for RNA recognition by PUF proteins. *Nature Chemical Biology, 7*, 425–427.

Finster, S., Legen, J., Qu, Y., & Schmitz-Linneweber, C. (2012). Land plant RNA editing or: don't be fooled by plant organellar DNA sequences. In R. Bock, & V. Knoop (Eds.), *Genomics of chloroplasts and mitochondria* (pp. 293–321). Netherlands: Dordrecht, NL: Springer.

Freyer, R., Kiefer-Meyer, M.-C., & Kössel, H. (1997). Occurrence of plastid RNA editing in all major lineages of land plants. *Proceedings of the National Academy of Sciences of the United States of America, 94*, 6285–6290.

Fujii, S., Bond, C. S., & Small, I. D. (2011). Selection patterns on restorer-like genes reveal a conflict between nuclear and mitochondrial genomes throughout angiosperm evolution. *Proceedings of the National Academy of Sciences of the United States of America, 108*, 1723–1728.

Fujii, S., Sato, N., & Shikanai, T. (2013). Mutagenesis of individual pentatricopeptide repeat motifs affects RNA binding activity and reveals functional partitioning of *Arabidopsis* PROTON gradient regulation3. *Plant Cell, 25*, 3079–3088.

Fujii, S., & Small, I. (2011). The evolution of RNA editing and pentatricopeptide repeat genes. *New Phytologist, 191*, 37–47.

Giegé, P., & Brennicke, A. (1999). RNA editing in *Arabidopsis* mitochondria effects 441 C to U changes in ORFs. *Proceedings of the National Academy of Sciences of the United States of America, 96*, 15324–15329.

Glass, F., Härtel, B., Zehrmann, A., Verbitskiy, D., & Takenaka, M. (2015). MEF13 requires MORF3 and MORF8 for RNA editing at eight targets in mitochondrial mRNAs in *Arabidopsis thaliana*. *Molecular Plant, 8*(10).

Grewe, F., Herres, S., Viehöver, P., Polsakiewicz, M., Weisshaar, B., & Knoop, V. (2011). A unique transcriptome: 1782 positions of RNA editing alter 1406 codon identities in mitochondrial mRNAs of the lycophyte *Isoetes engelmannii*. *Nucleic Acids Research, 39*, 2890–2902.

Grewe, F., Viehoever, P., Weisshaar, B., & Knoop, V. (2009). A trans-splicing group I intron and tRNA-hyperediting in the mitochondrial genome of the lycophyte *Isoetes engelmannii*. *Nucleic Acids Research, 37*, 5093–5104.

Groth-Malonek, M., Pruchner, D., Grewe, F., & Knoop, V. (2005). Ancestors of trans-splicing mitochondrial introns support serial sister group relationships of hornworts and mosses with vascular plants. *Molecular Biology and Evolution, 22*, 117–125.

Groth-Malonek, M., Wahrmund, U., Polsakiewicz, M., & Knoop, V. (2007). Evolution of a pseudogene: exclusive survival of a functional mitochondrial nad7 gene supports haplomitrium as the earliest liverwort lineage and proposes a secondary loss of RNA editing in marchantiidae. *Molecular Biology and Evolution, 24*, 1068–1074.
Gualberto, J. M., Lamattina, L., Bonnard, G., Weil, J. H., & Grienenberger, J. M. (1989). RNA editing in wheat mitochondria results in the conservation of protein sequences. *Nature, 341*, 660–662.
Gully, B. S., Cowieson, N., Stanley, W. A., Shearston, K., Small, I. D., Barkan, A., & Bond, C. S. (2015). The solution structure of the pentatricopeptide repeat protein PPR10 upon binding atpH RNA. *Nucleic Acids Research, 43*, 1918–1926.
Guo, W., Grewe, F., & Mower, J. P. (2015). Variable frequency of plastid RNA editing among ferns and repeated loss of uridine-to-cytidine editing from vascular plants. *PLoS One, 10*, e0117075.
Halter, C. P., Peeters, N. M., & Hanson, M. R. (2004). RNA editing in ribosome-less plastids of iojap maize. *Current Genetics, 45*, 331–337.
Hammani, K., Okuda, K., Tanz, S. K., Chateigner-Boutin, A. L., Shikanai, T., & Small, I. (2009). A study of new *Arabidopsis* chloroplast RNA editing mutants reveals general features of editing factors and their target sites. *Plant Cell, 21*, 3686–3699.
Handa, H. (2003). The complete nucleotide sequence and RNA editing content of the mitochondrial genome of rapeseed (*Brassica napus* L.): comparative analysis of the mitochondrial genomes of rapeseed and *Arabidopsis thaliana*. *Nucleic Acids Research, 31*, 5907–5916.
Härtel, B., Zehrmann, A., Verbitskiy, D., van der Merwe, J. A., Brennicke, A., & Takenaka, M. (2013). MEF10 is required for RNA editing at *nad2*-842 in mitochondria of *Arabidopsis thaliana* and interacts with MORF8. *Plant Molecular Biology, 81*, 337–346.
Hashimoto, M., Endo, T., Peltier, G., Tasaka, M., & Shikanai, T. (2003). A nucleus-encoded factor, CRR2, is essential for the expression of chloroplast ndhB in *Arabidopsis*. *Plant Journal, 36*, 541–549.
Hayes, M. L., Dang, K. N., Diaz, M. F., & Mulligan, R. M. (2015). A conserved Glutamate residue in the c-terminal deaminase domain of pentatricopeptide repeat proteins is required for RNA editing activity. *Journal of Biological Chemistry, 290*, 10136–10142.
Hayes, M. L., Giang, K., Berhane, B., & Mulligan, R. M. (2013). Identification of two pentatricopeptide repeat genes required for RNA editing and zinc binding by C-terminal cytidine deaminase-like domains. *Journal of Biological Chemistry, 288*, 36519–36529.
Hayes, M. L., & Hanson, M. R. (2007a). Assay of editing of exogenous RNAs in chloroplast extracts of *Arabidopsis*, maize, pea, and tobacco. *Methods in Enzymology, 424*, 459–482.
Hayes, M. L., & Hanson, M. R. (2007b). Identification of a sequence motif critical for editing of a tobacco chloroplast transcript. *RNA, 13*, 281–288.
Hayes, M. L., Reed, M. L., Hegeman, C. E., & Hanson, M. R. (2006). Sequence elements critical for efficient RNA editing of a tobacco chloroplast transcript *in vivo* and *in vitro*. *Nucleic Acids Research, 34*, 3742–3754.
Hecht, J., Grewe, F., & Knoop, V. (2011). Extreme RNA editing in coding islands and abundant microsatellites in repeat sequences of *Selaginella moellendorffii* mitochondria: the root of frequent plant mtDNA recombination in early tracheophytes. *Genome Biology and Evolution, 3*, 344–358.
Hegeman, C. E., Hayes, M. L., & Hanson, M. R. (2005). Substrate and cofactor requirements for RNA editing of chloroplast transcripts in *Arabidopsis in vitro*. *Plant Journal for Cell and Molecular Biology, 42*, 124–132.
Hein, A., Polsakiewicz, M., & Knoop, V. (2016). Frequent chloroplast RNA editing in early-branching flowering plants: pilot studies on angiosperm-wide co-existence of editing sites and their nuclear specificity factors. *BMC Evolutionary Biology, 16*(1). in press.

Hermann, M., & Bock, R. (1999). Transfer of plastid RNA-editing activity to novel sites suggests a critical role for spacing in editing-site recognition. *Proceedings of the National Academy of Sciences of the United States of America, 96*, 4856–4861.

Hiesel, R., Wissinger, B., Schuster, W., & Brennicke, A. (1989). RNA editing in plant mitochondria. *Science, 246*, 1632–1634.

Hirose, T., Kusumegi, T., Tsudzuki, T., & Sugiura, M. (1999). RNA editing sites in tobacco chloroplast transcripts: editing as a possible regulator of chloroplast RNA polymerase activity. *Molecular & General Genetics, 262*, 462–467.

Hirose, T., & Sugiura, M. (1997). Both RNA editing and RNA cleavage are required for translation of tobacco chloroplast ndhD mRNA: a possible regulatory mechanism for the expression of a chloroplast operon consisting of functionally unrelated genes. *EMBO Journal, 16*, 6804–6811.

Hirose, T., & Sugiura, M. (2001). Involvement of a site-specific trans-acting factor and a common RNA-binding protein in the editing of chloroplast mRNAs: development of a chloroplast in vitro RNA editing system. *EMBO Journal, 20*, 1144–1152.

Hiss, M., Laule, O., Meskauskiene, R. M., Arif, M. A., Decker, E. A., Erxleben, A. … Rensing, S. A. (2014). Large-scale gene expression profiling data for the model moss *Physcomitrella patens* aid understanding of developmental progression, culture and stress conditions. *Plant Journal, 79*, 530–539.

Hoch, B., Maier, R. M., Appel, K., Igloi, G. L., & Kössel, H. (1991). Editing of a chloroplast mRNA by creation of an initiation codon. *Nature, 353*, 178–180.

Hospattankar, A. V., Higuchi, K., Law, S. W., Meglin, N., & Brewer, H. B. J. (1987). Identification of a novel in-frame translational stop codon in human intestine apoB mRNA. *Biochemical and Biophysical Research Communications, 148*, 279–285.

Howard, M. J., Lim, W. H., Fierke, C. A., & Koutmos, M. (2012). Mitochondrial ribonuclease P structure provides insight into the evolution of catalytic strategies for precursor-tRNA 5′ processing. *Proceedings of the National Academy of Sciences of the United States of America, 109*, 16149–16154.

Ichinose, M., Sugita, C., Yagi, Y., Nakamura, T., & Sugita, M. (2013). Two DYW subclass PPR proteins are involved in RNA editing of ccmFc and atp9 transcripts in the moss *Physcomitrella patens*: first complete set of PPR editing factors in plant mitochondria. *Plant and Cell Physiology, 54*, 1907–1916.

Ichinose, M., Tasaki, E., Sugita, C., & Sugita, M. (2012). A PPR-DYW protein is required for splicing of a group II intron of *cox1* pre-mRNA in *Physcomitrella patens*. *Plant Journal for Cell and Molecular Biology, 70*, 271–278.

Ichinose, M., Uchida, M., & Sugita, M. (2014). Identification of a pentatricopeptide repeat RNA editing factor in *Physcomitrella patens* chloroplasts. *FEBS Letters, 588*, 4060–4064.

Inada, M., Sasaki, T., Yukawa, M., Tsudzuki, T., & Sugiura, M. (2004). A systematic search for RNA editing sites in pea chloroplasts: an editing event causes diversification from the evolutionarily conserved amino acid sequence. *Plant and Cell Physiology, 45*, 1615–1622.

Ishizaki, K., Nishihama, R., Yamato, K. T., & Kohchi, T. (2015). Molecular genetic tools and techniques for *Marchantia polymorpha* research. *Plant and Cell Physiology*.

Iyer, L. M., Zhang, D., Rogozin, I. B., & Aravind, L. (2011). Evolution of the deaminase fold and multiple origins of eukaryotic editing and mutagenic nucleic acid deaminases from bacterial toxin systems. *Nucleic Acids Research, 39*, 9473–9497.

Jonietz, C., Forner, J., Hölzle, A., Thuss, S., & Binder, S. (2010). RNA PROCESSING FACTOR2 is required for 5′ end processing of nad9 and cox3 mRNAs in mitochondria of *Arabidopsis thaliana*. *Plant Cell, 22*, 443–453.

Kakizaki, T., Yazu, F., Nakayama, K., Ito-Inaba, Y., & Inaba, T. (2012). Plastid signalling under multiple conditions is accompanied by a common defect in RNA editing in plastids. *Journal of Experimental Botany, 63*, 251–260.

Karcher, D., & Bock, R. (2002). Temperature sensitivity of RNA editing and intron splicing reactions in the plastid *ndhB* transcript. *Current Genetics, 41*, 48–52.

Karcher, D., & Bock, R. (2009). Identification of the chloroplast adenosine-to-inosine tRNA editing enzyme. *RNA, 15*, 1251–1257.

Karpenahalli, M. R., Lupas, A. N., & Soding, J. (2007). TPRpred: a tool for prediction of TPR-, PPR- and SEL1-like repeats from protein sequences. *BMC Bioinformatics, 8*, 2.

Ke, J., Chen, R. Z., Ban, T., Zhou, X. E., Gu, X., Tan, M. H. … Xu, H. E. (2013). Structural basis for RNA recognition by a dimeric PPR-protein complex. *Nature Structural & Molecular Biology, 20*, 1377–1382.

Kim, S. R., Yang, J. I., Moon, S., Ryu, C. H., An, K., Kim, K. M. … An, G. (2009). Rice OGR1 encodes a pentatricopeptide repeat-DYW protein and is essential for RNA editing in mitochondria. *Plant Journal, 59*, 738–749.

Kindgren, P., Yap, A., Bond, C. S., & Small, I. (2015). Predictable alteration of sequence recognition by RNA editing factors from *Arabidopsis*. *Plant Cell, 27*, 403–416.

Klein, R. R., Klein, P. E., Mullet, J. E., Minx, P., Rooney, W. L., & Schertz, K. F. (2005). Fertility restorer locus Rf1 of sorghum (*Sorghum bicolor* L.) encodes a pentatricopeptide repeat protein not present in the colinear region of rice chromosome 12. *Theoretical and Applied Genetics, 111*, 994–1012.

Knoop, V. (2011). When you can't trust the DNA: RNA editing changes transcript sequences. *Cellular and Molecular Life Sciences, 68*, 567–586.

Knoop, V., & Rüdinger, M. (2010). DYW-type PPR proteins in a heterolobosean protist: plant RNA editing factors involved in an ancient horizontal gene transfer? *FEBS Letters, 584*, 4287–4291.

Kobayashi, K., Kawabata, M., Hisano, K., Kazama, T., Matsuoka, K., Sugita, M., & Nakamura, T. (2012). Identification and characterization of the RNA binding surface of the pentatricopeptide repeat protein. *Nucleic Acids Research, 40*, 2712–2723.

Kobe, B., & Kajava, A. V. (2000). When protein folding is simplified to protein coiling: the continuum of solenoid protein structures. *Trends in Biochemical Sciences, 25*, 509–515.

Kotera, E., Tasaka, M., & Shikanai, T. (2005). A pentatricopeptide repeat protein is essential for RNA editing in chloroplasts. *Nature, 433*, 326–330.

Kugita, M., Yamamoto, Y., Fujikawa, T., Matsumoto, T., & Yoshinaga, K. (2003). RNA editing in hornwort chloroplasts makes more than half the genes functional. *Nucleic Acids Research, 31*, 2417–2423.

Kunzmann, A., Brennicke, A., & Marchfelder, A. (1998). 5′ end maturation and RNA editing have to precede tRNA 3′ processing in plant mitochondria. *Proceedings of the National Academy of Sciences of the United States of America, 95*, 108–113.

Lahmy, S., Barnèche, F., Derancourt, J., Filipowicz, W., Delseny, M., & Echeverria, M. (2000). A chloroplastic RNA-binding protein is a new member of the PPR family. *FEBS Letters, 480*, 255–260.

Legendre, P., Forstera, B., Juttner, R., & Meier, J. C. (2009). Glycine Receptors Caught between genome and Proteome - functional Implications of RNA editing and splicing. *Frontiers in Molecular Neuroscience, 2*, 23.

Lenz, H., & Knoop, V. (2013). PREPACT 2.0: predicting C-to-U and U-to-C RNA editing in organelle genome sequences with multiple references and curated RNA editing annotation. *Bioinformatics and Biology Insights, 7*, 1–19.

Lenz, H., Rüdinger, M., Volkmar, U., Fischer, S., Herres, S., Grewe, F., & Knoop, V. (2010). Introducing the plant RNA editing prediction and analysis computer tool PREPACT and an update on RNA editing site nomenclature. *Current Genetics, 56*, 189–201.

Li, Q., Yan, C., Xu, H., Wang, Z., Long, J., Li, W. … Yan, N. (2014). Examination of the dimerization states of the single-stranded RNA recognition protein pentatricopeptide repeat 10 (PPR10). *Journal of Biological Chemistry, 289*, 31503–31512.

Little, D. P., Moran, R. C., Brenner, E. D., & Stevenson, D. W. (2007). Nuclear genome size in *Selaginella*. *Genome, 50*, 351–356.

Liu, Y., Medina, R., & Goffinet, B. (2014). 350 My of mitochondrial genome stasis in mosses, an early land plant lineage. *Molecular Biology and Evolution, 31*, 2586–2591.

Lu, B., Wilson, R. K., Phreaner, C. G., Mulligan, R. M., & Hanson, M. R. (1996). Protein polymorphism generated by differential RNA editing of a plant mitochondrial rps12 gene. *Molecular and Cellular Biology, 16*, 1543–1549.

Lurin, C., Andrés, C., Aubourg, S., Bellaoui, M., Bitton, F., Bruyère, C. … Small, I. (2004). Genome-wide analysis of *Arabidopsis* pentatricopeptide repeat proteins reveals their essential role in organelle biogenesis. *Plant Cell, 16*, 2089–2103.

Mahendran, R., Spottswood, M. R., & Miller, D. L. (1991). RNA editing by cytidine insertion in mitochondria of *Physarum polycephalum*. *Nature, 349*, 434–438.

Malek, O., Lattig, K., Hiesel, R., Brennicke, A., Knoop, V., Lättig, K. … Knoop, V. (1996). RNA editing in bryophytes and a molecular phylogeny of land plants. *EMBO Journal, 15*, 1403–1411.

Manavski, N., Guyon, V., Meurer, J., Wienand, U., & Brettschneider, R. (2012). An essential pentatricopeptide repeat protein facilitates 5′ maturation and translation initiation of rps3 mRNA in Maize mitochondria. *Plant Cell, 24*, 3087–3105.

Mancebo, R., Zhou, X., Shillinglaw, W., Henzel, W., & Macdonald, P. M. (2001). BSF binds specifically to the bicoid mRNA 3′ untranslated region and contributes to stabilization of bicoid mRNA. *Molecular and Cellular Biology, 21*, 3462–3471.

Manna, S. (2015). An overview of pentatricopeptide repeat proteins and their applications. *Biochimie, 113*.

Marchfelder, A., Brennicke, A., & Binder, S. (1996). RNA editing is required for efficient excision of $tRNA^{Phe}$ from precursors in plant mitochondria. *Journal of Biological Chemistry, 271*, 1898–1903.

Maréchal-Drouard, L., Ramamonjisoa, D., Cosset, A., Weil, J. H., & Dietrich, A. (1993). Editing corrects mispairing in the acceptor stem of bean and potato mitochondrial phenylalanine transfer RNAs. *Nucleic Acids Research, 21*, 4909–4914.

Meier, J. C., Henneberger, C., Melnick, I., Racca, C., Harvey, R. J., Heinemann, U. … Grantyn, R. (2005). RNA editing produces glycine receptor alpha3(P185L), resulting in high agonist potency. *Nature Neuroscience, 8*, 736–744.

Meierhoff, K., Felder, S., Nakamura, T., Bechtold, N., & Schuster, G. (2003). HCF152, an *Arabidopsis* RNA binding pentatricopeptide repeat protein involved in the processing of chloroplast psbB-psbT-psbH-petB-petD RNAs. *Plant Cell, 15*, 1480–1495.

Miyamoto, T., Obokata, J., & Sugiura, M. (2002). Recognition of RNA editing sites is directed by unique proteins in chloroplasts: biochemical identification of *cis*-acting elements and trans-acting factors involved in RNA editing in tobacco and pea chloroplasts. *Molecular and Cellular Biology, 22*, 6726–6734.

Miyamoto, T., Obokata, J., & Sugiura, M. (2004). A site-specific factor interacts directly with its cognate RNA editing site in chloroplast transcripts. *Proceedings of the National Academy of Sciences of the United States of America, 101*, 48–52.

Miyata, Y., & Sugita, M. (2004). Tissue- and stage-specific RNA editing of *rps14* transcripts in moss (*Physcomitrella patens*) chloroplasts. *Journal of Plant Physiology, 161*, 113–115.

Miyata, Y., Sugiura, C., Kobayashi, Y., Hagiwara, M., & Sugita, M. (2002). Chloroplast ribosomal S14 protein transcript is edited to create a translation initiation codon in the moss *Physcomitrella patens*. *Biochimica et Biophysica Acta, 1576*, 346–349.

Moscou, M. J., & Bogdanove, A. J. (2009). A simple cipher governs DNA recognition by TAL effectors. *Science, 326*, 1501.

Mower, J. P. (2005). PREP-Mt: predictive RNA editor for plant mitochondrial genes. *BMC Bioinformatics, 6*, 96.

Mower, J. P. (2008). Modeling sites of RNA editing as a fifth nucleotide state reveals progressive loss of edited sites from angiosperm mitochondria. *Molecular Biology and Evolution, 25*, 52–61.
Mower, J. P. (2009). The PREP suite: predictive RNA editors for plant mitochondrial genes, chloroplast genes and user-defined alignments. *Nucleic Acids Research, 37*, W253–W259.
Mower, J. P., & Palmer, J. D. (2006). Patterns of partial RNA editing in mitochondrial genes of *Beta vulgaris*. *Molecular Genetics and Genomics, 276*, 285–293.
Nakajima, Y., & Mulligan, R. M. (2001). Heat stress results in incomplete C-to-U editing of maize chloroplast mRNAs and correlates with changes in chloroplast transcription rate. *Current Genetics, 40*, 209–213.
Nakajima, Y., & Mulligan, R. M. (2005). Nucleotide specificity of the RNA editing reaction in pea chloroplasts. *Journal of Plant Physiology, 162*, 1347–1354.
Nakamura, T., Meierhoff, K., Westhoff, P., & Schuster, G. (2003). RNA-binding properties of HCF152, an *Arabidopsis* PPR protein involved in the processing of chloroplast RNA. *European Journal of Biochemistry, 270*, 4070–4081.
Nakamura, T., & Sugita, M. (2008). A conserved DYW domain of the pentatricopeptide repeat protein possesses a novel endoribonuclease activity. *FEBS Letters, 582*, 4163–4168.
Navaratnam, N., Patel, D., Shah, R. R., Greeve, J. C., Powell, L. M., Knott, T. J., & Scott, J. (1991). An additional editing site is present in apolipoprotein B mRNA. *Nucleic Acids Research, 19*, 1741–1744.
Neuwirt, J., Takenaka, M., van der Merwe, J. A., & Brennicke, A. (2005). An in vitro RNA editing system from cauliflower mitochondria: editing site recognition parameters can vary in different plant species. *RNA, 11*, 1563–1570.
Notsu, Y., Masood, S., Nishikawa, T., Kubo, N., Akiduki, G., Nakazono, M.… Kadowaki, K. (2002). The complete sequence of the rice (*Oryza sativa* L.) mitochondrial genome: frequent DNA sequence acquisition and loss during the evolution of flowering plants. *Molecular Genetics and Genomics, 268*, 434–445.
Oda, K., Yamato, K., Ohta, E., Nakamura, Y., Takemura, M., Nozato, N.… Ohyama, K. (1992). Gene organization deduced from the complete sequence of liverwort *Marchantia polymorpha* mitochondrial DNA. A primitive form of plant mitochondrial genome. *Journal of Molecular Biology, 223*, 1–7.
Ohtani, S., Ichinose, M., Tasaki, E., Aoki, Y., Komura, Y., & Sugita, M. (2010). Targeted gene disruption identifies three PPR-DYW proteins involved in RNA editing for five editing sites of the moss mitochondrial transcripts. *Plant and Cell Physiology, 51*, 1942–1949.
Ohyama, K., Fukuzawa, H., Kohchi, T., Shirai, H., Sano, T., & Sano, S. (1986). Chloroplast gene organization deduced from complete sequence of liverwort *Marchantia polymorpha* chloroplast DNA. *Nature, 322*, 572–574.
Okuda, K., Chateigner-Boutin, A. L., Nakamura, T., Delannoy, E., Sugita, M., Myouga, F.… Shikanai, T. (2009). Pentatricopeptide repeat proteins with the DYW motif have distinct molecular functions in RNA Editing and RNA cleavage in *Arabidopsis* chloroplasts. *Plant Cell, 21*, 146–156.
Okuda, K., Habata, Y., Kobayashi, Y., & Shikanai, T. (2008). Amino acid sequence variations in Nicotiana CRR4 orthologs determine the species-specific efficiency of RNA editing in plastids. *Nucleic Acids Research, 36*, 6155–6164.
Okuda, K., Hammani, K., Tanz, S. K., Peng, L., Fukao, Y., Myouga, F.… Shikanai, T. (2010). The pentatricopeptide repeat protein OTP82 is required for RNA editing of plastid ndhB and ndhG transcripts. *Plant Journal, 61*, 339–349.
Okuda, K., Myouga, F., Motohashi, R., Shinozaki, K., & Shikanai, T. (2007). Conserved domain structure of pentatricopeptide repeat proteins involved in chloroplast RNA editing. *Proceedings of the National Academy of Sciences of the United States of America, 104*, 8178–8183.

Okuda, K., Nakamura, T., Sugita, M., Shimizu, T., & Shikanai, T. (2006). A pentatricopeptide repeat protein is a site recognition factor in chloroplast RNA editing. *Journal of Biological Chemistry, 281*, 37661–37667.
Okuda, K., & Shikanai, T. (2012). A pentatricopeptide repeat protein acts as a site-specificity factor at multiple RNA editing sites with unrelated *cis*-acting elements in plastids. *Nucleic Acids Research, 40*, 5052–5064.
Okuda, K., Shoki, H., Arai, M., Shikanai, T., Small, I., & Nakamura, T. (2014). Quantitative analysis of motifs contributing to the interaction between PLS-subfamily members and their target RNA sequences in plastid RNA editing. *Plant Journal, 80*, 870–882.
Oldenkott, B., Yamaguchi, K., Tsuji-Tsukinoki, S., Knie, N., & Knoop, V. (2014). Chloroplast RNA editing going extreme: more than 3400 events of C-to-U editing in the chloroplast transcriptome of the lycophyte *Selaginella uncinata*. *RNA, 20*, 1499–1506.
O'Toole, N., Hattori, M., Andres, C., Iida, K., Lurin, C., Schmitz-Linneweber, C.... Small, I. (2008). On the expansion of the pentatricopeptide repeat gene family in plants. *Molecular Biology and Evolution, 25*, 1120–1128.
Peeters, N. M., & Hanson, M. R. (2002). Transcript abundance supercedes editing efficiency as a factor in developmental variation of chloroplast gene expression. *RNA, 8*, 497–511.
Pfalz, J., Bayraktar, O. A., Prikryl, J., & Barkan, A. (2009). Site-specific binding of a PPR protein defines and stabilizes 5′ and 3′ mRNA termini in chloroplasts. *EMBO Journal, 28*, 2042–2052.
Phreaner, C. G., Williams, M. A., & Mulligan, R. M. (1996). Incomplete editing of rps12 transcripts results in the synthesis of polymorphic polypeptides in plant mitochondria. *Plant Cell, 8*, 107–117.
Powell, L. M., Wallis, S. C., Pease, R. J., Edwards, Y. H., Knott, T. J., & Scott, J. (1987). A novel form of tissue-specific RNA processing produces apolipoprotein-B48 in intestine. *Cell, 50*, 831–840.
Prikryl, J., Rojas, M., Schuster, G., & Barkan, A. (2011). Mechanism of RNA stabilization and translational activation by a pentatricopeptide repeat protein. *Proceedings of the National Academy of Sciences of the United States of America, 108*, 415–420.
Qiu, Y.-L., Li, L., Wang, B., Chen, Z., Knoop, V., Groth-Malonek, M.... Davis, C. C. (2006). The deepest divergences in land plants inferred from phylogenomic evidence. *Proceedings of the National Academy of Sciences of the United States of America, 103*, 15511–15516.
Rajasekhar, V. K., & Mulligan, R. M. (1993). RNA editing in plant mitochondria: [alpha]-phosphate is retained during C-to-U conversion in mRNAs. *Plant Cell, 5*, 1843–1852.
Reed, M. L. L., Peeters, N. M. M., & Hanson, M. R. R. (2001). A single alteration 20 nt 5′ to an editing target inhibits chloroplast RNA editing in vivo. *Nucleic Acids Research, 29*, 1507–1513.
Rensing, S. A., Lang, D., Zimmer, A. D., Terry, A., Salamov, A., Shapiro, H.... Boore, J. L. (2008). The *Physcomitrella* genome reveals evolutionary insights into the conquest of land by plants. *Science, 319*, 64–69.
Rice, D. W., Alverson, A. J., Richardson, A. O., Young, G. J., Sanchez-Puerta, M. V., Munzinger, J.... Palmer, J. D. (2013). Horizontal transfer of entire genomes via mitochondrial fusion in the angiosperm *Amborella*. *Science, 342*, 1468–1473.
Richardson, A. O., Rice, D. W., Young, G. J., Alverson, A. J., & Palmer, J. D. (2013). The "fossilized" mitochondrial genome of *Liriodendron tulipifera*: ancestral gene content and order, ancestral editing sites, and extraordinarily low mutation rate. *BMC Biology, 11*, 29.
Ringel, R., Sologub, M., Morozov, Y. I., Litonin, D., Cramer, P., & Temiakov, D. (2011). Structure of human mitochondrial RNA polymerase. *Nature, 478*, 269–273.
Robbins, J. C., Heller, W. P., & Hanson, M. R. (2009). A comparative genomics approach identifies a PPR-DYW protein that is essential for C-to-U editing of the *Arabidopsis* chloroplast *accD* transcript. *RNA, 15*, 1142–1153.

Rüdinger, M., Funk, H. T., Rensing, S. A., Maier, U. G., & Knoop, V. (2009). RNA editing: only eleven sites are present in the *Physcomitrella patens* mitochondrial transcriptome and a universal nomenclature proposal. *Molecular Genetics and Genomics, 281*, 473–481.

Rüdinger, M., Fritz-Laylin, L., Polsakiewicz, M., & Knoop, V. (2011). Plant-type mitochondrial RNA editing in the protist *Naegleria gruberi*. *RNA, 17*, 2058–2062.

Rüdinger, M., Polsakiewicz, M., & Knoop, V. (2008). Organellar RNA editing and plant-specific extensions of pentatricopeptide repeat (PPR) proteins in jungermanniid but not in marchantiid liverworts. *Molecular Biology and Evolution, 25*, 1405–1414.

Rüdinger, M., Szövényi, P., Rensing, S. A., & Knoop, V. (2011). Assigning DYW-type PPR proteins to RNA editing sites in the funariid mosses *Physcomitrella patens* and *Funaria hygrometrica*. *Plant Journal, 67*, 370–380.

Rüdinger, M., Volkmar, U., Lenz, H., Groth-Malonek, M., & Knoop, V. (2012). Nuclear DYW-type PPR gene families diversify with increasing RNA editing frequencies in liverwort and moss mitochondria. *Journal of Molecular Evolution, 74*, 37–51.

Ruwe, H., Castandet, B., Schmitz-Linneweber, C., & Stern, D. B. (2013). *Arabidopsis* chloroplast quantitative editotype. *FEBS Letters, 587*, 1429–1433.

Ruwe, H., & Schmitz-Linneweber, C. (2012). Short non-coding RNA fragments accumulating in chloroplasts: footprints of RNA binding proteins? *Nucleic Acids Research, 40*, 3106–3116.

Salone, V., Rüdinger, M., Polsakiewicz, M., Hoffmann, B., Groth-Malonek, M., Szurek, B. ... Lurin, C. (2007). A hypothesis on the identification of the editing enzyme in plant organelles. *FEBS Letters, 581*, 4132–4138.

Sasaki, T., Yukawa, Y., Miyamoto, T., Obokata, J., & Sugiura, M. (2003). Identification of RNA editing sites in chloroplast transcripts from the maternal and paternal progenitors of tobacco (*Nicotiana tabacum*): comparative analysis shows the involvement of distinct trans-factors for ndhB editing. *Molecular Biology and Evolution, 20*, 1028–1035.

Sasaki, T., Yukawa, Y., Wakasugi, T., Yamada, K., & Sugiura, M. (2006). A simple in vitro RNA editing assay for chloroplast transcripts using fluorescent dideoxynucleotides: distinct types of sequence elements required for editing of *ndh* transcripts. *Plant Journal, 47*, 802–810.

Schaap, P., Barrantes, I., Minx, P., Sasaki, N., Anderson, R. W., Bénard, M. ... Marwan, W. (2015). The *Physarum polycephalum* genome reveals extensive use of prokaryotic two-component and metazoan-type tyrosine kinase signaling. *Genome Biology and Evolution*. http://dx.doi.org/10.1093/gbe/evv237.

Schallenberg-Rüdinger, M., Kindgren, P., Zehrmann, A., Small, I., & Knoop, V. (2013). A DYW-protein knockout in *Physcomitrella* affects two closely spaced mitochondrial editing sites and causes a severe developmental phenotype. *Plant Journal, 76*, 420–432.

Schallenberg-Rüdinger, M., Lenz, H., Polsakiewicz, M., Gott, J. M., & Knoop, V. (2013). A survey of PPR proteins identifies DYW domains like those of land plant RNA editing factors in diverse eukaryotes. *RNA Biology, 10*, 1549–1556.

Schmitz-Linneweber, C., Williams-Carrier, R., & Barkan, A. (2005). RNA immunoprecipitation and microarray analysis show a chloroplast pentatricopeptide repeat protein to be associated with the 5′ region of mRNAs whose translation it activates. *Plant Cell, 17*, 2791–2804.

Schmitz-Linneweber, C., Williams-Carrier, R. E., Williams-Voelker, P. M., Kroeger, T. S., Vichas, A., & Barkan, A. (2006). A pentatricopeptide repeat protein facilitates the trans-splicing of the maize chloroplast rps12 pre-mRNA. *Plant Cell, 18*, 2650–2663.

Shi, X., Hanson, M. R., & Bentolila, S. (2015). Two RNA recognition motif-containing proteins are plant mitochondrial editing factors. *Nucleic Acids Research, 43*, 3814–3825.

Shikanai, T. (2006). RNA editing in plant organelles: machinery, physiological function and evolution. *Cellular and Molecular Life Sciences, 63*, 698–708.

Shikanai, T. (2015). RNA editing in plants: machinery and flexibility of the site recognition. *Biochimica et Biophysica Acta (BBA) — Bioenergetics, 1874*, 779—785.

Simpson, L., & Shaw, J. (1989). RNA editing and the mitochondrial cryptogenes of kinetoplastid protozoa. *Cell, 57*, 355—366.

Small, I., & Peeters, N. (2000). The PPR motif — a TPR-related motif prevalent in plant organellar proteins. *Trends in Biochemical Sciences, 25*, 46—47.

Smith, D. R. (2009). Unparalleled GC content in the plastid DNA of *Selaginella*. *Plant Molecular Biology, 71*, 627—639.

Smith, H. C., Gott, J. M., & Hanson, M. R. (1997). A guide to RNA editing. *RNA, 3*, 1105—1123.

Sosso, D., Mbelo, S., Vernoud, V., Gendrot, G., Dedieu, A., Chambrier, P. ... Rogowsky, P. M. (2012). PPR2263, a DYW-Subgroup pentatricopeptide repeat protein, is required for mitochondrial nad5 and cob transcript editing, mitochondrion biogenesis, and maize growth. *Plant Cell, 24*, 676—691.

Sper-Whitis, G. L., Moody, J. L., & Vaughn, J. C. (1996). Universality of mitochondrial RNA editing in cytochrome-c oxidase subunit I (*coxI*) among the land plants. *Biochimica et Biophysica Acta (BBA) — Gene Structure and Expression, 1307*, 301—308.

Staudinger, M., Bolle, N., & Kempken, F. (2005). Mitochondrial electroporation and in organello RNA editing of chimeric *atp6* transcripts. *Molecular Genetics and Genomics, 273*, 130—136.

Staudinger, M., & Kempken, F. (2003). Electroporation of isolated higher-plant mitochondria: transcripts of an introduced *cox2* gene, but not an *atp6* gene, are edited *in organello*. *Molecular Genetics and Genomics, 269*, 553—561.

Steinhauser, S., Beckert, S., Capesius, I., Malek, O., & Knoop, V. (1999). Plant mitochondrial RNA editing. *Journal of Molecular Evolution, 48*, 303—312.

Sturm, N. R., & Simpson, L. (1990). Kinetoplast DNA minicircles encode guide RNAs for editing of cytochrome oxidase subunit III mRNA. *Cell, 61*, 879—884.

Sugita, M., Ichinose, M., Ide, M., & Sugita, C. (2013). Architecture of the PPR gene family in the moss *Physcomitrella patens*. *RNA Biology, 10*, 1439—1445.

Sugita, M., Miyata, Y., Maruyama, K., Sugiura, C., Arikawa, T., & Higuchi, M. (2006). Extensive RNA editing in transcripts from the *psbB* operon and *rpoA* gene of plastids from the enigmatic moss *Takakia lepidozioides*. *Bioscience, Biotechnology, and Biochemistry, 70*, 2268—2274.

Sugiura, C., Kobayashi, Y., Aoki, S., Sugita, C., & Sugita, M. (2003). Complete chloroplast DNA sequence of the moss *Physcomitrella patens*: evidence for the loss and relocation of rpoA from the chloroplast to the nucleus. *Nucleic Acids Research, 31*, 5324—5331.

Sugiura, C., & Sugita, M. (2004). Plastid transformation reveals that moss tRNA(Arg)-CCG is not essential for plastid function. *Plant Journal, 40*, 314—321.

Sun, T., Germain, A., Giloteaux, L., Hammani, K., Barkan, A., Hanson, M. R., & Bentolila, S. (2013). An RNA recognition motif-containing protein is required for plastid RNA editing in *Arabidopsis* and maize. *Proceedings of the National Academy of Sciences of the United States of America, 110*, E1169—E1178.

Sun, T., Shi, X., Friso, G., Van Wijk, K., Bentolila, S., & Hanson, M. R. (2015). A zinc finger motif-containing protein is essential for chloroplast RNA editing. *PLoS Genetics, 11*, e1005028.

Takenaka, M. (2010). MEF9, an E-subclass pentatricopeptide repeat protein, is required for an RNA editing event in the *nad7* transcript in mitochondria of *Arabidopsis*. *Plant Physiology, 152*, 939—947.

Takenaka, M., & Brennicke, A. (2003). In vitro RNA editing in pea mitochondria requires NTP or dNTP, suggesting involvement of an RNA helicase. *Journal of Biological Chemistry, 278*, 47526—47533.

Takenaka, M., Neuwirt, J., & Brennicke, A. (2004). Complex *cis*-elements determine an RNA editing site in pea mitochondria. *Nucleic Acids Research, 32*, 4137–4144.

Takenaka, M., Verbitskiy, D., Zehrmann, A., & Brennicke, A. (2010). Reverse genetic screening identifies five E-class PPR proteins involved in RNA editing in mitochondria of *Arabidopsis thaliana*. *Journal of Biological Chemistry, 285*, 27122–27129.

Takenaka, M., Verbitskiy, D., Zehrmann, A., Härtel, B., Bayer-Csaszar, E., Glass, F., & Brennicke, A. (2014). RNA editing in plant mitochondria-connecting RNA target sequences and acting proteins. *Mitochondrion, 19*, 191–197.

Takenaka, M., Zehrmann, A., Brennicke, A., & Graichen, K. (2013). Improved computational target site prediction for pentatricopeptide repeat RNA editing factors. *PLoS One, 8*, e65343.

Takenaka, M., Zehrmann, A., Verbitskiy, D., Härtel, B., & Brennicke, A. (2013). RNA editing in plants and its evolution. *Annual Review of Genetics, 47*, 335–352.

Takenaka, M., Zehrmann, A., Verbitskiy, D., Kugelmann, M., Härtel, B., & Brennicke, A. (2012). Multiple organellar RNA editing factor (MORF) family proteins are required for RNA editing in mitochondria and plastids of plants. *Proceedings of the National Academy of Sciences of the United States of America, 109*, 5104–5109.

Tasaki, E., Hattori, M., & Sugita, M. (2010). The moss pentatricopeptide repeat protein with a DYW domain is responsible for RNA editing of mitochondrial ccmFc transcript. *Plant Journal for Cell and Molecular Biology, 62*, 560–570.

Tasaki, E., & Sugita, M. (2010). The moss *Physcomitrella patens*, a model plant for the study of RNA editing in plant organelles. *Plant Signaling and Behavior, 5*, 727–729.

Terasawa, K., Odahara, M., Kabeya, Y., Kikugawa, T., Sekine, Y., Fujiwara, M., & Sato, N. (2006). The mitochondrial genome of the moss *Physcomitrella patens* sheds new light on mitochondrial evolution in land plants. *Molecular Biology and Evolution, 24*, 699–709.

Tillich, M., Funk, H. T., Schmitz-Linneweber, C., Poltnigg, P., Sabater, B., Martin, M., & Maier, R. M. (2005). Editing of plastid RNA in *Arabidopsis thaliana* ecotypes. *Plant Journal, 43*, 708–715.

Tillich, M., Hardel, S. L., Kupsch, C., Armbruster, U., Delannoy, E., Gualberto, J. M.... Schmitz-Linneweber, C. (2009). Chloroplast ribonucleoprotein CP31A is required for editing and stability of specific chloroplast mRNAs. *Proceedings of the National Academy of Sciences of the United States of America, 106*, 6002–6007.

Tillich, M., Lehwark, P., Morton, B. R., & Maier, U. G. (2006). The evolution of chloroplast RNA editing. *Molecular Biology and Evolution, 23*, 1912–1921.

Tillich, M., Schmitz-Linneweber, C., Herrmann, R. G., & Maier, R. M. (2001). The plastid chromosome of maize (*Zea mays*): update of the complete sequence and transcript editing sites. *Maize Genetics Cooperation – Newsletter, 75*, 42–44.

Tseng, C.-C., Sung, T.-Y., Li, Y.-C., Hsu, S.-J., Lin, C.-L., & Hsieh, M.-H. (2010). Editing of *accD* and *ndhF* chloroplast transcripts is partially affected in the *Arabidopsis* vanilla cream1 mutant. *Plant Molecular Biology, 73*, 309–323.

Uchida, M., Ohtani, S., Ichinose, M., Sugita, C., & Sugita, M. (2011). The PPR-DYW proteins are required for RNA editing of rps14, cox1 and nad5 transcripts in *Physcomitrella patens* mitochondria. *FEBS Letters, 585*, 2367–2371.

Unseld, M., Marienfeld, J. R., Brandt, P., & Brennicke, A. (1997). The mitochondrial genome of *Arabidopsis thaliana* contains 57 genes in 366,924 nucleotides. *Nature Genetics, 15*, 57–61.

Van der Merwe, J. A., Takenaka, M., Neuwirt, J., Verbitskiy, D., & Brennicke, A. (2006). RNA editing sites in plant mitochondria can share *cis*-elements. *FEBS Letters, 580*, 268–272.

Vangerow, S., Teerkorn, T., & Knoop, V. (1999). Phylogenetic information in the mitochondrial nad5 gene of pteridophytes: RNA editing and intron sequences. *Plant Biology, 1*, 235–243.

Verbitskiy, D., van der Merwe, J. A., Zehrmann, A., Härtel, B., & Takenaka, M. (2012). The E-class PPR protein MEF3 of *Arabidopsis thaliana* can also function in mitochondrial RNA editing with an additional DYW domain. *Plant and Cell Physiology, 53*, 358–367.

Verbitskiy, D., Zehrmann, A., Härtel, B., Brennicke, A., & Takenaka, M. (2012). Two related RNA-editing proteins target the same sites in mitochondria of *Arabidopsis thaliana. Journal of Biological Chemistry, 287*, 38064–38072.

Verbitskiy, D., Zehrmann, A., van der Merwe, J. A., Brennicke, A., & Takenaka, M. (2010). The PPR protein encoded by the LOVASTATIN INSENSITIVE 1 gene is involved in RNA editing at three sites in mitochondria of *Arabidopsis thaliana. Plant Journal, 61*, 446–455.

Wagoner, J. A., Sun, T., Lin, L., & Hanson, M. R. (2015). Cytidine deaminase motifs within the DYW domain of two pentatricopeptide repeat-containing proteins are required for site-specific chloroplast RNA editing. *Journal of Biological Chemistry, 290*, 2957–2968.

Wickett, N. J., Mirarab, S., Nguyen, N., Warnow, T., Carpenter, E., Matasci, N. … Leebens-Mack, J. (2014). Phylotranscriptomic analysis of the origin and early diversification of land plants. *Proceedings of the National Academy of Sciences of the United States of America, 111*, 201323926.

Wolf, P. G., Rowe, C. A., & Hasebe, M. (2004). High levels of RNA editing in a vascular plant chloroplast genome: analysis of transcripts from the fern Adiantum capillus-veneris. *Gene, 339*, 89–97.

Yagi, Y., Hayashi, S., Kobayashi, K., Hirayama, T., & Nakamura, T. (2013). Elucidation of the RNA recognition code for pentatricopeptide repeat proteins involved in organelle RNA editing in plants. *PLoS One, 8*, e57286.

Yap, A., Kindgren, P., Colas des Francs-Small, C., Kazama, T., Tanz, S. K., Toriyama, K., & Small, I. (2015). AEF1/MPR25 is implicated in RNA editing of plastid *atpF* and mitochondrial *nad5* and also promotes *atpF* splicing in *Arabidopsis* and rice. *Plant Journal, 81*, 661–669.

Yin, P., Li, Q., Yan, C., Liu, Y., Liu, J., Yu, F. … Yan, N. (2013). Structural basis for the modular recognition of single-stranded RNA by PPR proteins. *Nature, 504*, 168–171.

Yoshinaga, K., Iinuma, H., Masuzawa, T., & Uedal, K. (1996). Extensive RNA editing of U to C in addition to C to U substitution in the *rbcL* transcripts of hornwort chloroplasts and the origin of RNA editing in green plants. *Nucleic Acids Research, 24*, 1008–1014.

Yu, Q. B., Jiang, Y., Chong, K., & Yang, Z. N. (2009). AtECB2, a pentatricopeptide repeat protein, is required for chloroplast transcript accD RNA editing and early chloroplast biogenesis in *Arabidopsis thaliana. Plant Journal, 59*, 1011–1023.

Yu, W., & Schuster, W. (1995). Evidence for a site-specific cytidine deamination reaction involved in C to U RNA editing of plant mitochondria. *Journal of Biological Chemistry, 270*, 18227–18233.

Zandueta-Criado, A., & Bock, R. (2004). Surprising features of plastid ndhD transcripts: addition of non-encoded nucleotides and polysome association of mRNAs with an unedited start codon. *Nucleic Acids Research, 32*, 542–550.

Zehrmann, A., Härtel, B., Glass, F., Bayer-Császár, E., Obata, T., Meyer, E. … Takenaka, M. (2015). Selective homo- and heteromer interactions between the multiple organellar RNA editing factor (MORF) proteins in *Arabidopsis thaliana. Journal of Biological Chemistry, 290*, 6445–6456.

Zehrmann, A., van der Merwe, J. A., Verbitskiy, D., Brennicke, A., & Takenaka, M. (2008). Seven large variations in the extent of RNA editing in plant mitochondria between three ecotypes of *Arabidopsis thaliana. Mitochondrion, 8*, 319–327.

Zehrmann, A., Verbitskiy, D., Härtel, B., Brennicke, A., & Takenaka, M. (2010). RNA editing competence of trans-factor MEF1 is modulated by ecotype-specific differences but requires the DYW domain. *FEBS Letters, 584*, 4181–4186.

Zehrmann, A., Verbitskiy, D., Hartel, B., Brennicke, A., Takenaka, M., & Härtel, B. (2011). PPR proteins network as site-specific RNA editing factors in plant organelles. *RNA Biology, 8*, 67–70.

Zehrmann, A., Verbitskiy, D., van der Merwe, J. A., Brennicke, A., & Takenaka, M. (2009). A DYW domain-containing pentatricopeptide repeat protein is required for RNA editing at multiple sites in mitochondria of *Arabidopsis thaliana*. *Plant Cell, 21*, 558–567.

Zeng, W. H., Liao, S. C., & Chang, C. C. (2007). Identification of RNA editing sites in chloroplast transcripts of *Phalaenopsis aphrodite* and comparative analysis with those of other seed plants. *Plant and Cell Physiology, 48*, 362–368.

Zhelyazkova, P., Hammani, K., Rojas, M., Voelker, R., Vargas-Suárez, M., Börner, T., & Barkan, A. (2012). Protein-mediated protection as the predominant mechanism for defining processed mRNA termini in land plant chloroplasts. *Nucleic Acids Research, 40*, 3092–3105.

Zhu, Q., Dugardeyn, J., Zhang, C., Takenaka, M., Kühn, K., Craddock, C.... Van der Straeten, D. (2012). SLO2, a mitochondrial pentatricopeptide repeat protein affecting several RNA editing sites, is required for energy metabolism. *Plant Journal, 71*, 836–849.

SECTION TWO

Genomes

CHAPTER THREE

The Genome of the Model Moss *Physcomitrella patens*

D. Lang*, N. van Gessel*, K.K. Ullrich[§], R. Reski*[,¶,†,1]
*University of Freiburg, Freiburg, Germany
[§]University of Marburg, Marburg, Germany
[¶]FRIAS – Freiburg Institute for Advanced Studies, Freiburg, Germany
[†]BIOSS – Centre for Biological Signalling Studies, Freiburg, Germany
[1]Corresponding author: E-mail: ralf.reski@biologie.uni-freiburg.de

Contents

Abstract

For more than two decades, the moss *Physcomitrella patens* has been developed and employed as a model species for comparative studies of plant biology as well as a safe production system for biotechnology.

Early on, the generation and dissemination of transcriptomic and genomic resources was an important focal point of *Physcomitrella* research, which, together with the ease

Advances in Botanical Research, Volume 78
ISSN 0065-2296
http://dx.doi.org/10.1016/bs.abr.2016.01.004

of genetic modification and flexibility of cultivation, over the years has attracted more and more research groups all around the world to use this moss as a model for basic and applied research. The establishment of genomic resources has culminated in the role of the *P. patens* genome as a reference genome for plant evolution. There are two main parts of this chapter: in the first part we provide an overview and history of the established genomic resources and in the second part we summarize the current biological knowledge about the genome structure, nature of nonprotein- and protein-coding genes including codon usage bias, repeats and transposable elements encoded by the *Physcomitrella* genome and, where applicable, discuss these attributes in an evolutionary context. Furthermore, we focus on the duplicated parts of the moss genome, like the paralogous genes that were retained after ancestral, large-scale to whole-genome duplication events and can be used to gain insights into the evolutionary history of *Physcomitrella*. Finally, we conclude this chapter by highlighting a special class of ancient paralogs in the *Physcomitrella* genome that have been actively retained as redundant copies and might act as pseudoalleles.

1. BRYOPHYTES AT THE FOREFRONT OF PLANT GENETICS AND GENOME RESEARCH

Given the arguably unmerited, very limited attention bryology has attracted in the scientific mainstream, the above statement seems rather bold if not presumptuous. But early on, the nuclear genomes of bryophytes were the target and focal points of genetic research, enabling some major breakthroughs in biology. For example, the search for the genetic basis of sex in plants was first crowned with success in 1917, when Charles E. Allen discovered plant sex chromosomes in the liverwort *Sphaerocarpos donnellii* Austin (Allen, 1917). In 1928 Emil Heitz discovered the longitudinal partitioning of chromosomes and was the first to coin the terms euchromatin and heterochromatin and furthermore suggested the importance of satellite DNA at chromosomal ends in the formation of the nucleolus when studying the liverworts *Pellia epiphylla* (L.) Corda and *P. fabroniana* Raddi (Heitz, 1928b). In the same year he confirmed the existence and further described the cytogenetical behaviour of heterochromatin in mosses by showing that 'with 70 species of true mosses from 20 families, always one chromosome behaves differently. It does not disappear in telophases as do the other chromosomes' (Heitz, 1928a; Passarge, 1979; Zacharias, 1995). In other words, he was the first to describe the continuity of chromosomes during the cell cycle. This research on bryophytes did not only pioneer major aspects in cytogenetic and chromatin biology, but also at about the same time was accompanied by other genetic discoveries like non-Mendelian inheritance in the moss *Funaria hygrometrica* (von Wettstein, 1928) or the application of X-ray mutagenesis in tetrad analysis in *Sphaerocarpos*, revealing nucleic acids and not proteins as the molecular basis of inheritable traits (Knapp, 1936). These studies have inspired multiple subsequent works that used mutagenesis to dissect the genetic basis of differentiation (reviewed by Reski, 1998).

A lot of time has passed since and most certainly a vast corpus of bryophyte research did not receive the proper attention it deserved, but ever since the publication and availability of the *Physcomitrella patens* genome [as the fourth genome of a land plant after *Arabidopsis thaliana* (The *Arabidopsis* genome initiative, 2000), *Oryza sativa* (Goff et al., 2002; Yu et al., 2002) and *Populus trichocarpa* (Tuskan et al., 2006), and as the first nonvascular plant genome (Rensing et al., 2008)] bryophytes, and mosses in particular, again serve as focal points of plant and genomic research. Given their phylogenetic position at the basis of the land plants, bryophytes are ideal for comparative studies frequently employed to infer the state of ancestral land plants. As we see in the following sections, today the moss *P. patens* is routinely employed as an evo-devo model organism and serves as a general plant reference genome helping to elucidate important questions about plant and genome evolution. Due to this important role as a comparator genome, it has been selected as one of few plant flagship genomes by the US Department of Energy.

At the same time, we also illustrate that these analyses have to be performed and interpreted with great caution, because bryophytes and mosses in particular are not frozen in time. With estimated median stem ages between 28 Ma (Mega-annum, ie, 1E + 06 years) and 41 Ma (Laenen et al., 2014), the genera of extant bryophytes are undoubtedly much younger than the assumed age of the last common ancestor (LCA) of land plants (500 Ma; median age of 15 studies eg, Lang et al., 2010) in the time-tree of life database (Hedges, Marin, Suleski, Paymer, & Kumar, 2015). *Physcomitrella* might be substantially younger (Laenen et al., 2014; McDaniel et al., 2010), likely representing a derived state, adapted to a specific life style and habitat (Beike et al., 2014). Thus, it should not be generalized blindly as an 'ancient' land plant, because it certainly does not represent the consensus state of other mosses and bryophytes, let alone the LCA of land plants.

Obviously this could be said for all model or flagship plants — nevertheless, if we consider these limitations and choose appropriate means to correct for them, the in-depth knowledge that can be acquired from the higher resolution, enabled by the concentrated focus on a smaller set of reference/model organisms, enables us to perform genome-scale comparative studies that can inform our view on almost any aspect of plant science. Dobzhanzky's omnipresent conclusion also applies in this case: *Nothing in Biology Makes Sense Except in the Light of Evolution* (Dobzhansky, 1973).

2. *PHYSCOMITRELLA*, THE MODEL PLANT

P. patens unites most of the favourable attributes for a plant model organism: a short generation time (4–8 weeks), small stature (1–5 mm), reduced morphological complexity, traceable cell lineage, high growth rate and simplicity of genetic transformation. These experimental traits, in combination with the phylogenetic position as part of the early diverging, paraphyletic group of bryophytes that harbour a haplo-dominant life cycle which is expressed through haploid tissues in most developmental stages, have attracted plant scientists from all fields. Over the last two decades, a growing community of researchers around the world has established *Physcomitrella* as a model organism with a well-developed molecular toolbox, including the uniquely efficient gene targeting by homologous recombination (Kamisugi et al., 2006; Schaefer & Zryd, 1997; Strepp, Scholz, Kruse, Speth, & Reski, 1998; Strotbek, Krinninger, & Frank, 2013). While these qualities and the availability of molecular tools lead to the role of the moss as a major model for evolutionary–developmental (evo-devo) studies (eg, Aya et al., 2011; Bartels, González Besteiro, Lang, & Ulm, 2010; Beike et al., 2015; Hirano et al., 2007; Landberg et al., 2013; Lind et al., 2015; Lindner et al., 2014; Paponov et al., 2009; Rensing et al., 2008; Sakakibara et al., 2013, 2014; Viaene et al., 2014; Zimmer et al., 2013), the moss also has become an important system for plant biotechnology (Bach, King, Zhan, Simonsen, & Hamberger, 2014; Decker & Reski, 2008; Lucumi, Posten, & Pons, 2005; Reski & Frank, 2005; Reski, Parsons, & Decker, 2015; Saidi et al., 2005; Weise et al., 2007, 2006).

3. HISTORY AND OVERVIEW OF *PHYSCOMITRELLA* GENOMICS

Early on, moss researches have dedicated substantial efforts to generate comprehensive genomics resources comprising genetic mapping (Kamisugi et al., 2008; Reski, Faust, Wang, Wehe, & Abel, 1994; von Stackelberg, Rensing, & Reski, 2006), Sanger-based expressed sequence tag (EST; Lang, Eisinger, Reski, & Rensing, 2005; Nishiyama et al., 2003; Rensing, Rombauts, Van de Peer, & Reski, 2002) and whole-genome shotgun

(WGS) sequencing (Rensing et al., 2008) and nowadays a plethora of 'next generation' sequencing projects targeting the moss transcriptome (Coruh et al., 2015; Frank & Scanlon, 2015; Nishiyama et al., 2012; O'Donoghue et al., 2013; Xiao, Wang, Wan, Kuang, & He, 2011), genome and epigenome (Widiez et al., 2014; Zemach, McDaniel, Silva, & Zilberman, 2010) greatly extending the taxonomic sampling from individual laboratory strain up to family (eg, Szövényi et al., 2013) and (in ongoing projects) to population level. These data have been made available using specific web resources like PHYSCObase (http://moss.nibb.ac.jp/) or the more general plant genome database Phytozome (Goodstein et al., 2012), but have been integrated and centralized early on in the *Physcomitrella* model organism database, bryophyte genomics and community resource cosmoss.org (http://www.cosmoss.org; Lang et al., 2005; Lang, Zimmer, Rensing, & Reski, 2008; Rensing, Fritzowsky, Lang, & Reski, 2005; Rensing et al., 2008; Zimmer et al., 2013).

3.1 From EST-Based Transcriptome to Genome

First studies of the *Physcomitrella* transcriptome date back to the late 1990s. cDNAs with sequence similarity to annotated genes of flowering plants, as well as putative species-specific transcripts without detectable homologs, were found by sequencing cDNAs isolated by subtractive hybridization (Reski, Reynolds, Wehe, Kleber-Janke, & Kruse, 1998). Sequencing of cDNAs from tissue treated with abscisic acid (ABA) yielded similar results and demonstrated that ABA leads to the induction of similar genes in moss and flowering plants (Machuka et al., 1999). Subsequently, several large-scale EST sequencing projects were initiated throughout the moss community worldwide: in the UK, the USA, Germany (Rensing, Rombauts, Hohe, et al., 2002; Rensing, Rombauts, Van de Peer, et al., 2002) and Japan (Nishiyama et al., 2003). Some of the libraries were normalized and/or subtracted, yielding a transcriptome representation of low redundancy and high coverage (Lang et al., 2005).

Although access to EST-based transcriptome databases provided valuable insights and greatly informed molecular and biotechnological research with the moss, it became clear that many questions can only be addressed with a sequenced genome. Thus, in an opinion article Rensing, Rombauts, Hohe, et al. (2002) suggested that 'this is a good time to establish an international moss-genome-sequencing project to exploit *Physcomitrella patens* to fullest as a model organism for functional and comparative genomics'. The International Moss Genome Consortium was founded by Brent Mishler

(UC Berkeley, USA), Ralph Quatrano (Washington University, St. Louis, USA), Ralf Reski and Stefan Rensing (University of Freiburg, Germany), Mitsuyasu Hasebe and Tomoaki Nishiyama (National Institute for Basic Biology, Japan) and David Cove and Andrew Cuming (Leeds University, UK) at the annual moss meeting in Freiburg in 2004. The plan to sequence the moss genome was realized together with the US Department of Energy's Joint Genome Institute (JGI). Since 2010, the *Physcomitrella* genome has been denoted a 'flagship' genome (http://jgi.doe.gov/our-science/science-programs/plant-genomics/plant-flagship-genomes/).

The draft genome sequence assembly of *Physcomitrella* (V1) was based on WGS Sanger sequencing and was published in 2008 (Rensing et al., 2008). The availability of the genomic sequence has proven an ideal foundation for extensive comparative and evo-devo studies. This is reflected in the publication record — more than 750 publications (about 95 per year) have cited the draft genome paper since its publication. An ever-growing community of researchers from all fields apply *Physcomitrella* as a model organism for comparative studies (eg, Alboresi, Caffarri, Nogue, Bassi, & Morosinotto, 2008; Axtell, Snyder, & Bartel, 2007; Aya et al., 2011; Bartels et al., 2010; Gitzinger, Parsons, Reski, & Fussenegger, 2009; Hirano et al., 2007; Lang et al., 2010; Lindner et al., 2014; Paponov et al., 2009; Pils & Heyl, 2009; Pitsch, Witsch, & Baier, 2010; Ranjan, Dickopf, Ullrich, Rensing, & Hoecker, 2014; Zimmer et al., 2013).

3.2 Evolution of Moss Genome Annotations

The digital representation of genomes (genome annotation) is usually comprised of three entities: (1) The (super)scaffolds representing the virtual assembly of overlapping raw reads into contiguous stretches of sequences which are combined with gaps inferred from large-insert library end sequencing or further scaffolded by genetic or physical mapping into the golden path of the pseudo-chromosomes. (2) The structural gene annotation usually inferred by a combination of *ab initio* gene finding utilizing Machine Learning approaches and extrinsic transcript and protein evidence derived from mapping of cDNA, EST and assembled RNA-Seq data and sequence homology comparisons with related taxa. (3) The functional gene annotation ideally derived from high-quality, curated, community-coordinated and large-scale integration of experimental evidences, but in reality often a semi- to fully automated, noncurated homology transfer of information gathered from multiple sources with varying quality and coverage.

The quality of genome annotation is the bottleneck for any downstream analysis. Especially conclusions from large-scale, high-throughput approaches like systems biology and comparative genomics are immensely affected by flaws of these data (Van den Berg, McCarthy, Lamont, & Burgess, 2010). Since its initial V1 release the *Physcomitrella* genome annotation has been iteratively improved. Like many other communities, the moss community records changes to the genome annotation using a numerical versioning system that provides information on both genome assembly and annotation. Versions are indicated using a *V* followed by the version of the assembly (changes once the genomic sequence — the assembly — is altered) and the annotation (changes of the structural annotation). This nomenclature is also reflected in the cosmoss.org gene ids (CGI; https://www.cosmoss.org/physcome_project/wiki/Cosmoss_Gene_IDs). CGIs of PpFtsZ1-1 (Martin et al., 2009) in different annotation versions are used as an example in the following paragraphs.

The nonpublic draft V1 assembly (Pp1s275_2V0; scaffold_275, locus 2) was based on WGS Sanger sequencing at 8.6x clone depth and comprised 2536 V1 scaffolds. This number was reduced to 2106 in the release V1.1 (Pp1s275_2V1) after the removal of bacterial contaminations (Rensing et al., 2008). After an additional round of scaffold filtering, released as V1.2 (Pp1s275_2V2), the genome sequence of the 27 chromosomes was still scattered over 1995 genomic scaffolds (Lang et al., 2008). The filtering and classification of transposable elements and other nonprotein-coding regions led to a catalogue of 27,966 protein-coding genes (Lang et al., 2008). Although clearly an improvement, V1.2 still had issues: there were cases of well-characterized moss genes that were present in the genomic sequence and were missing from the gene catalogue (Markmann-Mulisch et al., 2007). Furthermore, in V1.2 only 4515 (~16%) gene models had both 5′- and 3′-UTRs (untranslated regions) and over 23,000 genes lacked either 5′- or 3′-UTR annotation and thus were incomplete. Lastly, functional annotation was only available for 41% of the genes and hardly any of these annotations were backed by traceable experimental evidence.

In the meantime, the consortium set forth to further improve the genome assembly. The groups in Leeds and Freiburg joined forces to construct a first genetic map of the *Physcomitrella* genome (Kamisugi et al., 2008) based on two mapping populations derived from crosses of the standard laboratory strain based on the Gransden isolate initially collected by H. Whitehouse in Cambridgeshire, UK (Cuming, 2011) and the Villersexel K3 isolate collected near a pond close to Villersexel/Villers-la-Ville, Haute Saône, France by Michael Lüth. The joint map comprised 31 linkage groups based on the integration of 1420 AFLP (amplified fragment length polymorphisms) and SSR (short sequence repeat) markers. Although the insights into the genetic overall structure of the moss genome were immense, the coverage of the genome by sequence-anchored, physically locatable markers was not dense enough to reconstruct chromosome architecture and golden path to allow overall super-scaffolding of the V1 assembly based on the map.

The quality of the genome annotation was greatly improved with V1.6 (Zimmer et al., 2013; Pp1s275_2V6): the structural annotation comprised 32,275 protein-coding genes with 8387 additional loci as compared to V1.2, including 1456 loci with known protein domains or homologs in Plantae; 26,722 (~83%) loci were supported by transcript evidence (ie, Sanger ESTs or full-length cDNAs). Only 1582 gene models remained unchanged from V1.1 to V1.6. With V1.6, published genes and gene models released by the scientific community were mapped, manually curated and integrated into the gene catalogue. The number of protein-coding loci with both 5′- and 3′-UTRs increased from 4515 to 15,757. V1.6 was the first release to include information on transcript isoforms, suggesting alternative splicing events for at least 10.8% of the loci. Furthermore, this release also provided information on nonprotein-coding loci. Functional annotations were improved regarding quality and coverage, resulting in 58% annotated loci that comprised also 7200 additional loci with Gene Ontology term annotations.

3.3 Current Progress and Perspectives

In order to derive an optimal reference to infer SNP (single nucleotide polymorphism) markers used for high-density genetic linkage analysis using the Illumina Golden Gate assay, the group of Jeremy Schmutz at Hudson alpha reassembled the WGS traces to yield a nonreleased V2 assembly. Gene annotation was not performed for this version — the sole purpose of the assembly was to map reads from an Illumina-based resequencing of the Villersexel K3 isolate and call SNPs which could be used for high-throughput genetic screening. Doing so, for the to-be published V3 map and assembly of the genome, of >500,000 SNPs 6003 polymorphic markers where selected and utilized by Gerald Tuskan's group at Oak Ridge National Laboratory, USA to screen the Leeds and Freiburg mapping populations, the former of which was already employed for the V1 map.

The V3 sequence assembly and super-scaffolding into pseudo-chromosomes generated by Jerry Jenkins and Jeremy Schmutz at Hudson alpha was annotated in multiple rounds of gene prediction: (1) The initial V3.0 was generated with the Phytozome pipeline by David Goodstein's group (Pp3c22_4940V0; locus 4940 on chromosome 22). (2) V3.1 (Pp3c22_4940V1) was derived using the cosmoss.org gene prediction pipeline based on an extended training set, a new round of gene-caller training, extensive single-end RNA-Seq data and a novel machine learning approach for locus clustering and definition with subsequent variant selection (https://www.cosmoss.org/physcome_project/wiki/Genome_Annotation/V3.1). (3) Phytozome's V3.2 added splice variants based on transcript assemblies utilizing the upcoming, paired-end RNA-Seq libraries from JGI's Gene Atlas Project (http://jgi.doe.gov/doe-jgi-plant-flagship-gene-atlas/). (4) The latest version, V3.3 (Pp3c22_4940V3) again employed the cosmoss.org locus and variant selection pipeline to classify the representative, major isoform and alternative variants based on the gene models inferred for V3.1 and V3.2. V3.3 is available from Phytozome and cosmoss.org as of October, 2015; the publication describing the pseudo-chromosomal genome is work in progress.

3.4 The *Physcomitrella* MOD cosmoss.org

Model organism databases (MODs), like for example, Gramene (Liang et al., 2008), TAIR (Swarbreck et al., 2008), FlyBase (Tweedie et al., 2009) or PlantGenIE (Sundell et al., 2015), are an important prerequisite for the success and the quality of a reference genome and serve as integrated, web-accessible community resources for the respective models (Hirschman, Berardini, Drabkin, & Howe, 2010). MODs act as central repositories for all kinds of data and knowledge generated by the research community. They provide the necessary infrastructure for researchers working with a model species, serve as focal points for newcomers and act as primary interfaces for data exchange with more general data repositories, like for example, NCBI, UniProt (UniProt Consortium, 2012), Phytozome (Goodstein et al., 2012) and PLAZA (Van Bel et al., 2012). Thus, besides their direct importance for the respective research community, they play an important role in enabling comparative analyses for the wider research community and are crucial to ensure overall data quality.

Automated annotation without substantial manual curation is insufficient to ensure data quality and credible knowledge discovery (Howe et al., 2008). An actively annotating community is crucial in order to transfer and integrate available data, especially the biological knowledge contained in the literature or discussed at scientific conferences that otherwise remains 'hidden' from any form of large-scale analysis (Mazumder, Natale, Julio, Yeh, & Wu, 2010).

Since its initial launch in 2003, the cosmoss.org resource (http://www.cosmoss.org/) provides access to the *Physcomitrella* virtual transcriptome assemblies and annotation via BLAST service, keyword search and sequence retrieval (Lang et al., 2005). Subsequently, it was extended to provide services for splice site prediction (Rensing et al., 2005), for mining gene families of transcription-associated proteins (Richardt, Lang, Reski, Frank, & Rensing, 2007) and to predict dual protein targeting (Mitschke et al., 2009). Following the completion of the initial genome assembly V1, cosmoss.org introduced access to the draft genome sequence (Rensing et al., 2008), the genetic map (Kamisugi et al., 2008) and all genome annotation releases (Lang et al., 2008; Zimmer et al., 2013) including a complete history of all gene models ever called in the process from V1.0 to V3.3. More importantly, cosmoss.org serves as a central platform to coordinate the analysis and annotation of the moss genome sequence. Thus, a wiki and several mailing lists have been set up to report and discuss the results within the community. Additionally, an integrative genome browser serves as a main entry point for the exploration of the *Physcomitrella* genome and annotation. The cosmoss.org browser is based on the Gbrowse software (Stein et al., 2002) and provides base pair (bp) level resolution for large-scale annotation data covering predictions for all different kinds of genetic regions ranging from protein-coding genes, transposable elements and repeats to tRNA, rRNA, miRNAs and other nonprotein-coding RNAs.

We are continuously integrating external published data, for example, sRNAs, miRNAs and EST or short read data from the sequence read archive (SRA) and from collaborators around the world. In addition, the cosmoss.org gene annotation releases are shared and hosted at the NCBI and the comparative plant resources, Phytozome and PLAZA. In 2009, the *Physcomitrella* community annotation services were transferred from the JGI to the cosmoss.org website and the resource now functions as the central annotation repository for the moss *P. patens* (https://www.cosmoss.org/physcome_project/wiki/Downloads). Traceable version control implemented in the publicly accessible repository that also collects nightly dumps of the functional curations allows the maintenance of the complete history of moss annotations.

4. ECOLOGY AND PHYLOGENETIC CONTEXT OF *PHYSCOMITRELLA*

4.1 Habitat and Life Cycle of *P. patens*

Natural populations of *P. patens* (Fig. 1) are reported to be frequently selfing, annual opportunists growing in late summer to autumn in open, unshaded, limy, loamy, moist and disturbed, but nutrient-rich habitats, often close to the waterline (Goffinet, 2007; Nebel & Philippi, 2000). An example of such a habitat can be found regularly on a summer fallow near Reute in close vicinity to Freiburg, Germany (Fig. 1).

The heterophasic life cycle completes in 4–8 weeks (Collier & Hughes, 1982; Hohe, Rensing, Mildner, Lang, & Reski, 2002; Une & Tateishi, 1996). Sporophytes develop regularly, and there are no specialized tissues for asexual reproduction. Upon stress, the protonemata (ie, the moss' filamentous, tip-growing developmental stage) form vegetative diaspores called brachycytes and fragment at preformed breakpoints, the tmema cells (Goode, Stead, & Duckett, 1993; Schnepf & Reinhard, 1997).

In contrast to the diploid or polyploid seeds of flowering plants, the sexual propagation bodies of mosses are haploid spores, from which the dominant haploid gametophytic generation develops. The filamentous protonema forms buds from which the leafy gametophores develop (Fig. 1). On the latter, sexual organs (gametangia: antheridia and archegonia) develop on the same gametophore (monoicous), mixed on the same branch (synoicous). In *Physcomitrella*, like in most of its Funariaceae relatives, self-fertilization or intragametophytic reproduction seems to be the rule (Klips, 2015; Perroud, Cove, Quatrano, & McDaniel, 2011; Szövényi et al., 2014). The motile male gametes swim to the archegonia and fertilize the egg, from which the diploid sporophyte develops, growing in nutritionally dependent fashion on the gametophore. The life cycle closes with the production of several thousand spores per spore capsule (Kamisugi et al., 2008; Perroud et al., 2011).

4.2 The Funariaceae and the *Physcomitrium/Physcomitrella* Species Complex

Taxonomically, *P. patens* (Hedw.) Bruch & Schimp. (synonym: *Aphanorrhegma patens* (Hedw.) Lindb.) belongs to the family Funariaceae within the order Funariales that is part of the class Bryopsida (Goffinet, 2007). The Funariaceae are a family of monoicous, short-lived, minute to medium-sized, light to yellow-green and annual to biennial plants that grow

Figure 1 ***Habitus and natural habitat of the moss* P. patens.** Photographs of the natural habitus and habitat of *Physcomitrella* populations on a fallow acre near Freiburg, Germany discovered by Michael Lüth, that is forked by a small stream, visible as a green/red grass line in Fig. 1A. Fig. 1A–E demonstrate the habitat at different resolutions. Photographs (A–D) were taken by Daniel Lang in October 2008. *Arrows* indicate the location of a *Physcomitrella* plant with green-yellow, immature capsules. (E) was taken at the same location in 2006 by Mark von Stackelberg. The drawing in the lower right depicts the schematic habitus of an adult *Physcomitrella* gametophore with a mature sporophyte. Diploid sporophytic [2n] tissues are annotated in orange while haploid [1n] tissues are shown in green colour. (See colour plate)

gregarious to open tufts (McIntosh, 2007). The worldwide occurring family consists of about 15–31 genera containing 250–400 species (Liu, Budke, & Goffinet, 2012; as of 09/2015 the Global Biodiversity Information Facility lists 271 species; http://www.gbif.org/species/4645).

With regard to their mode of sexual reproduction, within the mosses, the Funariaceae belong to a minority: only 40% of all mosses are monoicous while 60% are dioicous (ie, have separate sexes; McDaniel, Atwood, & Burleigh, 2013). This is drastically different from angiosperms, where only 6% of all species have individuals of distinct sexes (Villarreal & Renner, 2013). While it is still unclear which of these represents the ancestral state, it has now become evident that transitions between sexual systems are frequent in all bryophytes (Devos et al., 2011; McDaniel et al., 2013; Villarreal & Renner, 2013). Apparently this has happened also within families or genera, for example, in liverworts in the large genus *Radula* (Devos et al., 2011) or within the moss genus *Encalypta* (Nebel & Philippi, 2000). In mosses the switch from mono- to dioicy is more common than the switch from di- to monoicy (McDaniel et al., 2013). Thus, the Funariaceae either represent an ancestral state or are the result of a reverse transition. Such a switch from dioicy to hermaphroditism in the Funariaceae could be the result of a genome duplication by autopolyploidization or hybridization (Rensing, Lang, & Zimmer, 2009).

Within the Funariaceae, *P. patens* belongs to a clade frequently called the *Physcomitrium/Physcomitrella* complex, comprised of species belonging to the genera *Physcomitrium* and *Physcomitrella*, clustering distinctly from the more basal *Funaria* clade (Beike et al., 2014; Liu et al., 2012; McDaniel et al., 2010; Medina et al., 2015).

The Funariaceae are characterized by a rather uniformly looking, vegetative, gametophytic body, but harbour substantial morphological and ecological diversity in terms of sporophyte architecture, ranging from indehiscent capsules almost entirely enclosed within the leaves, to sporangia elevated on a long seta with a complex annulus (ie, ring of cells involved in capsule opening) and double peristome (ie, tooth-like structure emerging from the annulus formed of dead cells with thickened cell walls that enables gradual spore release). One extreme is the cosmopolitan *F. hygrometrica* Hedwig, with a long seta and dehiscent capsules with a complex peristome adapted to humidity- and wind-dependent spore dispersal (Budke, Goffinet, & Jones, 2011; Shaw, 1991). At the opposing end of the spectrum is the cleistocarpous *Physcomitrella* (Fig. 1), whose large spores are thought to be dispersed locally, restricted by rupture and decay of the capsule, while mid-to-long-range dispersal is thought to occur by moving water or by random events like the adherence to the feathers or feet of migrant water birds (Beike et al., 2014). It has a very short seta and an indehiscent, emergent capsule lacking a peristome (Goffinet, 2007) — features proposed to represent the evolutionarily derived state, that is, comprising secondarily reduced features (Beike et al., 2014; Liu et al., 2012).

5. GENOMIC INSIGHTS INTO THE EVOLUTION OF *P. PATENS*, THE FUNARIACEAE AND LAND PLANTS

From a genomic perspective many questions arise considering the fundamental differences between diploid-dominant and haploid-dominant organisms: How is the undoubtedly lower morphological complexity of mosses as compared to angiosperms mirrored in their genomes? How is genomic integrity maintained in land plants where genomes are present in the haploid phase (ie, homozygous) for the larger part of the life cycle, where inbreeding and asexual reproduction is frequent and metabolic activity is largely suspended upon harsh environmental conditions? Population genetic theory predicts a reduction of effective population size and a lower rate of recombination for primarily self-fertilizing species (Wright, Kalisz, & Slotte, 2013), which should result in a stepwise accumulation of deleterious mutations and ultimately extinction (Muller, 1964). How did these mosses escape Muller's Ratchet?

While the above or similar questions likely are addressed in most chapters of this book, lineage- and species-specific trends should also be taken into consideration: How is the specific life style as a pioneer plant that is a fast-growing (for a moss), but short-lived opportunist reflected in the gene complement?

In the following sections we present our current knowledge regarding these questions and highlight future research avenues to further our knowledge about this species.

5.1 Global Genome Complexity: Genome Size

In terms of genome size, bryophytes still can be considered as under-investigated: as of September 2015, the Kew Royal Botanical Garden C-value database (Zonneveld, Leitch, & Bennett, 2005) lists 48 liverworts, 184 mosses and no hornworts (Greilhuber, Leitch, & Bennett, 2010). For mosses this only covers about 1.5% of the expected species diversity (Leitch & Leitch, 2013). With this limitation in mind, based on the data in the

C-value database, we can nevertheless come to some conclusions: Mosses, do not (yet) puzzle us with exceptionally enormous genome sizes like gymnosperms, ferns or some angiosperms and also do not show a great deal of variation in sizes, in contrast to angiosperms (Leitch & Leitch, 2013). Instead, they tend to have small genomes ranging from 170 to 2004 Mbp, with a median size of 433 Mbp and most values contained in the interval of 379–498 Mbp (25% and 75% quantiles). With a median genome size of 744 Mbp, ranging from 206 to 7791 Mbp, liverworts seem to have slightly larger genomes than mosses.

In the current C-value database release, usual angiosperm genomes are 3 (eudicot median: 1320 Mbp) to 14 (monocot median: 6030 Mbp) times larger, revealing a much larger variance comprising values ranging from 63 Mbp to 148,900 Mbp (quantiles: 25%: 880 Mbp, 75%: 6103 Mbp). Considering these data (Bennett & Leitch, 2012), which also represent only about 1.8% of the expected species diversity (Leitch & Leitch, 2013) of angiosperms, the generalizability of conclusions drawn from comparative analysis relying on the commonly used angiosperm model plants, *A. thaliana* (156 Mbp), *P. trichocarpa* (484 Mbp) and *O. sativa* (489 Mbp), needs to be assessed carefully.

Beike et al. (2014) have published DNA amounts for multiple isolates from different locations of representative Funariaceae species and particularly the *Physcomitrium/Physcomitrella* complex. As these data have not yet been integrated into the Kew Plant C-value database, we employed these measurements and the GC content of the assembled pseudo-chromosomes of *Physcomitrella* to infer bp-level genome-size estimates including 95% confidence intervals (Table 1). Based on these data we can infer a median genome size of 518 Mbp for the Funariaceae, ranging from the smallest genome in *F. hygrometrica* (238 Mbp) to the biggest genome in *Physcomitrium eurystomum* (815 Mbp). How can this 3.5x change be explained in a family that arose somewhere in between 30 and 172 Ma (Laenen et al., 2014; Newton, Wikström, Bell, Forrest, & Ignatov, 2006; Rensing, 2014)? We revisit this question in a later section of this chapter.

If we now reassess the genome size of the model moss *P. patens*, we can conclude that with a size of 518 Mbp, *Physcomitrella* surely is an excellent representative for mosses and Funariaceae, while it is only half the size of the average liverwort genome size measured so far.

Based on the assembled nuclear genome sequence of *Physcomitrella*, we now can ask how much sequence information is missing or how well the sequence data and the DNA amount measurements agree: Considering

Table 1 Estimated Genome Sizes for Selected Funariaceae

Species	FCM Measurements	Mean Relative 1C	Standard Deviation Relative 1C	Mean Absolute 1C (pg)	Lower Limit (Mbp)	Mean (Mbp)	Upper Limit (Mbp)
Aphanorrhegma serratum	1	0.9		0.50		485.98	
Funaria hygrometrica	2	0.44	0.03	0.24	215.08	237.59	260.10
Physcomitrella magdalenae	1	0.92		0.51		496.78	
Physcomitrella patens	8	0.96	0.15	0.53	496.78	518.38	539.97
Physcomitrella readeri	3	0.96	0.05	0.53	488.72	518.38	548.03
Physcomitrium collenchymatum	2	1.51	0.44	0.83	483.74	815.36	1146.98
Physcomitrium eurystomum	6	1.27	0.32	0.70	546.68	685.77	824.86
Physcomitrium pyriforme	15	1.33	0.42	0.73	603.17	718.17	833.17
Physcomitrium sphaericum	5	0.78	0.43	0.43	216.77	421.18	625.59

Inferred from mean and standard deviations of relative 1C values measured using flow cytometry (FCM) of 43 natural isolates published by Beike et al. (2014). Relative values were converted to absolute 1C values in relation to the published value of *Physcomitrella* (0.53 pg) identified by Schween, Gorr, Hohe, & Reski (2003). The GC content of the *P. patens* V3 pseudo-chromosomes (0.34) was used for all species to calculate genome sizes in megabase pairs (Mbp).

that genome-size estimation based on DNA content does have its limitations depending on the method, used reference and assumed GC content, we can conclude that the sequenced size of about 480 Mbp, which is comprised by the sequence data partitions classified to represent nuclear DNA in all genome assembly releases generated so far, largely agrees with the 95% confidence interval presented in Table 1. This suggests that most of the *P. patens* nuclear DNA is represented by the current sequencing data. Nevertheless the genome is not yet complete in terms of closure of the 1.2% gaps (length: 100 bp to 43 kbp) that still do remain in the assembly. This is also demonstrated by the known examples of genes that are missing from the sequence assembly, but have been identified in dedicated studies (eg, RAD51b in Rensing et al., 2008 or Aux/IAA1b in Prigge, Lavy, Ashton, & Estelle, 2010). We later discuss a possible source for these missing data.

5.2 Structural Genome Complexity and First Hints at Genome Evolution in Funariaceae: Chromosome Counts

Although one might expect to find similar genome sizes and chromosome complements in related species, due to interspecies variation in the rates of mechanisms driving genome expansion and shrinkage, these measures of global genome complexity usually do not carry sufficient phylogenetic signal and also cannot serve as proxies to measure organismic complexity (Lang et al., 2010), a phenomenon known as the C-value enigma (Gregory, 2005). This problem seems even more pronounced in chromosome number than on overall DNA content. Thus, we forgo a global comparative discussion of chromosome counts and simply review the status quo for *Physcomitrella* and its Funariaceae relatives.

The ancestral, base chromosome number in true mosses has been reported to be $n = 4-7$ (Frahm, 2001; Fritsch, 1991). As we can see in Fig. 2A, chromosome counts greatly vary in the genera of Funariaceae (descriptive statistics and graph based on chromosome count collections published in Fritsch, 1991; Rensing et al., 2007; Rice et al., 2015). The right-skewed distribution contains data from 40 species and culminates at 27–28 chromosomes (25–75% quantile range: 25–28) which encompass counts for all genera but *Funariella curviseta* (comprising only one count of 5). But the distribution is trimodal with a secondary peak at 14 chromosomes (25–75% quantile range: 14–17) which comprises species from the genera *Funaria* (*F. hygrometrica*: 18, *F. wallichii*: 1) and *Physcomitrella* (*P. patens*: 2). There is a tertiary peak at 52 chromosomes (25–75% quantile range: 51–54), which comprises counts reported for the genera *Physcomitrium*

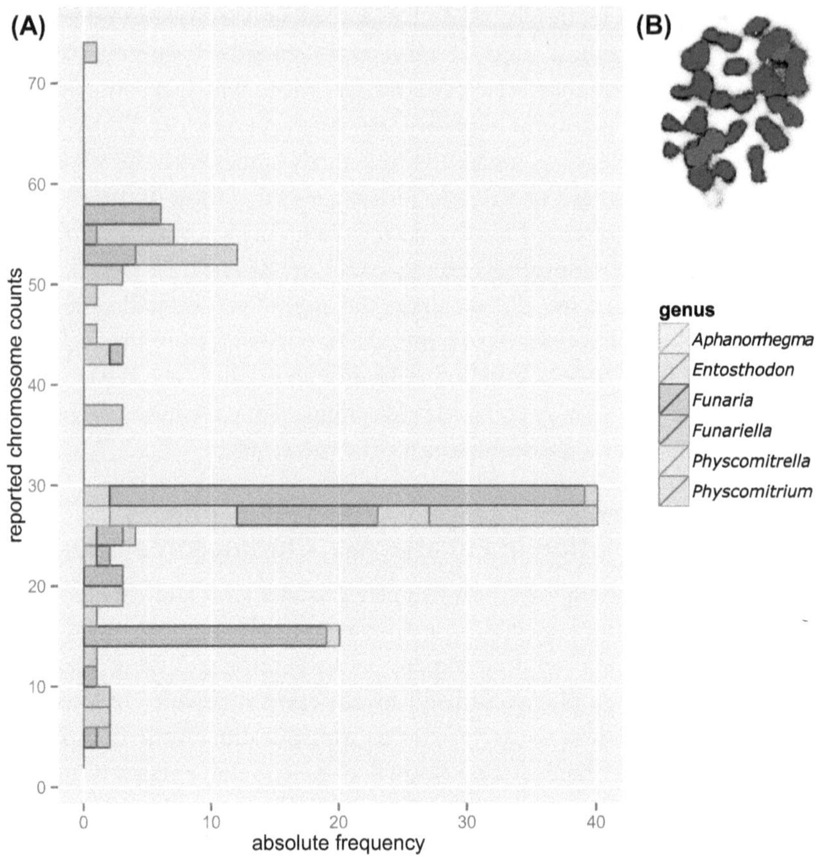

Figure 2 ***Chromosome numbers of selected Funariaceae.*** Fig. 2A depicts a stacked histogram (bin width = 2) of Funariaceae chromosome counts collected from Rensing et al. (2007) and the Chromosome Count Database (CCDB; http://ccdb.tau.ac.il/Bryophytes/Funariaceae/); 158 data points for 40 species were mapped to synonyms, curated and aggregated at genus level. Fig. 2B depicts an edited digital scan of an original meiotic chromosome squash published by Reski et al. (1994). Image processing was applied to select chromosomal boundaries and to highlight them in red. Putatively overlapping chromosomes are depicted in distinct red tones. (See colour plate)

(*P. pyriforme*: 6, *P. eurystomum*: 3, *P. coorgense*: 2, *P. cyathicarpum*: 2, *P. repandum*: 2, *P. immersum*: 1, *P. japonicum*: 1) and *Funaria* (*F. muhlenbergii*: 3, *F. hygrometrica*: 5, *F. hungarica*: 1).

These data also exhibit surprising levels of intraspecies variation. If we take *P. patens*, whose accepted chromosome complement is 27 as an example, we find counts of 14, 16, 26 and 27. This is similar for many of the 40 species in Fig. 2A. The observation can be explained either by an exceptional genomic instability in natural populations of Funariaceae, high rate of hybrid species with variable chromosome complements (Beike et al., 2014), misclassifications or by experimental difficulties in studying these tiny chromosomes. To illustrate this further, Fig. 2B reevaluates the original meiotic chromosome squashes (Reski et al., 1994) which lead to the manifestation of the chromosome number of 27 in the community. Contours of meiotic bivalents were identified by image processing and coloured to allow distinction of overlapping image layers, in order to demonstrate possible sources of ambiguity in such data.

Given the available Funariaceae chromosome counts and assuming an ancestral chromosome number of 7, two rounds of whole-genome duplications (WGDs) in the Funariaceae were previously hypothesized (Rensing et al., 2007, 2009): Beginning with an ancestral transition from dioicous ($n = 7$) to a monoicous ($n = 14$) state, which could have been the result of either hybridization (allopolyploidization) or autopolyploidization due to meiotic failure leading to the merger of two (male and female) parental chromosome sets in the LCA of the lineages leading to *Funaria* and the *Physcomitrium/Physcomitrella* species complex clades. According to this hypothesis, subsequently, at least the lineage leading to the extant *Physcomitrella/Physcomitrium* complex species underwent one to two additional, independent (auto- or allo-) polyploidization events ($n = 28$; $n = 56$), that in some cases were accompanied or followed by aneuploidization, resulting in the loss of one or several chromosomes ($n = 27$; $n = 52-54$; hypothesis suggested in Beike et al., 2014; McDaniel et al., 2010; Rensing et al., 2007, 2009; chromosome counts corresponding to the peaks in Fig. 2A were obtained from Fritsch, 1991; Rensing et al., 2007; Rice et al., 2015). This hypothesis requires molecular validation based on genomic or genome-scale data.

5.3 Gene Density and Repeat Content

Based on the data of the V3.1 genome annotation (https://www.cosmoss.org/physcome_project/wiki/Genome_Annotation/V3.1), the *P. patens* chromosomes have a gene density of 149 genes/Mbp. This is substantially smaller (1.7x) than the chromosomal gene density of *A.s thaliana* (248 genes/Mbp, TAIR 10; https://www.arabidopsis.org/portals/genAnnotation/gene_structural_annotation/annotation_data.jsp), similar to that of *O. sativa* (116 genes/Mbp, 1.4x smaller than *A. thaliana*, MSU 7 http://rice.plantbiology.msu.edu/annotation_pseudo_current.shtml) and higher than that of *P. trichocarpa* chromosomes (100 genes/Mbp, 2.4x smaller than *A. thaliana*,

JGI Ptr V2 ftp://ftp.jgi-psf.org/pub/JGI_data/phytozome/v5.0/). The *A. thaliana* genome represents a secondarily reduced genome (Hu et al., 2011). Which factors contribute to the differences in gene densities and genome size? The obvious answer to this question is twofold: (a) gene number and (b) repeat content.

The analysis of the draft genome sequence of *P. patens* revealed that about half of the genome is repetitive (Lang et al., 2008; Rensing et al., 2008). As illustrated in the cosmoss.org genome browser snapshot of the V1.6 genome annotation in Fig. 3, protein-coding genes are encoded on islands surrounded by areas mostly comprised of long terminal—repeat retrotransposons (LTR-Rs), most of which are fragmented or nested. In the V1 assembly only 4795 LTR-Rs were predicted by the MIPS ANGELA pipeline to be intact. Gypsy-like LTR-Rs are the predominant class (46%) while only 2% are Copia-like LTR-Rs. Nested insertions of LTR elements into other LTR-Rs are common (14%). The activity of these LTR-Rs seems to be tightly controlled by posttranscriptional gene silencing via sRNA-mediated epigenetic silencing (Coruh et al., 2015), as indicated by the overlapping data in the genome browser tracks comprising the mapping of sRNA reads, the CG/CHG/CHH methylation histograms (Zemach et al., 2010) and the repressive histone marks (red colour) of peak regions derived from the analysis of H3 ChIP-Seq data (Widiez et al., 2014) in Fig. 3. Up to 3% of a given Sanger-based EST library (see above) originate from regions annotated as LTR-Rs. Timing of insertion ages of LTR-Rs suggests multiple possibly distinct waves of activity of the two LTR-R classes, with the most recent peaks dating back longer than 1—3 Ma (Fig. 4).

In contrast to other eukaryotic genomes, *P. patens* contains only one family of Helitron rolling-circle DNA transposons (Rensing et al., 2008). Further, class II DNA-transposable elements could not be detected in large-scale analyses (Lang et al., 2008). Transposons are not the only type of genomic intruders that *Physcomitrella* had to cope with in its evolutionary past — like several algae and the lycophyte *Selaginella moellendorfii*, the moss genome harbours evidence of genomic integrations of nucleocytoplasmic large DNA viruses (Maumus, Epert, Nogué, & Blanc, 2014).

Comparing the repeat or transposon content (Lang et al., 2008) of *Arabidopsis* (10%), rice (30%) and poplar (37%), we see that the *Physcomitrella* genome has a substantially larger repeat content, providing part of the answer to the differences in genome size. Then again, why isn't this pattern directly reflected in the chromosomal gene density? A possible answer might be given considering the ages of the most recent insertion waves (Fig. 4), the

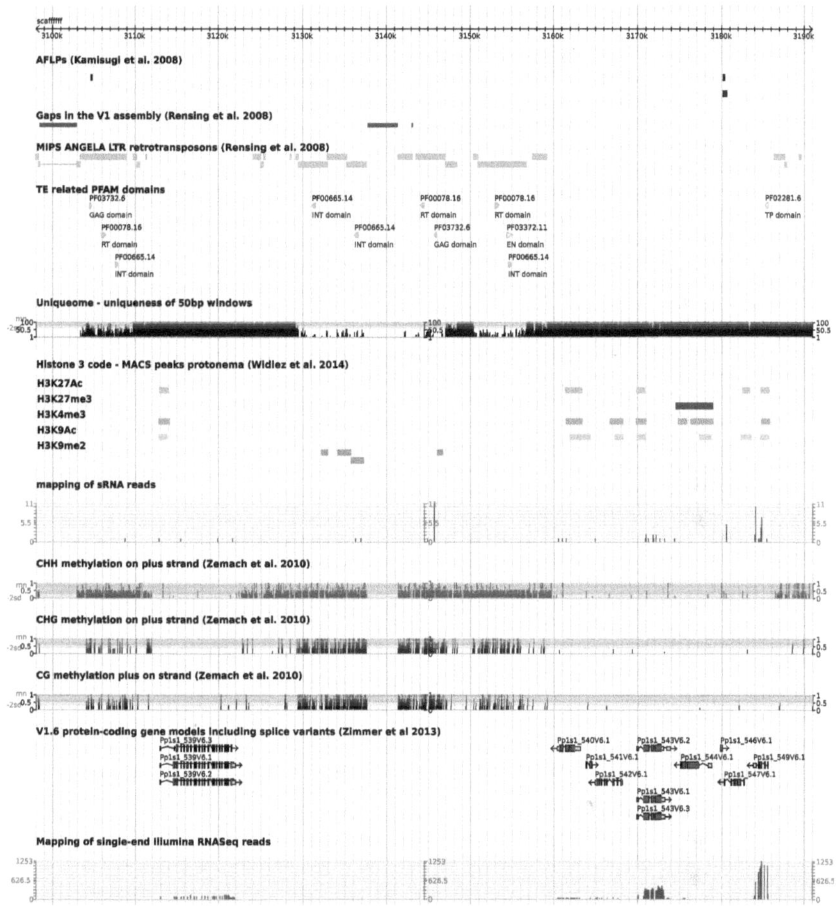

Figure 3 ***Genomic region of*** **Physcomitrella** ***in the cosmoss.org genome browser.*** Snapshot of the cosmoss.org genome browser displaying available V1.6 annotation as tracks comprising chromosomally localized features. Tracks represent (from top to bottom): (1) Mappings of AFLP markers used for genetic map (blue segments). (2) Assembly gaps (red segments). (3) LTR-Rs (brown segments). (4) PFAM protein domains specific for LTR-Rs (brown segments). (5) Uniquenome: histogram of sliding-window analysis of uniqueness of 50 bp windows (black). (6) Peak regions of histone 3 ChIP-Seq data derived from protonema cultures (red = repressive mark; green = activating mark). (7) Histogram of spliced alignments of sRNA Illumina reads. (8–10) Histograms of CHH (red), CHG (blue) and CG (black) methylation on the plus strand. (11) V1.6 gene models for splice variants (red segments = coding sequence; light-grey segments = UTR). (12) Mapping of RNA-Seq reads derived by Trueseq single-end Illumina sequencing. (See colour plate)

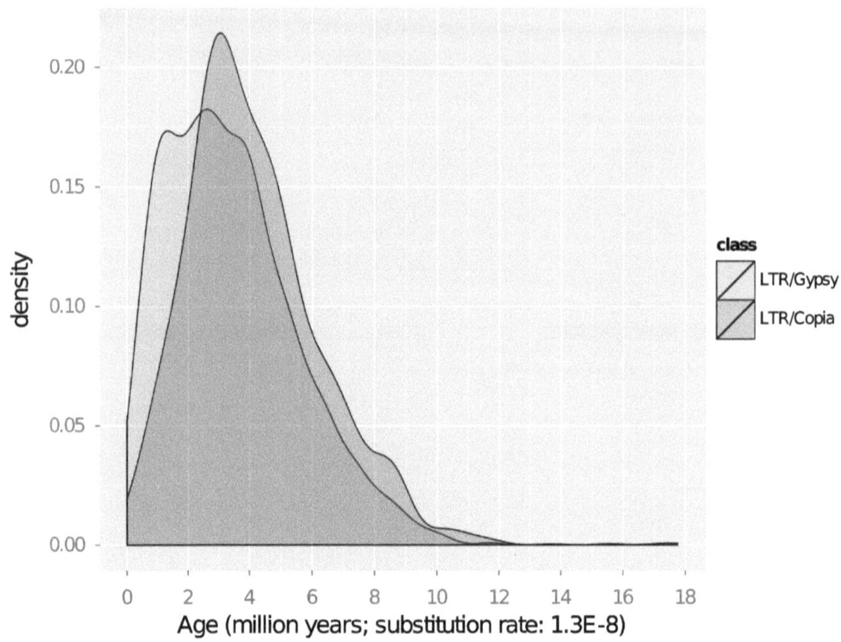

Figure 4 ***Insertion ages of LTR-Rs in* P. patens.** Density plot of estimated insertion ages (in Ma; estimated based on substitution rate described in Rensing et al., 2007) of Copia-like (purple (grey in print versions)) and Gypsy-like (orange (light grey in print versions)) LTR-Rs in the *P. patens* genome. Copia- and Gypsy-like LTR-Rs are autonomous retro-transcribing mobile genetic elements, which are usually comprised of ORFs for Gag and Pol precursor proteins that in return are flanked on both sides by the LTRs and variable-sized noncoding regions. Both classes are distinguished by the structure of the Pol precursor protein, namely the position of the integrase among the protease, reverse transcriptase and ribonuclease H. Depending on the subfamily other protein domains like the chromodomain are also observed in *Physcomitrella* and other plants. The distributions suggest recurring cycles of LTR-Rs activity that partly overlap for the two types. Like in most plant genomes, Gypsy-like are more abundant. *Reanalysis of raw data from Rensing, S. A., Lang, D., Zimmer, A. D., Terry, A., Salamov, A., Shapiro, H., … Boore, J. L. (2008). The* Physcomitrella *genome reveals evolutionary insights into the conquest of land by plants.* Science, 319*(5859), 64–69. http://doi.org/10.1126/science.1150646.*

level of fragmentation and low abundance of full-length LTR-Rs (Fig. 3), the relatively high uniqueness of LTR-R regions (Uniqueome track in Fig. 3) and also the almost complete absence of class II transposons. Leitch and Leitch (2013) previously suggested that this might be the result of an efficient silencing machinery and elimination of DNA by recombination. Thus, this could be the result of the prevalence of the moss to repair DNA double-strand breaks (DSBs) by homologous recombination rather than nonhomologous end joining (Kamisugi et al., 2012), yielding a higher degree of fragmentation of repeats and reshuffling of gene and transposon regions than in seed plants and thus resulting in a higher than expected gene density. Possibly this could also be related to the efficient purging of deleterious mutations observed in the primarily selfing Funariaceae *Physcomitrella* and *Funaria* (Szövényi et al., 2014).

5.4 Complements of Protein-Coding and Nonprotein-Coding Genes

The V1.6 release of the *P. patens* genome annotation (Zimmer et al., 2013) comprises 32,275 protein-coding genes, 432 tRNA loci, 798 rDNA regions, 229 miRNA precursors (108 families), 213 snRNA genes and 6 SRP (signal recognition particle) loci.

While the accuracy and completeness of gene structures is greatly improved due to the increased experimental support from comprehensive RNA-Seq data, the global number of protein-coding genes remains largely stable comparing the published V1.6 to the latest, yet unpublished V3.3 release (adding about 2000 genes).

A comprehensive annotation of *Physcomitrella* sRNA loci based on 10 sRNA-Seq libraries from 10-day-old protonemata (Coruh et al., 2015) identifies 1462 loci sRNA, 1090 of which are annotated as heterochromatic siRNAs (23–24 bp). These siRNA loci are localized mostly in 'intergenic regions with dense DNA methylation': that is, LTR-Rs and other repetitive regions (Fig. 3). Employing very stringent filtering criteria, from the remaining loci the authors deduce 130 miRNA-encoding genes (114 of which overlap with miRBase release 21, also incorporated into V1.6). Coruh et al. (2015) suggest that the remainder of the miRBase annotations should not be considered miRNA *sensu strictu*, attributing these potential misannotations to a lower sequencing depth in previous studies and less strict parameters for identifying miRNA hairpins. Depending on which sRNA mapping tool and parameters are employed, the mere number of sRNA-overlapping loci in our experience varies at least tenfold. From these candidate loci, again the choice of program and parameters used to define miRNA hairpins and true miRNAs ('Dicer-calls') largely affect how many miRNAs are annotated subsequently. Some of these parameters are based on results obtained in other species. What is missing so far is an unbiased selection of attributes determining the action and precision of moss DCL proteins. The efficiency of gene targeting in *Physcomitrella* should enable an experimental setup to identify 'true' miRNAs and their characteristics, for example, based on mutant analysis. Until these results are available the topic remains open for debate, and we can only conclude that the *P. patens* genome harbours about 114–229 miRNA-encoding genes.

The comparative analysis of gene complement and structural evolution along the green plant lineage (Zimmer et al., 2013) demonstrated that overall (excluding recent polyploids) gene complements are consistent in size, and gene structures – especially positioning and length of introns in coding sequences – are highly conserved among land plants. There might be a trend towards reduction of intron lengths in land plants. In contrast, the algae of the Volvocales seem to have acquired introns independently, resulting in longer genes with more and longer introns. There was one striking feature that distinguishes *P. patens* genes – a comparatively high number of loci with more and longer 5′-UTR introns in the moss than in any other of the studied Viridiplantae. Considering the influence of 5′-UTR introns on gene expression level, regulation, translation and nonsense-mediated mRNA decay and the fact that almost 50% of all genes encode for transcripts with 5′-UTR introns, Zimmer et al. (2013) hypothesized that *Physcomitrella* frequently relies on these mechanisms of co- and posttranscriptional gene regulation. Another possible explanation could be that the additional and longer introns arose by the recombination-driven reshuffling of DNA to eliminate transposable elements (see above). Possibly only the observed, highly efficient purging of deleterious mutations in the primarily selfing haploid species (Szövényi et al., 2014) prevents an additional accumulation and expansion of such introns in the coding sequences.

Compared to the flowering plant *Arabidopsis*, the open reading frames (ORFs) of protein-coding genes in *Physcomitrella* display low to no codon usage bias (Rensing et al., 2005; Zimmer et al., 2013, Fig. 5). This is mirrored in the ORFs of moss *Ceratodon purpureus* and the liverwort *Marchantia polymorpha* which also peak at 57 effective codons (Fig. 5). Further evaluation and taxonomic sampling is necessary to allow conclusions about the generalizability of this trend.

Comparative analysis of the functional annotation of the *Physcomitrella* genome to that of other land plants consistently revealed expansions of moss housekeeping and metabolic genes (Lang et al., 2005; Zimmer et al., 2013) and further possibly adaptive, lineage-specific expansions and gains including at least 13% orphan genes (ie, species- or lineage-specific genes; Zimmer et al., 2013).

Possible functions for some of these orphan genes have been suggested by a time series analysis of the cold response in *Physcomitrella* (Beike et al.,

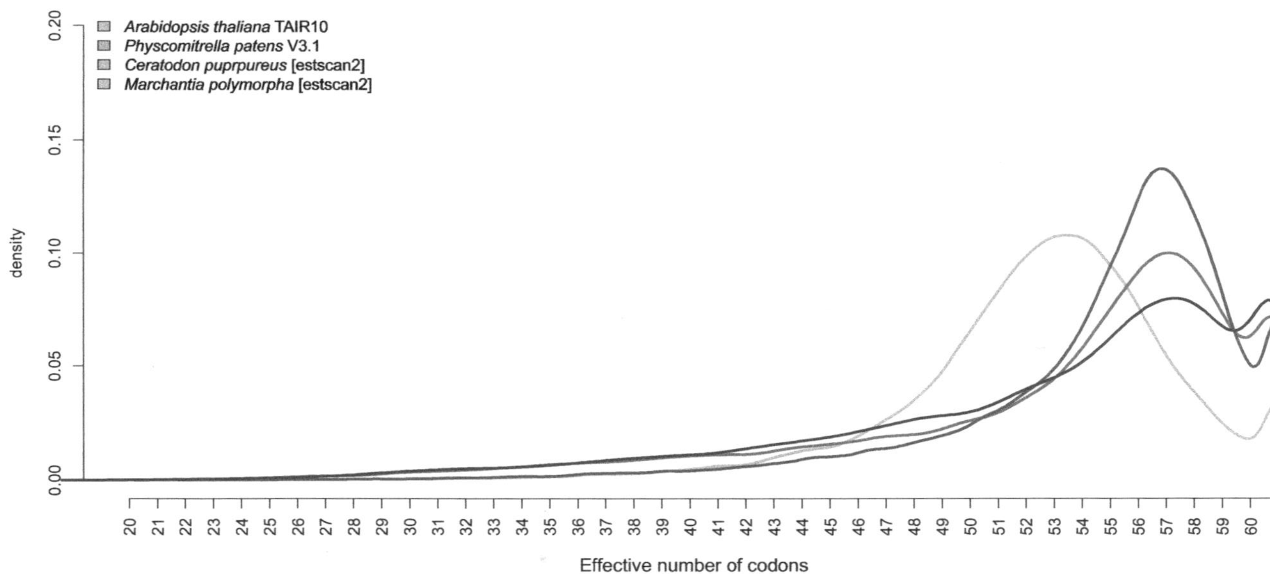

Figure 5 ***Comparison of the effective number of codons (ENC) suggests low codon usage bias.*** The ENC was calculated with CodonW v1.4.4 (http://codonw.sourceforge.net/) for nuclear encoded primary transcripts of *P. patens* V3.1 and *A. thaliana* vTAIR10. In addition, transcriptome assemblies of two other bryophyte species, namely *M. polymorpha* (Sharma, Jung, Bhalla, & Singh, 2014) and *C. purpureus* (Szövényi et al., 2015), were used to infer coding sequences with estscan2 (http://sourceforge.net/projects/estscan/files/ESTScan2/) and subsequently cluster them using CD-HIT-EST v4.6.1 reducing 99% identical isoforms to a representative sequence. Compared to *A. thaliana* (3.0%), *P. patens* and two other bryophyte species show a high proportion of transcripts with no codon bias (*P. patens*: 7.9%; *M. polymorpha*: 11.1%; *C. purpureus*: 5.3%). While *A. thaliana* peaks at 53.3 ENC, the studied bryophyte species peak at about57 ENC (*P. patens*: 57.0; *M. polymorpha*: 57.2; *C. purpureus*: 56.8).

2015). Orphan genes are among the earliest loci responding to cold treatment. Nevertheless, the analysis also suggested that most of transcriptional regulation and physiological/developmental response to this abiotic stress is largely conserved. For example, a consistent global specialization of transcriptional regulators was observed resulting in up-regulation of conserved stress-responsive regulators and a down-regulation of conserved developmental regulators. The results are consistent with a conserved regulatory toolkit of land plants (Floyd & Bowman, 2007; Frank & Scanlon, 2015; Hiss et al., 2014; Richardt et al., 2007) and supported by the observation that the majority of transcriptional regulators either were acquired or expanded in the LCA of land plants (Lang et al., 2010).

A study (Yue, Hu, Sun, Yang, & Huang, 2012) of horizontal gene transfer in the green linage identified 57 families of nuclear genes that were putatively acquired from prokaryotes, fungi or viruses. While 18 of these gene families could be detected in all studied Viridiplantae, 39 were only identified in embryophytes. From the latter set 19 families were restricted to *Physcomitrella* and might represent linage-specific acquisitions, for example, like the insertions of nucleocytoplasmic large DNA viruses discussed earlier (Maumus et al., 2014).

Utilizing the wealth of expression data that now is available for various conditions, developmental stages and even tissue or cell types (Beike et al., 2015; Cooper et al., 2013; Cuming, Cho, Kamisugi, Graham, & Quatrano, 2007; Frank & Scanlon, 2015; Hiss et al., 2014; Nishiyama et al., 2012; O'Donoghue et al., 2013; Shinde, Behpouri, McElwain, & Ng, 2015; Wolf, Rizzini, Stracke, Ulm, & Rensing, 2010; Xiao et al., 2011), we now can obtain a better understanding of how the more than 30,000 protein-coding genes in the *Physcomitrella* genome are expressed along the gametophytic and sporophytic generations of the moss.

According to microarray data (Hiss et al., 2014), at least 74% of the protein-coding genes show evidence of expression in the gametophytic generation. About 13% of these genes appear only to be active in gametophores, while about 3% appear to be specifically expressed in protonemal filaments. O'Donoghue et al. (2013) found 12% of all protein-coding genes to be affected by the transition from gametophyte to sporophyte; 7% of all genes were found to be active specifically in the sporophytic tissues. A lower number of generation-biased genes were reported for the close relative *F. hygrometrica* (Szövényi, Rensing, Lang, Wray, & Shaw, 2011; Szövényi et al., 2013). In their studies comparing RNA-Seq data from gametophores and sporophytes, Szövényi et al. found only 2–3% of all expressed genes to be generation biased. They found the majority of genes (97%) in *Funaria* to be expressed in both generations. Comparative analysis of *Arabidopsis* and *Funaria* revealed only limited conservation of generation-biased gene expression, supporting the view of an ancestral regulatory toolkit that operates in the respective dominant generation of the life cycle (Frank & Scanlon, 2015).

O'Donoghue et al. (2013) attributed the observed difference in generation-biased genes between the two Funariaceae (*Physcomitrella* and *Funaria*) to experimental problems in the *Funaria* study, like a lower number of sampled genes and the exclusion of protonemal tissues. But if we consider that numbers of specifically expressed genes in the latter are only relatively low and most genes are detectable in both gametophytic stages, we can raise the question whether the observed difference has other, biologically founded sources? If so, why does the cleistocarpous *P. patens* with a possibly secondarily reduced sporophyte harbour a larger genome with more generation-biased genes than *F. hygrometrica*, which harbours one of the most complex sporophyte morphologies found in mosses (Budke, Goffinet, & Jones, 2012)?

5.5 Remnants of Ancestral Funariaceae Speciation Events in the *P. patens* Paranome?

As discussed in the previous sections, the suggested direction of the evolution of sexuality from an ancestral dioicous to a monoicous state resulting from the merger of unisexual parental genomes and the inspection of available DNA content and chromosome count data early on implicated the occurrence of one or multiple genome duplication events in the lineage of the Funariaceae (Beike et al., 2014; McDaniel et al., 2010; Rensing et al., 2007, 2009, 2008).

Genome duplication as the result of autopolyploidization or hybridization seems to be a widespread phenomenon among land plants (Blanc, Hokamp, & Wolfe, 2003; Bowers, Chapman, Rong, & Paterson, 2003; Cheng et al., 2013; Chester et al., 2012; Cui et al., 2006; Freeling & Thomas, 2006; Jiao & Paterson, 2014; Maere et al., 2005; Panaud, Jackson, & Wendel, 2014; Rensing et al., 2007, 2008; Soltis & Soltis, 2012; Tang et al., 2008; Tuskan et al., 2006; Vandepoele, Simillion, & Van de Peer, 2003; Vanneste, Baele, Maere, & Van de Peer, 2014; Vanneste, Sterck, Myburg, Van de Peer, & Mizrachi, 2015; Wu et al., 2013). Large-scale or WGD events are commonly thought to have acted as driving forces behind radiation, diversification and speciation of plant lineages that shaped the evolution of organismal complexity (Carroll, 2001; Crow & Wagner, 2006; Freeling & Thomas, 2006). The latter hypothesis was corroborated by phylogenetic comparative analysis of land plants suggesting a strong correlation of WGD with morphological complexity (Lang et al., 2010).

In evolutionary terms, the impact of genome duplication can be condensed to two major aspects: (1) The facilitation of biological innovation resulting from the expansion of gene families with subsequent neo- and subfunctionalization of duplicates (paralogs). (2) The role as a motor of speciation by establishment of reproductive isolation of polyploids due to post-duplication chromosome loss, rearrangements of the genome and biased fractionation of paralogs, for example, in the (re)diploidization of tetrapolyploids (Freeling & Thomas, 2006).

The analysis of synonymous, silent substitution (K_S) rates of codons across all *P. patens* paralogs (ie, the paranome) based on assembled EST data provided initial, molecular confirmation of the suggested paleopolyploid history (Rensing et al., 2007). The distribution plot of pairwise K_s values revealed a peak at 0.85 (0.6–1.1). Based on molecular clock analysis of linearized phylogenetic trees inferred for the moss, *Arabidopsis*, poplar and rice, this peak was dated to have occurred 30–60 Ma (mean 45 Ma). Duplicates which arose from such a paleoploidy are frequently termed paleologs.

The distribution pattern was largely confirmed by subsequent analyses based on genomic data (Rensing et al., 2008), including the analysis with an alternative method that particularly focuses on transversion rates of fourfold degenerate codons (4DTv; Kumar & Subramanian, 2002) shown in Fig. 6. The distribution of pairwise 4DTv distances derived for the V1.2 gene predictions mirrors that inferred from K_s rates on both transcriptomic and genomic data (Fig. 6). There is one broad peak comprising 4342 paralog pairs involving 5586 unique loci in the 4DTv range between 0.1 and 0.4 (blue vertical lines). This broad pattern could be either the result of multiple, subsequent events in 'short' evolutionary time resulting in overlapping distributions of mutation rates or could be due to a large variation of synonymous substitution rates of paleologs from the same event. Additional, possibly older peaks cannot be reliably detected by these methods due to the problem of multiple substitution (black line Fig. 6; 4DTv > 0.5; $K_s > 1$).

The existence of further genome duplication events that are either contained in the broad peak or cannot be detected due to the problem arising from multiple substitution so far can only be assumed by the analysis of independent gene trees. Phylogenetic trees of *P. patens* gene families

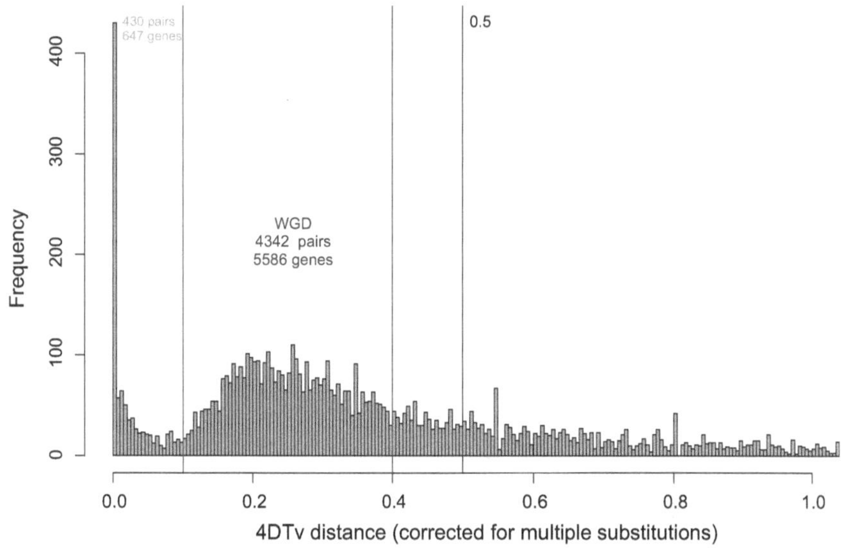

Figure 6 ***The analysis of paralogs in the* Physcomitrella *genome provides evidence for at least one large-scale genome duplication event.*** Histogram of pairwise synonymous substitution rates among pairs of paralogs in the V1.2 genome measured using the 4DTv method demonstrates at least one broad peak comprising 4342 paralog pairs involving 5586 unique loci. This peak is consistent with what was observed using assembled EST data (Rensing et al., 2007). A higher, sharper peak is observed at bin 0, comprising 647 loci. Further, possibly older peaks cannot reliably be detected by this method due to the problem of multiple substitution (4DTv > 0.5). (See colour plate)

frequently display a peculiar pattern in the clustering of paralogs that arose subsequent to speciation (inparalogs). They usually comprise 2–3 times nested clades of closely related inparalogs, hinting at additional polyploidization events. An example of this can be observed in genes of the *Physcomitrella* MADS-box superfamily (Barker & Ashton, 2013). The parsimonious model of lineage-specific expansions MADS-box type I and II families of Elizabeth Barker and Neil Ashton assume three segmental duplications, of which at least two are suspected to be WGD events. Indeed, if we analyse 4DTv distances for the nodes in the phylogeny of the two type II subfamilies MIKC^{C} and MIKC⋆ by median-condensing pairwise rates, we can observe at least two distinct signals (MIKC^{C}/MIKC⋆: 0.2/0.27, 0.37/0.46).

Due to the status of the V1 genome assembly these data can only be considered as preliminary. For higher resolution and confidence, identification of syntenic blocks and structural analysis of chromosome evolution is required. The analysis of the now available V3 pseudo-chromosomes will shed further light on this topic. For the time being, we can conclude that there is convincing evidence for one WGD and initial evidence for one or two older WGD events in the genomic sequence of *P. patens*.

In light of this data, it is tempting to speculate which of these events is shared with other Funariaceae. Preliminary insight again can be found in a study focussing on the MADS-box type 2 MIKC* gene family (Zobell, Faigl, Saedler, & Münster, 2010). Zobell et al. inferred phylogenetic trees of the MIKC* family in the flowering plant *A. thaliana*, the liverwort *M. polymorpha* and the mosses *Sphagnum subsecundum*, *P. patens* and *F. hygrometrica*. The phylogenetic clustering of the Funaria and Physcomitrella MIKC* proteins clearly suggests 1:1 orthologous relationships and does not suggest the existence of additional paleologs in the lineage of Physcomitrella. Thus, combining the data from these two studies and considering the phylogenetic relationship of the two species, we can formulate the working hypothesis that the WGD event(s) detectable in the Physcomitrella patens genome happened in the lineage leading to the LCA of Funaria and Physcomitrella.

There is evidence for more recent, subsequent genome duplication events in the Funariaceae based on the analysis of gene trees and SSR markers in multiple isolates (Beike et al., 2014; Klips, 2015; Liu et al., 2012; McDaniel et al., 2010). These data suggest hybridizations within the *Physcomitrium/Physcomitrella* species complex that gave rise to recent allopolyploid species with increased DNA content and chromosome complement: *Physcomitrium pyriforme*, *P. collenchymatum* and *P. eurystomum* (Beike et al., 2014).

5.6 Initial Evidence for Concerted Pseudoalleles

The analysis of the moss paranome highlights an additional peculiarity (Fig. 6). Strikingly, all plots of synonymous substitution rates of paralogs in *Physcomitrella* generated so far reveal a higher, sharper peak at bin 0. While paralogs clustering in this bin are usually assumed to result from recent, local and small-scale duplications which are observable in most plants, there seems to be an additional, unique source for the low mutation rates observed in these paralogs in *Physcomitrella*.

Closer inspection of the paralogs included in bin 0 of histograms inferred from pairwise synonymous substitution rates revealed that many of them represent tandem arrayed genes (TAGs) residing physically close with no or only few genes in between them (Rensing et al., 2008). TAGs are thought to originate from small-scale duplication events like unequal cross-over, that is, the reciprocal transfer of DNA between sister chromatids in mitosis or homologous chromosomes in meiosis. Their genomic distribution pattern has been shown to be positively correlated with the recombination rate (Rizzon, Ponger, & Gaut, 2006; Zhang & Gaut, 2003), for example, resulting in low abundance of TAGs in the centromeric regions.

TAGs found in the *Physcomitrella* genome significantly deviate from those observed in di- and polyploid flowering plants in terms of frequency, orientation and conservation. Most analysed plant and animal genomes contain about 10–15% of all genes in tandem arrays (Pan & Zhang, 2008; Rizzon et al., 2006; Semple & Wolfe, 1999; Zhang & Gaut, 2003), and these TAGs account for about one-third (30–34%) of all paralogs in the genomes.

The *Physcomitrella* genome contains significantly less genes in tandem arrays (Rensing et al., 2008). Only 1% of all protein-coding genes in V1.1 where identified to be tandemly arrayed.

The distribution of tandem array sizes corroborates what has been reported for other organisms (Pan & Zhang, 2008; Rizzon et al., 2006): the majority of tandem arrays detected in *Physcomitrella* comprise only two members that are separated by short, unrelated, mostly gene-free, spacer regions. An example for a genomic region comprising such a TAG paralog pair is shown in Fig. 7, which depicts the loci encoding the LHCB1N6 protein, a component of the LHCII major antenna (Rensing et al., 2008; Lang et al., 2008). As indicated by the green, interconnected segments in Fig. 7, the entire coding region, most of the UTRs and the promoter regions are part of inverted, perfectly identical repeat regions. Only some bases differ in the exon-encoding 3′-UTR region, resulting in polymorphic ESTs which can be used to confirm that indeed both loci are expressed. Additional, nonconnected green segments indicate repetitive blocks that are identical to a remote location in the genome. These perfectly identical regions frequently lead to difficulties in genome assembly. This was also the case for the missing second Aux/IAA locus mentioned earlier, which is part of a TAG that could not be assembled properly. As these cases can usually only be resolved via primer-walking, the genome-wide extent of this problem cannot be assessed easily (eg, in the analysis of the CHS family performed by Wolf et al., 2010).

This high level of sequence conservation is reflected on the functional level – while flowering plant paralogs derived from small-scale duplications (SSD) like TAGs harbour higher sequence and functional divergence rates than those retained after WGD, the contrary was observed in *Physcomitrella* (Carretero-Paulet & Fares, 2012; Rensing, 2014).

Unsurprisingly, these perfectly identical regions are found among the paralogs clustering in bin 0. Do these paralogs really represent recent local

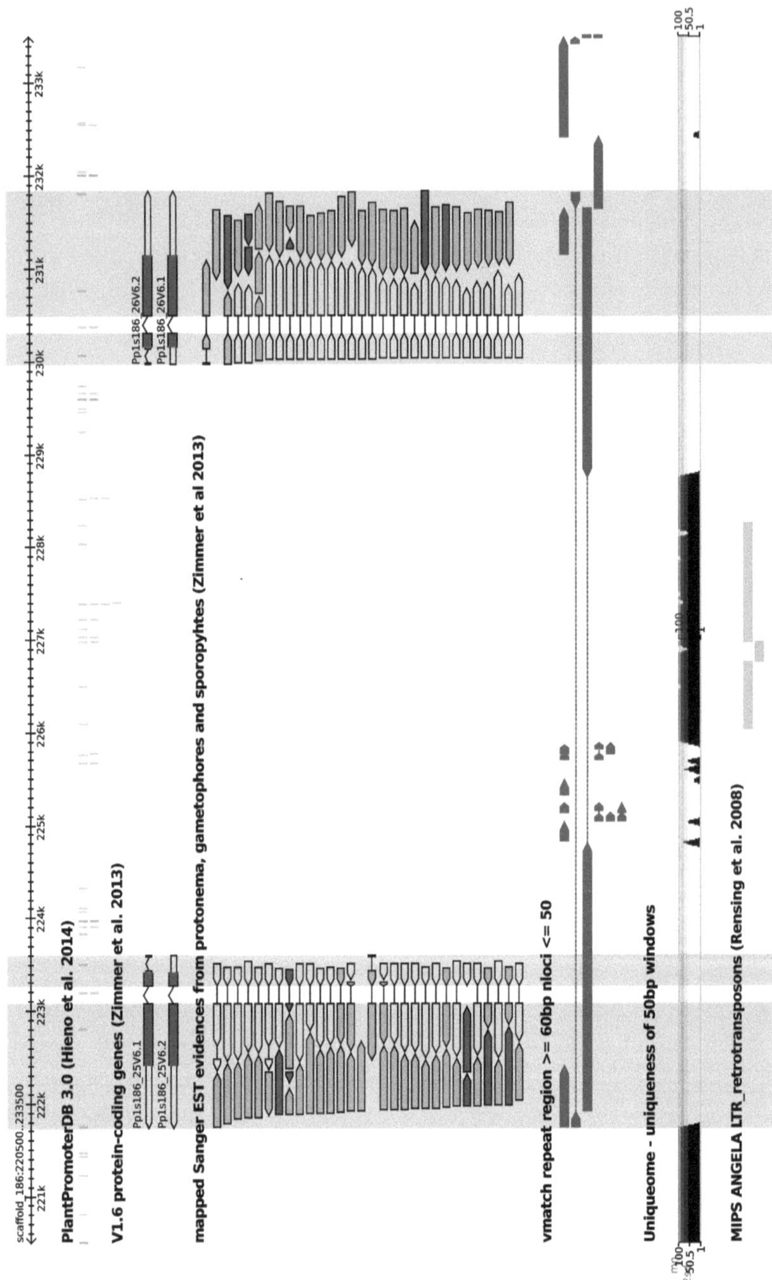
scaffold_186:220500..233500
221k
222k
223k
224k
225k
226k
227k
228k
229k
230k
231k
232k
233k
PlantPromoterDB 3.0 (Hieno et al. 2014)
V1.6 protein-coding genes (Zimmer et al. 2013)
Pp1s186_25V6.1
Pp1s186_25V6.2
Pp1s186_26V6.2
Pp1s186_26V6.1
mapped Sanger EST evidences from protonema, gametophores and sporopyhtes (Zimmer et al 2013)
vmatch repeat region >= 60bp nloci <= 50
Uniqueome - uniqueness of 50bp windows
MIPS ANGELA LTR_retrotransposons (Rensing et al. 2008)
100
50.5
1

Figure 7 ***Example of a putative pseudoallele with identical coding sequences in the* P. patens *genome.*** cosmoss.org genome browser snapshot displaying the inverted, head-to-head tandem duplication pair encoding the LHCB1N6 protein, a component of the LHCII major antenna (Rensing et al., 2008; Lang et al., 2008). Coding regions of two splice variants at each locus are shown as red boxes, while UTR regions are displayed as light grey boxes. In order to demonstrate overlaps with other features, the respective exonic regions are highlighted in grey in all tracks. Additional tracks (top to bottom) are putative promoter elements from the plant promoter database (Hieno et al., 2014), spliced alignments of Sanger EST evidences used for V1.6 predictions from multiple tissues (dark green sporophytes, bright green gametophores, green protonema; original number reduced for brevity), identical repeats longer than 60 bp with less than 50 loci in the genome identified by Vmatch (dark green; pairs connected by *dashed* lines; http://www.vmatch.de/), Uniqueome (black histogram; level of uniqueness in a sliding-window approach of 50 bp), full-length and fragmented LTR-Rs (brown segments; Rensing et al., 2008).

duplication events? An alternative source for such homogenization could be concerted evolution of these regions due to unequal crossover and gene conversion. The latter describes the unidirectional, nonreciprocal transfer of genetic information between duplicated regions (Chen, Cooper, Chuzhanova, Férec, & Patrinos, 2007) and is thought to be mainly the result of DSB repair via homologous recombination.

Thus, this pattern might be related to the high rate of DNA repair of moss by homologous recombination (Kamisugi et al., 2012; Markmann-Mulisch et al., 2007), resulting in concerted evolution of these loci via gene conversion allowing the haploid-dominant organism to maintain 'pseudoalleles' (Lewis, 1951) of highly expressed and other dosage-sensitive genes (Lang et al., 2008).

6. CONCLUDING REMARKS AND OUTLOOK

Over the last 20 years, the dedication and hard work of the international moss community has amounted in substantial genomic resources for the moss *P. patens*. These efforts have led us from the early beginnings starting with a few sequenced cDNA clones, via EST-based transcriptome representations to the present, tentatively cumulating in the first pseudo-chromosome assembly of a nonvascular plant, which give us genome-scale, single bp resolution of genetic and epigenetic regulation of protein- and nonprotein-coding genes, transposable elements and other genetic regions in the moss.

Building on these resources, in combination with the comprehensive molecular toolbox which has been developed in parallel, the model *Physcomitrella* continues to attract an increasing community of researchers from all fields. The accumulated knowledge and genomic resources serve as flagship and reference frameworks to extend our genomic resolution of bryophytes beyond the *P. patens* genome. As illustrated by some of the other chapters in this book, the community has begun to extend genomic and transcriptomic resolution to other bryophyte taxa and even to population level. Bryophyte research has arrived in the genomic and post-genomic age.

In order to serve as a reliable reference, we need more than a few sequencing runs and gigabytes on our hard drives — the backflow of information inferred from the genomic resources and a continued, community-driven effort to improve and extend the structural and functional annotation of the *Physcomitrella* and other bryophyte genomes are vital to the success of these endeavours.

Most certainly, mosses and other bryophytes still present us with many riddles about their ancestry, biology and the evolutionary mechanisms that shaped their genomes and morphology. As demonstrated with the discovery of heterochromatin (Heitz, 1928a), solutions to these puzzles can lead to important biological insights with an impact beyond the taxonomic level of bryophytes. The answers to the questions that attracted the early pioneers of genetic research to mosses and other bryophytes are now within our reach. Genomic, phylogenomic and population genomic insights into the biology and evolutionary history of *Physcomitrella* and the Funariaceae surely will continue to contribute to this progress.

REFERENCES

Alboresi, A., Caffarri, S., Nogue, F., Bassi, R., & Morosinotto, T. (2008). *In silico* and biochemical analysis of *Physcomitrella patens* photosynthetic antenna: identification of subunits which evolved upon land adaptation. *PLoS One, 3*(4), e2033. http://doi.org/10.1371/journal.pone.0002033.

Allen, C. E. (1917). A chromosome difference correlated with sex differences in *Sphaerocarpos*. *Science, 46*(1193), 466–467. http://doi.org/10.1126/science.46.1193.466.

Axtell, M. J., Snyder, J. A., & Bartel, D. P. (2007). Common functions for diverse small RNAs of land plants. *The Plant Cell, 19*(6), 1750–1769. http://doi.org/10.1105/tpc.107.051706.

Aya, K., Hiwatashi, Y., Kojima, M., Sakakibara, H., Ueguchi-Tanaka, M., Hasebe, M., & Matsuoka, M. (2011). The Gibberellin perception system evolved to regulate a pre-existing GAMYB-mediated system during land plant evolution. *Nature Communications, 2*, 544. http://doi.org/10.1038/ncomms1552.

Bach, S. S., King, B. C., Zhan, X., Simonsen, H. T., & Hamberger, B. (2014). Heterologous stable expression of terpenoid biosynthetic genes using the moss *Physcomitrella patens*. *Methods in Molecular Biology (Clifton, NJ), 1153*, 257–271. http://doi.org/10.1007/978-1-4939-0606-2_19.

Barker, E. I., & Ashton, N. W. (2013). A parsimonious model of lineage-specific expansion of MADS-box genes in *Physcomitrella patens*. *Plant Cell Reports, 32*(8), 1161–1177. http://doi.org/10.1007/s00299-013-1411-8.

Bartels, S., González Besteiro, M. A., Lang, D., & Ulm, R. (2010). Emerging functions for plant MAP kinase phosphatases. *Trends in Plant Science, 15*(6), 322–329.

Beike, A. K., Lang, D., Zimmer, A. D., Wüst, F., Trautmann, D., Wiedemann, G.… Reski, R. (2015). Insights from the cold transcriptome of *Physcomitrella patens*: global specialization pattern of conserved transcriptional regulators and identification of orphan genes involved in cold acclimation. *The New Phytologist, 205*(2), 869–881. http://doi.org/10.1111/nph.13004.

Beike, A. K., von Stackelberg, M., Schallenberg-Rüdinger, M., Hanke, S. T., Follo, M., Quandt, D.… Rensing, S. A. (2014). Molecular evidence for convergent evolution and allopolyploid speciation within the *Physcomitrium-Physcomitrella* species complex. *BMC Evolutionary Biology, 14*(158). http://doi.org/10.1186/1471-2148-14-158.

Bennett, M. D., & Leitch, I. J. (2012). *Angiosperm DNA c-values database* (Release 8.0, December 2012).

Blanc, G., Hokamp, K., & Wolfe, K. H. (2003). A recent polyploidy superimposed on older large-scale duplications in the *Arabidopsis* genome. *Genome Research, 13*(2), 137–144.

Bowers, J. E., Chapman, B. A., Rong, J., & Paterson, A. H. (2003). Unravelling angiosperm genome evolution by phylogenetic analysis of chromosomal duplication events. *Nature, 422*(6930), 433–438.

Budke, J. M., Goffinet, B., & Jones, C. S. (2011). A hundred-year-old question: is the moss calyptra covered by a cuticle? A case study of *Funaria hygrometrica. Annals of Botany, 107*(8), 1279–1286. http://doi.org/10.1093/aob/mcr079.

Budke, J. M., Goffinet, B., & Jones, C. S. (2012). The cuticle on the gametophyte calyptra matures before the sporophyte cuticle in the moss *Funaria hygrometrica* (Funariaceae). *American Journal of Botany, 99*(1), 14–22. http://doi.org/10.3732/ajb.1100311.

Carretero-Paulet, L., & Fares, M. A. (2012). Evolutionary dynamics and functional specialization of plant paralogs formed by whole and small-scale genome duplications. *Molecular Biology and Evolution, 29*(11), 3541–3551. http://doi.org/10.1093/molbev/mss162.

Carroll, S. B. (2001). Chance and necessity: the evolution of morphological complexity and diversity. *Nature, 409*(6823), 1102–1109. http://doi.org/10.1038/35059227.

Chen, J.-M., Cooper, D. N., Chuzhanova, N., Férec, C., & Patrinos, G. P. (2007). Gene conversion: mechanisms, evolution and human disease. *Nature Reviews. Genetics, 8*(10), 762–775. http://doi.org/10.1038/nrg2193.

Cheng, S., van den Bergh, E., Zeng, P., Zhong, X., Xu, J., Liu, X. … Schranz, M. E. (2013). The *Tarenaya hassleriana* genome provides insight into reproductive trait and genome evolution of crucifers. *The Plant Cell, 25*(8), 2813–2830. http://doi.org/10.1105/tpc.113.113480.

Chester, M., Gallagher, J. P., Symonds, V. V., Cruz da Silva, A. V., Mavrodiev, E. V., Leitch, A. R. … Soltis, D. E. (2012). Extensive chromosomal variation in a recently formed natural allopolyploid species, *Tragopogon miscellus* (Asteraceae). *Proceedings of the National Academy of Sciences of the United States of America, 109*(4), 1176–1181. http://doi.org/10.1073/pnas.1112041109.

Collier, P. A., & Hughes, K. W. (1982). Life cycle of the moss, *Physcomitrella patens*, in culture. *Journal of Tissue Culture Methods,* 7(1), 19–22. http://doi.org/10.1007/BF01666875.

Cooper, L., Walls, R. L., Elser, J., Gandolfo, M. A., Stevenson, D. W., Smith, B. … Jaiswal, P. (2013). The plant ontology as a tool for comparative plant anatomy and genomic analyses. *Plant & Cell Physiology, 54*(2), e1. http://doi.org/10.1093/pcp/pcs163.

Coruh, C., Cho, S. H., Shahid, S., Liu, Q., Wierzbicki, A., & Axtell, M. J. (2015). Comprehensive annotation of *Physcomitrella patens* small RNA loci reveals that the heterochromatic short interfering RNA pathway is largely conserved in land plants. *The Plant Cell, 27*(8), 2148–2162. http://doi.org/10.1105/tpc.15.00228.

Crow, K. D., & Wagner, G. P. (2006). Proceedings of the SMBE Tri-national young investigators' workshop 2005. What is the role of genome duplication in the evolution of complexity and diversity? *Molecular Biology and Evolution, 23*(5), 887–892. http://doi.org/10.1093/molbev/msj083.

Cui, L., Wall, P. K., Leebens-Mack, J. H., Lindsay, B. G., Soltis, D. E., Doyle, J. J. … Depamphilis, C. W. (2006). Widespread genome duplications throughout the history of flowering plants. *Genome Research, 15*, 15.

Cuming, A. C. (2011). Molecular bryology: mosses in the genomic era. *Field Bryology, 103*, 9–13.

Cuming, A. C., Cho, S. H., Kamisugi, Y., Graham, H., & Quatrano, R. S. (2007). Microarray analysis of transcriptional responses to abscisic acid and osmotic, salt, and drought stress in the moss, *Physcomitrella patens. The New Phytologist, 176*(2), 275–287. http://doi.org/10.1111/j.1469-8137.2007.02187.x.

Decker, E. L., & Reski, R. (2008). Current achievements in the production of complex biopharmaceuticals with moss bioreactors. *Bioprocess and Biosystems Engineering, 31*(1), 3–9. http://doi.org/10.1007/s00449-007-0151-y.

Devos, N., Renner, M. A. M., Gradstein, R., Shaw, A. J., Laenen, B., & Vanderpoorten, A. (2011). Evolution of sexual systems, dispersal strategies and habitat selection in the liverwort genus *Radula*. *The New Phytologist, 192*(1), 225–236. http://doi.org/10.1111/j.1469-8137.2011.03783.x.

Dobzhansky, T. (1973). Nothing in biology makes sense except in the light of evolution. *The American Biology Teacher, 35*(3), 125–129.

Floyd, S. K., & Bowman, J. L. (2007). The ancestral developmental tool kit of land plants. *International Journal of Plant Sciences, 168*(1), 1–35.

Frahm, J.-P. (2001). *Biologie der Moose*. Heidelberg, Berlin: Spektrum Akademischer Verlag.

Frank, M. H., & Scanlon, M. J. (2015). Transcriptomic evidence for the evolution of shoot meristem function in sporophyte-dominant land plants through concerted selection of ancestral gametophytic and sporophytic genetic programs. *Molecular Biology and Evolution, 32*(2), 355–367. http://doi.org/10.1093/molbev/msu303.

Freeling, M., & Thomas, B. C. (2006). Gene-balanced duplications, like tetraploidy, provide predictable drive to increase morphological complexity. *Genome Research, 16*(7), 805–814.

Fritsch, R. (1991). *Index to bryophyte chromosome counts* (Vol. 40). Berlin, Stuttgart: J. Cramer/Gebrueder Borntraeger.

Gitzinger, M., Parsons, J., Reski, R., & Fussenegger, M. (2009). Functional cross-kingdom conservation of mammalian and moss (*Physcomitrella patens*) transcription, translation and secretion machineries. *Plant Biotechnology Journal, 7*(1), 73–86. http://doi.org/10.1111/j.1467-7652.2008.00376.x.

Goff, S. A., Ricke, D., Lan, T. H., Presting, G., Wang, R., Dunn, M. … Briggs, S. (2002). A draft sequence of the rice genome (*Oryza sativa* L. ssp. japonica). *Science, 296*(5565), 92–100.

Goffinet, B. (2007). *Physcomitrella* bruch & schimper. *Flora of North America, 27*, 194–195.

Goode, J. A., Stead, A. D., & Duckett, J. G. (1993). Redifferentiation of moss protonemata: an experimental and immunofluorescence study of brood cell formation. *Canadian Journal of Botany, 71*(11), 1510–1519. http://doi.org/10.1139/b93-183.

Goodstein, D. M., Shu, S., Howson, R., Neupane, R., Hayes, R. D., Fazo, J. … Rokhsar, D. S. (2012). Phytozome: a comparative platform for green plant genomics. *Nucleic Acids Research, 40*(Database issue), D1178–D1186. http://doi.org/10.1093/nar/gkr944.

Gregory, T. R. (2005). The C-value enigma in plants and animals: a review of parallels and an appeal for partnership. *Annals of Botany, 95*(1), 133–146. http://doi.org/10.1093/aob/mci009.

Greilhuber, J., Leitch, I. J., & Bennett, M. D. (2010). *Bryophyte DNA c-values database (Release 3.0, December 2010).*

Hedges, S. B., Marin, J., Suleski, M., Paymer, M., & Kumar, S. (2015). Tree of life reveals clock-like speciation and diversification. *Molecular Biology and Evolution, 32*(4), 835–845. http://doi.org/10.1093/molbev/msv037.

Heitz, E. (1928a). Das Heterochromatin der Moose: I. In *Jahrbuch für wissenschaftliche Botanik, 69* pp. 762–818).

Heitz, E. (1928b). Der bilaterale Bau der Geschlechtschromosomen und Autosomen bei *Pellia fabbroniana, P. epiphylla* und einigen anderen Iungermanniaceen. *Planta, 5*, 725–768.

Hieno, A., Naznin, H. A., Hyakumachi, M., Sakurai, T., Tokizawa, M., Koyama, H. … Yamamoto, Y. Y. (2014). Ppdb: plant promoter database version 3.0. *Nucleic Acids Research, 42*(D1). http://doi.org/10.1093/nar/gkt1027.

Hirano, K., Nakajima, M., Asano, K., Nishiyama, T., Sakakibara, H., Kojima, M. … Matsuoka, M. (2007). The GID1-Mediated gibberellin perception mechanism is conserved in the lycophyte *Selaginella moellendorffii* but not in the bryophyte *Physcomitrella patens*. *Plant Cell, 26*, 26.

Hirschman, J., Berardini, T. Z., Drabkin, H. J., & Howe, D. (2010). A MOD(ern) perspective on literature curation. *Molecular Genetics and Genomics, 283*(5), 415–425. http://doi.org/10.1007/s00438-010-0525-8.

Hiss, M., Laule, O., Meskauskiene, R. M., Arif, M. A., Decker, E. L., Erxleben, A.… Rensing, S. A. (2014). Large-scale gene expression profiling data for the model moss *Physcomitrella patens* aid understanding of developmental progression, culture and stress conditions. *The Plant Journal : For Cell and Molecular Biology, 79*(3), 530–539. http://doi.org/10.1111/tpj.12572.

Hohe, A., Rensing, S. A., Mildner, M., Lang, D., & Reski, R. (2002). Day length and temperature strongly influence sexual reproduction and expression of a novel MADS-Box gene in the moss *Physcomitrella patens. Plant Biology, 4*(6), 762.

Howe, D., Costanzo, M., Fey, P., Gojobori, T., Hannick, L., Hide, W.… Rhee, S. Y. (2008). Big data: the future of biocuration. *Nature, 455*(7209), 47–50. http://doi.org/10.1038/455047a.

Hu, T. T., Pattyn, P., Bakker, E. G., Cao, J., Cheng, J.-F., Clark, R. M.… Guo, Y.-L. (2011). The *Arabidopsis lyrata* genome sequence and the basis of rapid genome size change. *Nature Genetics, 43*(5), 476–481. http://doi.org/10.1038/ng.807.

Jiao, Y., & Paterson, A. H. (2014). Polyploidy-associated genome modifications during land plant evolution. *Philosophical Transactions of the Royal Society of London. Series B, Biological Sciences, 369*(1648), 20130355. http://doi.org/10.1098/rstb.2013.0355.

Kamisugi, Y., Schaefer, D. G., Kozak, J., Charlot, F., Vrielynck, N., Holá, M.… Nogué, F. (2012). MRE11 and RAD50, but not NBS1, are essential for gene targeting in the moss *Physcomitrella patens. Nucleic Acids Research, 40*(8), 3496–3510. http://doi.org/10.1093/nar/gkr1272.

Kamisugi, Y., Schlink, K., Rensing, S. A., Schween, G., von Stackelberg, M., Cuming, A. C.… Cove, D. J. (2006). The mechanism of gene targeting in *Physcomitrella patens*: homologous recombination, concatenation and multiple integration. *Nucleic Acids Research, 34*(21), 6206–6214. http://doi.org/10.1093/nar/gkl832.

Kamisugi, Y., Von Stackelberg, M., Lang, D., Care, M., Reski, R., Rensing, S. A., & Cuming, A. C. (2008). A sequence-anchored genetic linkage map for the moss, *Physcomitrella patens. The Plant Journal, 56*(5), 855–866. http://doi.org/10.1111/j.1365-313X.2008.03637.x.

Klips, R. A. (2015). DNA microsatellite analysis of sporophytes of the short-lived moss *Physcomitrium pyriforme* reveals a predominantly self-fertilizing mating pattern. *The Bryologist, 118*(2), 200–211. http://doi.org/10.1639/0007-2745-118.2.200.

Knapp, E. (1936). Zur Genetik von *Sphaerocarpus. Berichte der Deutschen Botanischen Gesellschaft, 54*(11), 58–69.

Kumar, S., & Subramanian, S. (2002). Mutation rates in mammalian genomes. *Proceedings of the National Academy of Sciences of the United States of America, 99*(2), 803–808. http://doi.org/10.1073/pnas.022629899.

Laenen, B., Shaw, B., Schneider, H., Goffinet, B., Paradis, E., Désamoré, A.… Shaw, A. J. (2014). Extant diversity of bryophytes emerged from successive post-Mesozoic diversification bursts. *Nature Communications, 5*, 5134. http://doi.org/10.1038/ncomms6134.

Landberg, K., Pederson, E. R. A., Viaene, T., Bozorg, B., Friml, J., Jönsson, H.… Sundberg, E. (2013). The moss *Physcomitrella patens* reproductive organ development is highly organized, affected by the two SHI/STY genes and by the level of active auxin in the SHI/STY expression domain. *Plant Physiology, 162*(3), 1406–1419. http://doi.org/10.1104/pp.113.214023.

Lang, D., Eisinger, J., Reski, R., & Rensing, S. A. (2005). Representation and high-quality annotation of the *Physcomitrella patens* transcriptome demonstrates a high proportion of proteins involved in metabolism in mosses. *Plant Biology, 7*(3), 238–250. http://doi.org/10.1055/s-2005-837578.

Lang, D., Weiche, B., Timmerhaus, G., Richardt, S., Riaño-Pachón, D. M., Corrêa, L. G. ... Rensing, S. A. (2010). Genome-wide phylogenetic comparative analysis of plant transcriptional regulation: a timeline of loss, gain, expansion and correlation with complexity. *Genome Biology and Evolution, 2*, 488–503. http://doi.org/10.1093/gbe/evq032.

Lang, D., Zimmer, A. D., Rensing, S. A., & Reski, R. (2008). Exploring plant biodiversity: the *Physcomitrella* genome and beyond. *Trends in Plant Science, 13*(10), 542–549.

Leitch, I. J., & Leitch, A. R. (2013). Genome size diversity and evolution in land plants. In *Plant genome diversity* (Vol. 2, pp. 307–322). Vienna: Springer Verlag.

Lewis, E. B. (1951). Pseudoallelism and gene evolution. *Cold Spring Harbor Symposia on Quantitative Biology, 16*, 159–174.

Liang, C., Jaiswal, P., Hebbard, C., Avraham, S., Buckler, E. S., Casstevens, T. ... Stein, L. (2008). Gramene: a growing plant comparative genomics resource. *Nucleic Acids Research, 36*(Database issue), D947–D953. http://doi.org/10.1093/nar/gkm968.

Lind, C., Dreyer, I., López-Sanjurjo, E. J., von Meyer, K., Ishizaki, K., Kohchi, T. ... Hedrich, R. (2015). Stomatal guard cells co-opted an ancient ABA-dependent desiccation survival system to regulate stomatal closure. *Current Biology : CB, 25*(7), 928–935. http://doi.org/10.1016/j.cub.2015.01.067.

Lindner, A.-C., Lang, D., Seifert, M., Podlešáková, K., Novák, O., Strnad, M. ... von Schwartzenberg, K. (2014). Isopentenyltransferase-1 (IPT1) knockout in *Physcomitrella* together with phylogenetic analyses of IPTs provide insights into evolution of plant cytokinin biosynthesis. *Journal of Experimental Botany, 65*(9), 2533–2543. http://doi.org/10.1093/jxb/eru142.

Liu, Y., Budke, J. M., & Goffinet, B. (2012). Phylogenetic inference rejects sporophyte based classification of the Funariaceae (Bryophyta): rapid radiation suggests rampant homoplasy in sporophyte evolution. *Molecular Phylogenetics and Evolution, 62*(1), 130–145. http://doi.org/10.1016/j.ympev.2011.09.010.

Lucumi, A., Posten, C., & Pons, M.-N. (2005). Image analysis supported moss cell disruption in photo-bioreactors. *Plant Biology, 7*(3), 276–282. http://doi.org/10.1055/s-2005-865638.

Machuka, J., Bashiardes, S., Ruben, E., Spooner, K., Cuming, A., Knight, C. ... Spooner, K. (1999). Sequence analysis of expressed sequence tags from an ABA-treated cDNA library identifies stress response genes in the moss *Physcomitrella patens*. *Plant Cell Physiology, 40*(4), 378–387.

Maere, S., De Bodt, S., Raes, J., Casneuf, T., Van Montagu, M., Kuiper, M., & Van de Peer, Y. (2005). Modeling gene and genome duplications in eukaryotes. *Proceedings of the National Academy of Sciences of the United States of America, 102*(15), 5454–5459.

Markmann-Mulisch, U., Wendeler, E., Zobell, O., Schween, G., Steinbiss, H.-H. H., & Reiss, B. (2007). Differential requirements for RAD51 in *Physcomitrella patens* and *Arabidopsis thaliana* development and DNA damage repair. *The Plant Cell, 19*(10), 3080–3089. http://doi.org/10.1105/tpc.107.054049.

Martin, A., Lang, D., Hanke, S. T., Mueller, S. J. X., Sarnighausen, E., Vervliet-Scheebaum, M., & Reski, R. (2009). Targeted gene knockouts reveal overlapping functions of the five *Physcomitrella patens* FtsZ isoforms in chloroplast division, chloroplast shaping, cell patterning, plant development, and gravity sensing. *Molecular Plant, 2*(6), 1359–1372. http://doi.org/10.1093/mp/ssp076.

Maumus, F., Epert, A., Nogué, F., & Blanc, G. (2014). Plant genomes enclose footprints of past infections by giant virus relatives. *Nature Communications, 5*, 4268. http://doi.org/10.1038/ncomms5268.

Mazumder, R., Natale, D. A., Julio, J. A. E., Yeh, L.-S., & Wu, C. H. (2010). Community annotation in biology. *Biology Direct, 5*, 12. http://doi.org/10.1186/1745-6150-5-12.

McDaniel, S. F., von Stackelberg, M., Richardt, S., Quatrano, R. S., Reski, R., & Rensing, S. A. (2010). The speciation history of the *Physcomitrium-Physcomitrella* species complex. *Evolution; International Journal of Organic Evolution, 64*(1), 217–231. http://doi.org/10.1111/j.1558-5646.2009.00797.x.

McDaniel, S. F., Atwood, J., & Burleigh, J. G. (2013). Recurrent evolution of dioecy in bryophytes. *Evolution; International Journal of Organic Evolution, 67*(2), 567–572. http://doi.org/10.1111/j.1558-5646.2012.01808.x.

McIntosh, T. T. (2007). 12. Funariaceae Schwägrichen. *Flora of North America, 27*, 180.

Medina, R., Liu, Y., Li-Song, W., Shuiliang, G., Hylander, K., & Goffinet, B. (2015). DNA based revised geographic circumscription of species of Physcomitrella S.l. (Funariaceae): *P. Patens* new to East Asia and *P. Magdalenae* new to East Africa. *The Bryologist, 118*(1), 22–31. http://dx.doi.org/10.1639/0007-2745-118.1.022. The American Bryological and Lichenological Society, Inc.

Mitschke, J., Fuss, J., Blum, T., Höglund, A., Reski, R., Kohlbacher, O., & Rensing, S. A. (2009). Prediction of dual protein targeting to plant organelles. *The New Phytologist, 183*(1), 224–235. http://doi.org/10.1111/j.1469-8137.2009.02832.x.

Muller, H. J. (1964). The relation of recombination to mutational advance. *Mutation Research/Fundamental and Molecular Mechanisms of Mutagenesis, 1*(1), 2–9. http://doi.org/10.1016/0027-5107(64)90047-8.

Nebel, M., & Philippi, G. (2000). *Die Moose Baden-Württembergs Band 1*. Stuttgart: Eugen Ulmer.

Newton, A. E., Wikström, N., Bell, N., Forrest, L. L., & Ignatov, M. S. (2006). Dating the diversification of the pleurocarpous mosses. In Newton, & Tangney (Eds.), *Pleurocarpous mosses: Systematics and evolution*. Boca Raton: CRC Press (Taylor & Francis Books, Inc).

Nishiyama, T., Fujita, T., Shin, I. T., Seki, M., Nishide, H., Uchiyama, I. … Hasebe, M. (2003). Comparative genomics of *Physcomitrella patens* gametophytic transcriptome and *Arabidopsis thaliana*: implication for land plant evolution. *Proceedings of the National Academy of Sciences of the United States of America, 100*(13), 8007–8012.

Nishiyama, T., Miyawaki, K., Ohshima, M., Thompson, K., Nagashima, A., Hasebe, M., & Kurata, T. (2012). Digital gene expression profiling by 5′-end sequencing of cDNAs during reprogramming in the moss *Physcomitrella patens*. *PLoS One*, 7(5), e36471. http://doi.org/10.1371/journal.pone.0036471.

O'Donoghue, M.-T., Chater, C., Wallace, S., Gray, J. E., Beerling, D. J., & Fleming, A. J. (2013). Genome-wide transcriptomic analysis of the sporophyte of the moss *Physcomitrella patens*. *Journal of Experimental Botany, 64*(12), 3567–3581. http://doi.org/10.1093/jxb/ert190.

Pan, D., & Zhang, L. (2008). Tandemly arrayed genes in vertebrate genomes. *Comparative and Functional Genomics*, 545269. http://doi.org/10.1155/2008/545269.

Panaud, O., Jackson, S. A., & Wendel, J. F. (2014). Drivers and dynamics of diversity in plant genomes. *The New Phytologist, 202*(1), 15–18. http://doi.org/10.1111/nph.12633.

Paponov, I. A., Teale, W., Lang, D., Paponov, M., Reski, R., Rensing, S. A., & Palme, K. (2009). The evolution of nuclear auxin signalling. *BMC Evolutionary Biology, 9*(1), 126. http://doi.org/10.1186/1471-2148-9-126.

Passarge, E. (1979). Emil Heitz and the concept of heterochromatin: longitudinal chromosome differentiation was recognized fifty years ago. *American Journal of Human Genetics, 31*(2), 106–115.

Perroud, P.-F., Cove, D. J., Quatrano, R. S., & McDaniel, S. F. (2011). An experimental method to facilitate the identification of hybrid sporophytes in the moss *Physcomitrella patens* using fluorescent tagged lines. *The New Phytologist, 191*(1), 301–306. http://doi.org/10.1111/j.1469-8137.2011.03668.x.

Pils, B., & Heyl, A. (2009). Unraveling the evolution of cytokinin signaling. *Plant Physiology, 151*(2), 782–791. http://doi.org/10.1104/pp.109.139188.

Pitsch, N. T., Witsch, B., & Baier, M. (2010). Comparison of the chloroplast peroxidase system in the chlorophyte *Chlamydomonas reinhardtii*, the bryophyte *Physcomitrella patens*, the

lycophyte *Selaginella moellendorffii* and the seed plant *Arabidopsis thaliana. BMC Plant Biology, 10*(133). http://doi.org/10.1186/1471-2229-10-133.
Prigge, M. J., Lavy, M., Ashton, N. W., & Estelle, M. (2010). *Physcomitrella patens* auxin-resistant mutants affect conserved elements of an auxin-signaling pathway. *Current Biology, 20*(21), 1907–1912. http://doi.org/10.1016/j.cub.2010.08.050.
Ranjan, A., Dickopf, S., Ullrich, K. K., Rensing, S. A., & Hoecker, U. (2014). Functional analysis of COP1 and SPA orthologs from *Physcomitrella* and rice during photomorphogenesis of transgenic *Arabidopsis* reveals distinct evolutionary conservation. *BMC Plant Biology, 14*(1), 178. http://doi.org/10.1186/1471-2229-14-178.
Rensing, S. A. (2014). Gene duplication as a driver of plant morphogenetic evolution. *Current Opinion in Plant Biology, 17*, 43–48. http://doi.org/10.1016/j.pbi.2013.11.002.
Rensing, S. A., Fritzowsky, D., Lang, D., & Reski, R. (2005). Protein encoding genes in an ancient plant: analysis of codon usage, retained genes and splice sites in a moss, *Physcomitrella patens. BMC Genomics, 6*(1), 43. http://doi.org/10.1186/1471-2164-6-43.
Rensing, S. A., Ick, J., Fawcett, J. A., Lang, D., Zimmer, A., Van de Peer, Y., & Reski, R. (2007). An ancient genome duplication contributed to the abundance of metabolic genes in the moss *Physcomitrella patens. BMC Evolutionary Biology,* 7(1), 130. http://doi.org/10.1186/1471-2148-7-130.
Rensing, S. A., Lang, D., & Zimmer, A. D. (2009). Comparative genomics. In C. D. Knight, P.-F. Perroud, & D. J. Cove (Eds.), *The Moss* Physcomitrella patens. Oxford, UK: Wiley-Blackwell. http://doi.org/10.1002/9781444316070.
Rensing, S. A., Lang, D., Zimmer, A. D., Terry, A., Salamov, A., Shapiro, H. … Boore, J. L. (2008). The *Physcomitrella* genome reveals evolutionary insights into the conquest of land by plants. *Science, 319*(5859), 64–69. http://doi.org/10.1126/science.1150646.
Rensing, S. A., Rombauts, S., Hohe, A., Lang, D., Duwenig, E., Rouze, P. … Reski, R. (2002). *The transcriptome of the moss* Physcomitrella patens*: Comparative analysis reveals a rich source of new genes.* http://www.plant-biotech.net/Rensing_et_al_transcriptome2002.pdf.
Rensing, S. A., Rombauts, S., Van de Peer, Y., & Reski, R. (2002). Moss transcriptome and beyond. *Trends in Plant Science,* 7(12), 535–538.
Reski, R. (1998). Development, genetics and molecular biology of mosses. *Botanica Acta, 111*(1), 1–15.
Reski, R., Faust, M., Wang, X. H., Wehe, M., & Abel, W. O. (1994). Genome analysis of the moss *Physcomitrella patens* (Hedw.) B.S.G. *Molecular and General Genetics, 244*(4), 352–359.
Reski, R., & Frank, W. (2005). Moss (*Physcomitrella patens*) functional genomics – gene discovery and tool development, with implications for crop plants and human health. *Briefings in Functional Genomics & Proteomics, 4*(1), 48–57.
Reski, R., Parsons, J., & Decker, E. L. (2015). Moss-made pharmaceuticals: from bench to bedside. *Plant Biotechnology Journal, 13*(8), 1191–1198. http://doi.org/10.1111/pbi.12401.
Reski, R., Reynolds, S., Wehe, M., Kleber-Janke, T., & Kruse, S. (1998). Moss (*Physcomitrella patens*) expressed sequence tags include several sequences which are novel for plants. *Botanica Acta, 111*(2), 143–149.
Rice, A., Glick, L., Abadi, S., Einhorn, M., Kopelman, N. M., Salman-Minkov, A. … Mayrose, I. (2015). The Chromosome Counts Database (CCDB) – a community resource of plant chromosome numbers. *The New Phytologist, 206*(1), 19–26. http://doi.org/10.1111/nph.13191.
Richardt, S., Lang, D., Reski, R., Frank, W., & Rensing, S. A. (2007). PlanTAPDB, a phylogeny-based resource of plant transcription-associated proteins. *Plant Physiology, 143*(4), 1452–1466. http://doi.org/10.1104/pp.107.095760.
Rizzon, C., Ponger, L., & Gaut, B. S. (2006). Striking similarities in the genomic distribution of tandemly arrayed genes in *Arabidopsis* and rice. *PLoS Computational Biology, 2*(9), e115.
Saidi, Y., Finka, A., Chakhporanian, M., Zryd, J. P., Schaefer, D. G., & Goloubinoff, P. (2005). Controlled expression of recombinant proteins in *Physcomitrella patens* by a

conditional heat-shock promoter: a tool for plant research and biotechnology. *Plant Molecular Biology, 59*(5), 697–711.

Sakakibara, K., Ando, S., Yip, H. K., Tamada, Y., Hiwatashi, Y., Murata, T. ... Bowman, J. L. (2013). KNOX2 genes regulate the haploid-to-diploid morphological transition in land plants. *Science, 339*(6123), 1067–1070. http://doi.org/10.1126/science.1230082.

Sakakibara, K., Reisewitz, P., Aoyama, T., Friedrich, T., Ando, S., Sato, Y. ... Laux, T. (2014). WOX13-like genes are required for reprogramming of leaf and protoplast cells into stem cells in the moss *Physcomitrella patens. Development, 141*(8), 1660–1670. http://doi.org/10.1242/dev.097444.

Schaefer, D. G., & Zryd, J.-P. (1997). Efficient gene targeting in the moss *Physcomitrella patens. The Plant Journal, 11*(6), 1195–1206. http://doi.org/10.1046/j.1365-313X.1997.11061195.x.

Schnepf, E., & Reinhard, C. (1997). Brachycytes in *Funaria* protonemate: Induction by abscisic acid and fine structure. *Journal of Plant Physiology, 151*(2), 166–175. http://doi.org/10.1016/S0176-1617(97)80149-6.

Schween, G., Gorr, G., Hohe, A., & Reski, R. (2003). Unique tissue-specific cell cycle in physcomitrella. *Plant Biology, 5*(1), 50–58.

Semple, C., & Wolfe, K. H. (1999). Gene duplication and gene conversion in the *Caenorhabditis elegans* genome. *Journal of Molecular Evolution, 48*(5), 555–564.

Sharma, N., Jung, C.-H., Bhalla, P. L., & Singh, M. B. (2014). RNA sequencing analysis of the gametophyte transcriptome from the liverwort, *Marchantia polymorpha. PLoS One, 9*(5), e97497. http://doi.org/10.1371/journal.pone.0097497.

Shaw, A. J. (1991). The genetic structure of sporophytic and gametophytic populations of the Moss, *Funaria hygrometrica* hedw. *Evolution, 45*(5), 1260–1274.

Shinde, S., Behpouri, A., McElwain, J. C., & Ng, C. K.-Y. (2015). Genome-wide transcriptomic analysis of the effects of sub-ambient atmospheric oxygen and elevated atmospheric carbon dioxide levels on gametophytes of the moss, *Physcomitrella patens. Journal of Experimental Botany, 66*(13), 4001–4012. http://doi.org/10.1093/jxb/erv197.

Soltis, P., & Soltis, D. (2012). In P. Soltis, & D. Soltis (Eds.), *Polyploidy and genome evolution.* Berlin Heidelberg: Springer. http://doi.org/10.1007/978-3-642-31442-1.

von Stackelberg, M., Rensing, S. A., & Reski, R. (2006). Identification of genic moss SSR markers and a comparative analysis of twenty-four algal and plant gene indices reveal species-specific rather than group-specific characteristics of microsatellites. *BMC Plant Biology, 6*, 9. http://doi.org/10.1186/1471-2229-6-9.

Stein, L. D., Mungall, C., Shu, S., Caudy, M., Mangone, M., Day, A. ... Lewis, S. (2002). The generic genome browser: a building block for a model organism system database. *Genome Research, 12*(10), 1599–1610. http://doi.org/10.1101/gr.403602.

Strepp, R., Scholz, S., Kruse, S., Speth, V., & Reski, R. (1998). Plant nuclear gene knockout reveals a role in plastid division for the homolog of the bacterial cell division protein FtsZ, an ancestral tubulin. *Proceedings of the National Academy of Sciences of the United States of America, 95*(8), 4368–4373.

Strotbek, C., Krinninger, S., & Frank, W. (2013). The moss *Physcomitrella patens*: methods and tools from cultivation to targeted analysis of gene function. *The International Journal of Developmental Biology, 57*(6–8), 553–564. http://doi.org/10.1387/ijdb.130189wf.

Sundell, D., Mannapperuma, C., Netotea, S., Delhomme, N., Lin, Y.-C., Sjödin, A. ... Street, N. R. (2015). The plant genome integrative explorer resource: *New Phytologist.* n/a–n/a PlantGenIE.org http://doi.org/10.1111/nph.13557.

Swarbreck, D., Wilks, C., Lamesch, P., Berardini, T. Z., Garcia-Hernandez, M., Foerster, H. ... Huala, E. (2008). The *Arabidopsis* information resource (TAIR): gene structure and function annotation. *Nucleic Acids Research, 36*(Database issue), D1009–D1014. http://doi.org/10.1093/nar/gkm965.

Szövényi, P., Devos, N., Weston, D. J., Yang, X., Hock, Z., Shaw, J. A.... Wagner, A. (2014). Efficient purging of deleterious mutations in plants with haploid selfing. *Genome Biology and Evolution, 6*(5), 1238–1252. http://doi.org/10.1093/gbe/evu099.

Szövényi, P., Perroud, P.-F., Symeonidi, A., Stevenson, S., Quatrano, R. S., Rensing, S. A.... McDaniel, S. F. (2015). De novo assembly and comparative analysis of the *Ceratodon purpureus* transcriptome. *Molecular Ecology Resources, 15*(1), 203–215. http://doi.org/10.1111/1755-0998.12284.

Szövényi, P., Rensing, S. A., Lang, D., Wray, G. A., & Shaw, A. J. (2011). Generation-biased gene expression in a bryophyte model system. *Molecular Biology and Evolution, 28*(1), 803–812. http://doi.org/10.1093/molbev/msq254.

Szövényi, P., Ricca, M., Hock, Z., Shaw, J. A., Shimizu, K. K., & Wagner, A. (2013). Selection is no more efficient in haploid than in diploid life stages of an angiosperm and a moss. *Molecular Biology and Evolution, 30*(8), 1929–1939. http://doi.org/10.1093/molbev/mst095.

Tang, H., Wang, X., Bowers, J. E., Ming, R., Alam, M., & Paterson, A. H. (2008). Unraveling ancient hexaploidy through multiply-aligned angiosperm gene maps. *Genome Research, 18*(12), 1944–1954. http://doi.org/10.1101/gr.080978.108.

The Arabidopsis Genome Initiative. (2000). Analysis of the genome sequence of the flowering plant *Arabidopsis thaliana*. *Nature, 408*(6814), 796–815.

Tuskan, G. A., Difazio, S., Jansson, S., Bohlmann, J., Grigoriev, I., Hellsten, U.... Rokhsar, D. (2006). The genome of black cottonwood, *Populus trichocarpa* (Torr. & Gray). *Science, 313*(5793), 1596–1604.

Tweedie, S., Ashburner, M., Falls, K., Leyland, P., McQuilton, P., Marygold, S.... Zhang, H. (2009). FlyBase: enhancing *Drosophila* gene ontology annotations. *Nucleic Acids Research, 37*(Database issue), D555–D559. http://doi.org/10.1093/nar/gkn788.

Une, K., & Tateishi, Y. (1996). Life cycle of *Physcomitrella patens* (Hedw.) B.S.G. subsp. Californica (Crum & Anderson) Tan in Japan. *Hikobia, 12*, 151–156.

UniProt Consortium. (2012). Reorganizing the protein space at the Universal Protein resource (UniProt). *Nucleic Acids Research, 40*(Database issue), D71–D75. http://doi.org/10.1093/nar/gkr981.

Van Bel, M., Proost, S., Wischnitzki, E., Movahedi, S., Scheerlinck, C., Van de Peer, Y., & Vandepoele, K. (2012). Dissecting plant genomes with the PLAZA comparative genomics platform. *Plant Physiology, 158*(2), 590–600. http://doi.org/10.1104/pp.111.189514.

Van den Berg, B. H. J., McCarthy, F. M., Lamont, S. J., & Burgess, S. C. (2010). Reannotation is an essential step in systems biology modeling of functional genomics data. *PLoS One, 5*(5), e10642. http://doi.org/10.1371/journal.pone.0010642.

Vandepoele, K., Simillion, C., & Van de Peer, Y. (2003). Evidence that rice and other cereals are ancient aneuploids. *The Plant Cell, 15*(9), 2192–2202.

Vanneste, K., Baele, G., Maere, S., & Van de Peer, Y. (2014). Analysis of 41 plant genomes supports a wave of successful genome duplications in association with the Cretaceous-Paleogene boundary. *Genome Research, 24*(8), 1334–1347. http://doi.org/10.1101/gr.168997.113.

Vanneste, K., Sterck, L., Myburg, A. A., Van de Peer, Y., & Mizrachi, E. (2015). Horsetails are ancient polyploids: evidence from *Equisetum giganteum*. *The Plant Cell, 27*(6), 1567–1578. http://doi.org/10.1105/tpc.15.00157.

Viaene, T., Landberg, K., Thelander, M., Medvecka, E., Pederson, E., Feraru, E.... Friml, J. (2014). Directional auxin transport mechanisms in early diverging land plants. *Current Biology, 24*(23), 2786–2791. http://doi.org/10.1016/j.cub.2014.09.056.

Villarreal, J. C., & Renner, S. S. (2013). Correlates of monoicy and dioicy in hornworts, the apparent sister group to vascular plants. *BMC Evolutionary Biology, 13*, 239. http://doi.org/10.1186/1471-2148-13-239.

Von Wettstein, F. (1928). Über plasmatische Vererbung und über das Zusammenwirken von Genen und Plasma. *Berichte der Deutschen Botanischen Gesellschaft, 46*(11), 32–49.

Weise, A., Altmann, F., Rodriguez-Franco, M., Sjoberg, E. R., Bäumer, W., Launhardt, H. … Gorr, G. (2007). High-level expression of secreted complex glycosylated recombinant human erythropoietin in the *Physcomitrella* Delta-fuc-t Delta-xyl-t mutant. *Plant Biotechnology Journal, 5*(3), 389–401. http://doi.org/10.1111/j.1467-7652.2007.00248.x.

Weise, A., Rodriguez-Franco, M., Timm, B., Hermann, M., Link, S., Jost, W., & Gorr, G. (2006). Use of *Physcomitrella patens* actin 5′ regions for high transgene expression: importance of 5′ introns. *Applied Microbiology and Biotechnology, 70*(3), 337–345. http://doi.org/10.1007/s00253-005-0087-6.

Widiez, T., Symeonidi, A., Luo, C., Lam, E., Lawton, M., & Rensing, S. A. (2014). The chromatin landscape of the moss *Physcomitrella patens* and its dynamics during development and drought stress. *The Plant Journal, 79*(1), 67–81. http://doi.org/10.1111/tpj.12542.

Wolf, L., Rizzini, L., Stracke, R., Ulm, R., & Rensing, S. A. (2010). The molecular and physiological responses of *Physcomitrella patens* to ultraviolet-B radiation. *Plant Physiology, 153*(3), 1123–1134. http://doi.org/10.1104/pp.110.154658.

Wright, S. I., Kalisz, S., & Slotte, T. (2013). Evolutionary consequences of self-fertilization in plants. *Proceedings of Biological Sciences/The Royal Society, 280*(1760), 20130133. http://doi.org/10.1098/rspb.2013.0133.

Wu, J., Wang, Z., Shi, Z., Zhang, S., Ming, R., Zhu, S. … Zhang, S. (2013). The genome of the pear (*Pyrus bretschneideri* Rehd.). *Genome Research, 23*(2), 396–408. http://doi.org/10.1101/gr.144311.112.

Xiao, L., Wang, H., Wan, P., Kuang, T., & He, Y. (2011). Genome-wide transcriptome analysis of gametophyte development in *Physcomitrella patens*. *BMC Plant Biology, 11*(1), 177. http://doi.org/10.1186/1471-2229-11-177.

Yu, J., Hu, S., Wang, J., Wong, G. K., Li, S., Liu, B. … Yang, H. (2002). A draft sequence of the rice genome (*Oryza sativa* L. ssp. indica). *Science, 296*(5565), 79–92.

Yue, J., Hu, X., Sun, H., Yang, Y., & Huang, J. (2012). Widespread impact of horizontal gene transfer on plant colonization of land. *Nature Communications, 3*, 1152. http://doi.org/10.1038/ncomms2148.

Zacharias, H. (1995). Emil Heitz (1892–1965): chloroplasts, heterochromatin, and polytene chromosomes. *Genetics, 141*(1), 7–14.

Zemach, A., McDaniel, I. E., Silva, P., & Zilberman, D. (2010). Genome-wide evolutionary analysis of eukaryotic DNA methylation. *Science, 328*(5980), 916–919. http://doi.org/10.1126/science.1186366.

Zhang, L., & Gaut, B. S. (2003). Does recombination shape the distribution and evolution of tandemly arrayed genes (TAGs) in the *Arabidopsis thaliana* genome? *Genome Research, 13*(12), 2533–2540.

Zimmer, A. D., Lang, D., Buchta, K., Rombauts, S., Nishiyama, T., Hasebe, M. … Reski, R. (2013). Reannotation and extended community resources for the genome of the non-seed plant *Physcomitrella patens* provide insights into the evolution of plant gene structures and functions. *BMC Genomics, 14*(1), 498. http://doi.org/10.1186/1471-2164-14-498.

Zobell, O., Faigl, W., Saedler, H., & Münster, T. (2010). MIKC* MADS-box proteins: conserved regulators of the gametophytic generation of land plants. *Molecular Biology and Evolution, 27*(5), 1201–1211. http://doi.org/10.1093/molbev/msq005.

Zonneveld, B. J. M., Leitch, I. J., & Bennett, M. D. (2005). First nuclear DNA amounts in more than 300 angiosperms. *Annals of Botany, 96*(2), 229–244. http://doi.org/10.1093/aob/mci170.

CHAPTER FOUR

The *Ceratodon purpureus* Transcriptome Ushers in the Era of Moss Comparative Genomics

S.F. McDaniel[*,1], P.-F. Perroud[§], A.C. Cuming[¶], P. Szövényi[†]
[*]University of Florida, Gainesville, FL, United States
[§]University of Marburg, Marburg, Germany
[¶]University of Leeds, Leeds, United Kingdom
[†]University of Zurich, Zurich, Switzerland
[1]Corresponding author: E-mail: stuartmcdaniel@ufl.edu

Contents

Abstract

This chapter outlines the scope of the ongoing *Ceratodon purpureus* genome project and provides an overview of the *C. purpureus* transcriptome, the evolution of the *C. purpureus* UV sex chromosomes, and the patterns of polymorphism in the species. Comparative analyses of the transcriptomes of a male and a female isolate showed that *C. purpureus* and the moss model *Physcomitrella patens* had highly overlapping gene sets, and that most of the genes shared between these two species evolve under strong purifying selection. However, the differences between the *C. purpureus* and *P. patens* genomes refined our understanding of the timing of gene family gain and loss across the land plants and the heterogeneity in rate of molecular evolution across the genome of these two species. *Ceratodon purpureus* showed a slightly greater codon usage bias compared to *P. patens*, which may be explained by the contrasting mating system of the two species. The *C. purpureus* transcriptomes also showed evidence of a genome doubling

Advances in Botanical Research, Volume 78
ISSN 0065-2296
http://dx.doi.org/10.1016/bs.abr.2016.02.003

event ~65–76 MYA that was independent of the contemporaneous polyploidy event inferred for *P. patens*. These data also suggested considerable physiological and developmental divergence between the two species. Genetic mapping and molecular evolutionary analysis showed that the nonrecombining UV chromosomes of *C. purpureus* are actively capturing new genes, illustrating that at least this part of the genome is highly dynamic. Moreover, patterns of polymorphism were highly variable across the genome, suggesting that sexual recombination in other parts of the genome decouples even genes on the same chromosome, and they experience different patterns of natural selection. The forthcoming *C. purpureus* genome will build on these existing resources and enable us to answer definitively many questions regarding the evolution of land plant gene families, genome structure, and the genetic basis of adaptive variation.

1. AN INTRODUCTION TO THE *CERATODON PURPUREUS* GENOME PROJECT

In this chapter we outline the scope of the *Ceratodon purpureus* genome project and provide an overview of the *C. purpureus* transcriptome, the evolution of the *C. purpureus* UV sex chromosomes, and the patterns of polymorphism in the species. *Ceratodon purpureus* is the second moss, after the laboratory model *Physcomitrella patens*, chosen for whole genome sequencing. The genome of *C. purpureus* is estimated to be smaller than that of *P. patens* (~360 Mbs vs 514 Mbs; Voglmayr, 2000), and the chromosome number is approximately half as well (n = 13 vs n = 27; Crum & Anderson, 1981). The analysis of homology between the transcriptomes of *C. purpureus* and *P. patens* has already begun to provide insights into the ancestral gene content of more than 90% of all mosses (Szövényi et al., 2015). *Ceratodon purpureus* is a very widespread species that is common in disturbed sites in temperate regions of the northern and southern hemispheres, as well as at high elevations in the cool montane tropics. It is found growing under a variety of extreme conditions. As a consequence, the potential for gene-discovery projects using natural variation is considerable. However, what sets *C. purpureus* apart from other mosses of similar evolutionary or ecological interest is its long history as a model system in genetics, physiology and development. Heitz (1928) coined the term heterochromatin based in part on his studies of chromosome behaviour in *C. purpureus*. More recently, *C. purpureus* has served as a model for studies of cell biology and physiology (Knight, Cove, Cuming, & Quatrano, 2002; Cove et al., 2009; Cove & Quatrano, 2006), the evolution of metal tolerance (Jules & Shaw, 1994, reviewed in McDaniel, 2009), and associations between mosses and sperm-dispersing microarthropods (Rosenstiel, Shortlidge,

Melnychenko, Pankow, & Eppley, 2012). Moreover, the powerful transgenic tools developed for *P. patens* also function in *C. purpureus* (Brücker, Mittmann, Hartmann, & Lamparter, 2005; Mittmann, Dienstbach, Weisert, & Forreiter, 2009; Trouiller, Charlot, Choinard, Schaefer, & Nogue, 2007).

The US Department of Energy Joint Genome Institute Community Science Program initiated the *C. purpureus* genome project by sequencing a single haploid female lab isolate (GG1, from Gross Gerungs, Austria). Currently the genome assembly consists of approximately 5600 scaffolds. This assembly will soon be superseded by a chromosome-scale assembly, based on resequencing data from a mapping population generated from a cross between GG1 and a male isolate (R40, from Rensselaer, New York, USA). The genetic mapping approach, however, cannot be used to order scaffolds on the nonrecombining sex chromosomes. To accomplish this, a BAC library was generated for the GG1 isolate. This library was end-sequenced and ends mapping to sex-linked scaffolds, based on the genetic map, were identified. A minimum spanning path of BACs to cross the GG1 female sex chromosome were identified and sequenced. These genetic mapping and physical mapping in parallel will be used to generate a version 1 of the *C. purpureus* genome.

To develop genomic tools to understand the genetic basis of phenotypic variation in *C. purpureus*, the R40 isolate and 23 other isolates from across the distribution of *C. purpureus* were resequenced. In parallel to the DNA sequencing efforts, total RNA sequencing has been conducted on gametophores from the GG1 and R40 laboratory isolates, and on protonema from ten male—female sibling pairs from seven localities throughout the distribution of the species. The potential to use these genomic resources to combine studies of natural variation with the well-developed transgenic methods developed for *P. patens* gives a powerful set of tools to connect genotype to phenotype. Collectively these data will enable researchers to use *C. purpureus* to tackle several long-term problems concerning fundamental evolutionary and developmental processes.

2. LIFE CYCLE FACTORS THAT INFLUENCE THE GENOME STRUCTURE OF *CERATODON PURPUREUS*

Mosses are unusual among eukaryotes, in that approximately 60% of the species are dioecious and 40% are hermaphroditic, and that sex is expressed in the dominant haploid phase (McDaniel, Neubig, Payton, Quatrano, & Cove, 2013). The difference in mating system has profound consequences for several aspects of genome evolution. It is these consequences that we examine in this chapter Importantly, *C. purpureus* and *P. patens* have contrasting mating systems, providing us with an instructive comparison (although one that must be critically evaluated with data from additional species). The unisexual gametophytes of *C. purpureus* bear either male or female gametangia but never both. By contrast, *P. patens* gametophytes bear both male and female gametangia on one single shoot. Therefore, mating in *C. purpureus* can occur between gametophytes originating from different sporophytes (equivalent to outcrossing in the angiosperm sense) or between male and female gametophytes originating from the same sporophyte (intergametophytic selfing). This is comparable to selfing in diploid plants, like angiosperms, and results in the halving of heterozygosity in every generation. *Physcomitrella patens,* in addition to intergametophytic selfing, can also undergo selfing involving the fusion of genetically identical male and female gametes produced by a single genetic individual (termed intragametophytic selfing). This kind of selfing is not possible in angiosperms. The fusion of gametes produced by a single haploid gametophyte results in a completely homozygous sporophyte that at meiosis produces genetically homogeneous spores. Experimental crossing studies suggest that *P. patens* indeed undergoes frequent intragametophytic selfing (Perroud, Cove, Quatrano, & McDaniel, 2011). As such, mosses like *P. patens* use the sexual cycle to produce essentially clonal offspring.

Several lines of evidence suggest that these contrasting life histories may drive differences in genome architecture (Szövényi et al., 2014; Wright, Kalisz, & Slotte, 2013). First, evolutionary theory suggests that frequent intergametophytic selfing (ie, inbreeding in angiosperms) leads to a reduced efficacy of selection (Wright et al., 2013). This is because inbreeding leads to a decrease in the effective population size, which increases the relative importance of genetic drift against natural selection. That is, the frequency of weakly deleterious or beneficial mutations in a population will be largely driven by genetic drift and not by natural selection. Interestingly, intragametophytic selfing (what is possible in a hermaphroditic moss, like *P. patens*) reduces the chance that recessive deleterious mutations will persist in heterozygotes, because heterozygotes are largely absent. Transcriptomic data indicate that some deleterious mutations may be more rapidly purged in (intragametophytic) selfing *P. patens* relative to (intergametophytic) selfing *Arabidopsis thaliana* (Szövényi et al., 2014).

The second major genomic consequence of breeding system variation is the possession of heteromorphic sex chromosomes (Bachtrog et al., 2011).

In dioecious species, like *C. purpureus*, sex determination happens at the haploid (gametophyte) stage in the life cycle (Heitz, 1932; McDaniel, Neubig, et al., 2013; McDaniel, Willis, & Shaw, 2007). The diploid part of the life cycle (the sporophyte) is always heterozygous at the sex-determining locus. At meiosis the female- and male-determining factors (U and V, following Bachtrog et al., 2011) segregate to haploid female and male spores, respectively. These spores ultimately develop into multicellular haploid female and male gametophytes, the dominant part of the moss life cycle. Sporophytes are only produced when a U-bearing egg on a female gametophyte is fertilized by a V-bearing sperm from a male gametophyte; sporophytes homozygous at the sex-determining locus are never produced in dioecious mosses. Thus, each sex possesses a chromosomal region that is passed only through one sex (ie, from fathers to sons or from mothers to daughters). This pattern of transmission has important consequences for the evolution of gene content on the U and V chromosomes, as well as the developmental pathways shaping sporophyte development and the outcome of meiosis.

Finally, these contrasting mating systems influence the levels of polymorphism that are expected within populations and the degree of variation in polymorphism across the genome. Outcrossers, like *C. purpureus*, typically harbour more segregating genetic variation than inbreeding species, like *P. patens,* which commonly selfs (Eppley, Taylor, & Jesson, 2006; Perroud et al., 2011). However, the independent assortment and crossing-over associated with outbreeding means that different regions of the genome may have vastly different evolutionary histories and therefore patterns of polymorphism (McDaniel, Atwood, & Burleigh, 2013; McDaniel & Shaw, 2005; McDaniel, van Baren, Jones, Payton, & Quatrano, 2013). We discuss below our current understanding of the evolution of the gene content of *C. purpureus*, relative to *P. patens*, the evolution of the *C. purpureus* sex chromosomes, and the patterns of polymorphism across the *C. purpureus* genome.

3. THE *CERATODON PURPUREUS* TRANSCRIPTOME

Szövényi et al. (2015) analysed protonemal transcriptomes from a male and female genotype of *C. purpureus* (R40 and GG1, respectively) and discuss this resource in the context of the evolution of land plant genomes. The assembly of high-quality sequences produced 26,270 contigs which were further classified into 15,506 genes. Using the proteome of *P. patens* as reference (Zimmer et al., 2013), approximately 80% of the *C. purpureus* transcripts covered at least 80% of the length of the *P. patens* protein sequences, indicating that the coding sequence length was well conserved between *P. patens* and *C. purpureus*. Notably, the number of unique transcripts assembled in male R40 was always lower than in female GG1, suggesting that R40 possesses a less complex transcriptome. How much of this is due to sex-difference or simply individual differences requires analyses of replicates within males and females. Nevertheless, the R40 assembly contained 346 putative homologues of the 357 *A. thaliana* single copy genes conserved across all eukaryotes, while the GG1 assembly contained only six more (352), implying that both transcriptome assemblies are largely complete.

The *C. purpureus* assembled transcripts were enriched in some particular functions, compared to the genomic average of the *P. patens* genome, and whether the same functional categories are enriched in chloronemata in both *P. patens* (based on RPKM >1, in SRR072918) and *C. purpureus* assemblies (based on RPKM >1 in *C. purpureus* control library, see Szövényi et al., 2015). Interestingly, genes that were expressed in chloronemata of *P. patens* showed no particular enrichment compared to the genomic average. In contrast, genes of the *C. purpureus* assemblies showed significant enrichment for 17 Gene Ontology (GO) terms. This suggests that the functional distribution of the genes in the present *C. purpureus* assemblies significantly differs from the gene set expressed in a comparable sample of the *P. patens* protonemata, potentially reflecting a difference in developmental timing and protonemal growth between the two species.

3.1 Refining the Bryophyte-Specific Gene Set

Szövényi et al. (2015) reported that approximately 85% (GG1) and 87% (R40) of the assembled transcripts could be assigned to gene families in the plaza version 2.5 database (Van Bel et al., 2012) based on standard measures of homology (blastx e-value 10^{-6}; 6356 genes for GG1 and 6197 genes for R40). Of the *C. purpureus* families, 97% (GG1) and 96% (R40) contained one or more *P. patens* members. The highly overlapping gene sets of *C. purpureus* and *P. patens* suggest that the common ancestor of *C. purpureus* and *P. patens* (the common ancestor of all of the arthrodontous mosses, >90% of all moss species) had a similar set of gene families. A considerable proportion of the *C. purpureus* families (10% in GG1 and 11% in R40) were moss specific (ie, contained only *P. patens* sequences) indicating that a significant proportion of what were previously thought to be *P. patens*-specific families (1046 in the plaza database) are likely to represent moss-specific genes (Zimmer et al., 2013). The strongly supported moss-specific families were enriched for biosynthetic processes, cell adhesion, ion transport, signalling, and DNA repair.

The *C. purpureus* transcripts additionally extended the set of gene families that is shared with other groups of plants, including green algae, *Selaginella moellendorffii*, and the angiosperms. About 3% (both in GG1 and R40) of the assigned *C. purpureus* assembled transcripts hit 324 (GG1) and 300 (R40) gene families that contained no *P. patens* sequences. Szövényi et al. (2015) found 137 (GG1) and 124 (R40) families that were matched by *C. purpureus* sequences lacked *P. patens* sequences but contained one or more chlorophyte sequences. These extend the list of families that are likely shared with green algae and were putatively present in the common ancestor of Viridiplantae. These families were functionally enriched for a highly interconnected network of various basic molecular processes. These genes were presumably lost or replaced in *P. patens* but remained active and functional in *C. purpureus*. Another 104 (GG1) and 88 (R40) *C. purpureus* families were shared exclusively with *S. moellendorffii* sequences. Thus, these were not truly *S. moellendorffii*-specific but were shared with bryophytes and were lost in or along the branch leading to *P. patens*. These families were only weakly enriched for specific biological process terms. Finally, *C. purpureus* sequences were also assigned to 97 (GG1) and 99 (R40) gene families that were previously believed to be angiosperm specific. These families may have been present in the common ancestor of embryophytes, but were lost from *P. patens* and *S. moellendorffii*, or experienced some more complex history. These families were highly enriched for processes involved in DNA metabolism; repair and methylation; biosynthetic, metabolic processes and response to stimulus, potentially suggesting that *C. purpureus* differs from *P. patens* in these respects. Thus, the *P. patens* gene set is substantially representative of the mosses, but the upcoming genome sequence of *C. purpureus* will significantly improve the information on the gene sets specific to mosses and shared with other clades of land plants.

3.2 Genomic Divergence Between *Physcomitrella patens* and *Ceratodon purpureus*

Consistent with the phylogenetic profiling analysis conducted by Szövényi et al. (2015), the *C. purpureus* transcripts were similar to the *P. patens* proteome (blastx, e-value threshold 10^{-6}), with 22,288 of the 25,373 genes in the GG1 assembly and 18,257 of the 20,277 genes in the R40 assembly having significant matches. The average identity between the *C. purpureus* and

P. patens proteomes peaked at around 80%, with some proteins showing 100% identity. Szövényi et al. (2015) used these data to derive an estimate for the time since divergence between *P. patens* and *C. purpureus*, based on the number of synonymous substitutions per synonymous site for their pairwise orthologs. Since synonymous sites are thought to evolve under relaxed evolutionary constraints, this means that the number of mutations is nearly proportional to the time since divergence (Hurst, 2002). In the 7213 (GG1) and 7034 (R40) pairwise orthologs between the proteomes, the Ks distribution peaked around 0.8–1.0. Assuming that the species diverged at the modal Ks value, this corresponds to a divergence time of c. 40–60 myr (1.9×10^{-8} per synonymous site/year, Rensing et al., 2007; the rate reported in the paper was incorrectly calculated by a factor of two).

The ratio of nonsynonymous changes per nonsynonymous sites (Ka) to the synonymous changes per synonymous site (Ks) estimates the number of genes that may have been affected by positive or negative selection. Ka/Ks ratios between *P. patens* and *C. purpureus* were <1 (mean ratios 0.1 and 0.14 for GG1 and R40, respectively), showing that nonsynonymous mutations at nonsynonymous sites occur with a considerably lower rate than that expected under neutrality (Ks). This suggests that evolution of proteins is primarily driven by strong purifying selection. These estimates are in the range of values recorded in other genome-wide studies in plant species such as in *A. thaliana* (Yang & Gaut, 2011). Genes with an elevated Ka/Ks value were enriched for biological functions including response to metal ions, secondary and pigment metabolic processes, protein break down mechanisms, intracellular transport, nucleotide biosynthesis, alcohol catabolism, and glucose catabolism. In short, genes with diverse basic biological processes show accelerated evolution since the split of *P. patens* and *C. purpureus* suggesting further comparative analysis of the two species may uncover new discoveries in physiology and potentially biology, in general.

3.3 *Ceratodon purpureus* Codon Usage

To evaluate the strength of selection of synonymous codon usage, Szövényi et al. (2015) identified preferred codons using correspondence analysis on relative synonymous codon usage values of genes. In the GG1 and R40 assemblies, the first two axes of the correspondence analysis on RSCU values explained in total 27.8% (19.8% and 8.03% GG1) and 27.4% (19.3% and 8.16% R40) of the synonymous codon usage variation in the data. Although the first two axes explain a relatively low proportion of the variation in the data, the frequency of optimal codons is significantly positively correlated with the expression level of genes (RPKM in the control library vs the frequency of optimal codons (FOP) Spearman's rho GG1 = 0.0832, $P < 2.2e-16$; Spearman's rho R40 = 0.1540, $P < 2.2e-16$). The list of preferred codons was very similar between the GG1 and R40 assemblies, and included 20 preferred codons. This set of codons was very similar in *C. purpureus* and in *P. patens*, all the inferred preferred codons are C- or G-ending (15 and 5 codons, respectively). Only one codon was preferred in *C. purpureus* but nonpreferentially used in *P. patens*. This similarity extends to larger evolutionary timescales as well. Of the 20 preferred codons, 15 and 16 preferred codons were shared with those identified in *A. thaliana* and *Silene latifolia*, and only eight or four were different, respectively (Qiu, Bergero, Zeng, & Charlesworth, 2011; Wright, Yau, Looseley, & Meyers, 2004;). Whether differences in the remaining four preferred codons are biologically relevant or are just a result of sampling bias (only protonemal tissues were used to generate these transcriptomes) is not clear.

One predicted consequence of reduced selection efficacy is less optimized synonymous codon usage in hermaphroditic species (like *P. patens*) than species possessing separate sexes (like *C. purpureus*). After identifying the set of preferred codons, Szövényi et al. (2015) analysed the difference in the codon bias (using both FOP and the effective number of codons, Nc) between *P. patens* the *C. purpureus* GG1 and the R40 strains. They found that the FOP values were slightly higher, whereas Nc values lower in the GG1 and R40 strains than in *P. patens*. The absolute difference between *P. patens* and *C. purpureus* FOP and Nc values was small but turned out to be statistically highly significant. Post hoc test showed that FOP values were significantly greater in both strains of *C. purpureus* than in *P. patens*. Furthermore, FOP values of the GG1 transcripts were higher than that of the R40 strain. The effective number of codons was very slightly but nonetheless statistically significantly lower in both strains of *C. purpureus* suggesting more effective selection on codon usage compared to *P. patens*. Furthermore, similar to the FOP values, the GG1 strain had a significantly smaller effective number of codons than the R40 strain. Altogether, both statistics (FOP and Nc) indicated that codon bias was somewhat higher in *C. purpureus* than in *P. patens*, as predicted by population genetic theory (Wright et al., 2013).

This difference between *P. patens* and *C. purpureus* could also be caused by biased gene conversion (Beletskii & Bhagwat, 2001). The effect of neutral and selective processes can be dissected by comparing base composition of coding and noncoding genomic regions, such as introns, that should not be under selection for synonymous codon usage. If the GC content of third codon positions (GC3) content covaries with biased codon usage, compositional biases alone may be sufficient to generate the codon usage patterns. As not enough intron data was available at that time, Szövényi et al. (2015) used GC3 as an indicator of compositional biases (Qiu et al., 2011). They found that both GC3 content and codon bias were lower in *P. patens*, indicating that mutational biases could potentially explain the codon bias difference between *P. patens* and *C. purpureus*. However, this conclusion did not apply for the difference between GG1 and R40 as GC3 content and codon bias were inversely related in this comparison. That is, R40 had a lower GC3 content, but a greater codon bias than the GG1 strain. Szövényi et al. (2015) additionally calculated the ΔRSCU+ index, which is less sensitive to base and amino acid compositional biases (Cutter, Wasmuth, & Blaxter, 2006). ΔRSCU+ values were greatest in the GG1 (mean: 1.0367), smallest in the R40 strain (mean: 0.7786) and intermediate in *P. patens* (mean: 0.9680). ΔRSCU+ values among the three data sets were, however, not significantly different. In combination with the FOB and Nc analyses, these data suggest that codon usage may be slightly more biased in *C. purpureus* than in *P. patens* as predicted by evolutionary theory, potentially more in males. Nevertheless, whether this difference is driven by natural selection or rather by mutational biases remains to be seen.

3.4 An Ancient Genome Duplication Event in *Ceratodon purpureus*

To test for evidence of large-scale genome duplications, Szövényi et al. (2015) aligned the both GG1 and R40 transcriptome to itself. The distribution of the synonymous substitutions (Ks) for both transcriptome assemblies was very similar, suggesting that the lineage containing *C. purpureus* experienced its most prominent duplication approximately 32–38 MYA (Ks = 0.6–0.7). This is nearly contemporaneous with the whole genome duplication described for *P. patens* (Rensing et al., 2007), although both occurred well after the divergence of the *P. patens* and *C. purpureus* lineages from one another. The over- or under-represented GO terms in GG1 and R40 were enriched for response to metal ions, consistent with earlier work suggesting that *C. purpureus* has the capacity to evolve resistance to heavy-metal containing substrates (Jules & Shaw 1994). The preferential retention of these genes may be related to the tolerance of a broad range of substrate chemistries in *C. purpureus*.

4. *CERATODON PURPUREUS* HAS A UV SEX CHROMOSOME SYSTEM

One of the most obvious differences between the genomes of *C. purpureus* and *P. patens* is that *C. purpureus* possesses large, nonrecombining sex chromosomes. This is a direct consequence of having separate sexes at the gametophyte stage. In dioecious mosses, like *C. purpureus*, sex determination happens at the haploid phase (following meiosis), rather than at the diploid phase (following fertilization) as in dioecious angiosperms and animals. This difference has important consequences for the evolution of bryophyte sex chromosomes. First, female and male gametophytes each carry a sex-specific chromosome, which, following Bachtrog et al. (2011), we refer to as U and V, respectively. The diploid part of the life cycle (the sporophyte) is always heterozygous (ie, UV) for the sex-determining chromosomes (see Fig. 1). At meiosis, the U and V chromosomes segregate to haploid female and male spores, respectively, which develop into multicellular haploid female and male gametophytes, the dominant part of the moss life cycle. The sex-specific portions of the bryophyte U and V sex chromosomes are genetically analogous to the mating type loci of some fungi (eg, *Microbotryum* or *Neurospora*) and algae (eg, *Chlamydomonas* or *Ectocarpus*), in that mating type is determined at meiosis (Ferris & Goodenough, 1994; Hood, 2002; Hood, Antonovics, & Koskella, 2004; Menkis, Jacobson, Gustafsson, & Johannesson, 2008). However, unlike the best-studied fungi and algae, mosses possess the gamete-size differences (and often extensive gametophyte sexual dimorphism) that generally are associated with male–female systems, which may influence the long-term evolution of gene content on the sex chromosomes.

The evolution of genetically determined males and females sets the stage for the evolution of differentiated sex chromosomes (Fig. 2). Theory suggests that mutations with sex-specific fitness effects have a greater chance of increasing in frequency when they are linked to the sex that they benefit. For example, a male-beneficial allele has a greater chance of fixing in the population when it is linked to a male-determining region than when it is found on an autosome where half the time it will be expressed in a female background (Rice, 1987). This in turn generates selection for modifiers that suppress recombination between male-beneficial alleles and the male-determining region. The repeated fixation of alleles that have sex-specific fitness effects and the subsequent suppression of recombination between these loci and the sex-determining region is thought to be the major process

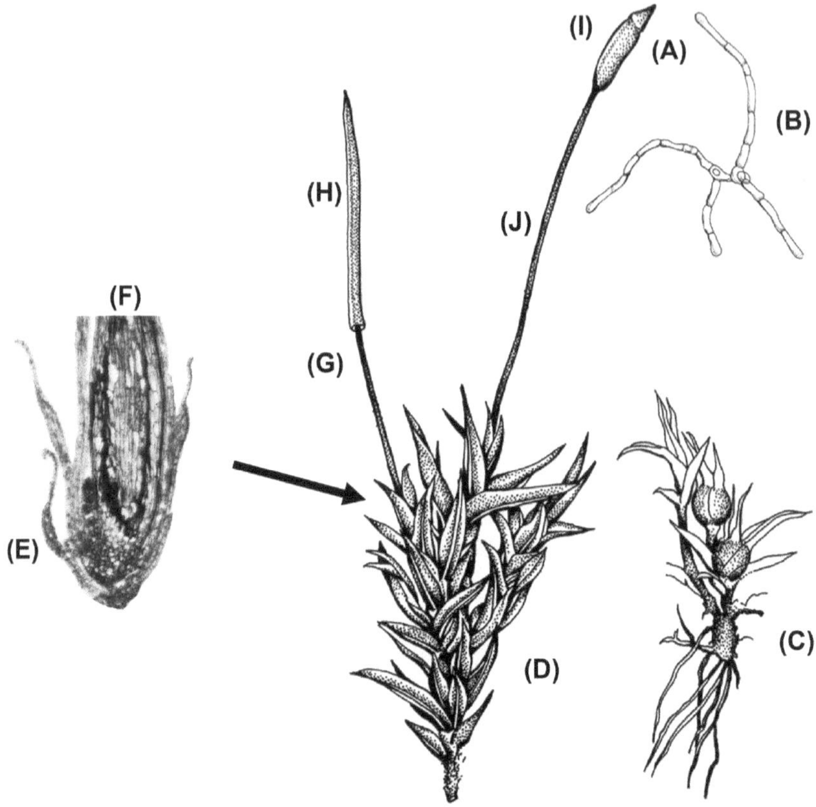

Figure 1 The life cycle of mosses, like all plants, begins with a spore produced by meiosis (A). In mosses the spore germinates to form a gametophyte (a gamete-making plant) initially consisting of filamentous protonema (B) that grows by serial division. Protonemata grow indeterminately, unlike the gametophytes of seed plants that reach maturity as pollen and ovules after relatively few cell divisions. Some protonema cells receive an age- and position-dependent signal causing them to differentiate into buds which establish a 3D division pattern, although the protonema continues to grow by tip growth. 3D division development leads to the production of phyllids characteristic of the mature gametophore (C and D). At sexual maturity, the gametophore produces gametangia that produce either sperm (antheridia) or eggs (archegonia). The sperm are broadcast in semiaqueous circumstances and transported through the terrestrial environment to fertilize the egg, which remains inside the archegonia. As in all eukaryotes, fertilization results in a zygote, which in plants develops into the sporophyte (the spore-making plant). The zygote undergoes an asymmetric cell division. The lower cell will ultimately differentiate into transfer cells (E) that carry nutrients from the maternal gametophyte to the developing embryo (F). In the mature sporophyte this tissue is called the foot. The upper cell in the zygote differentiates into the embryo (G) which remains covered by archegonia (now called a calyptra, H) as the sporophyte matures. The mature sporophyte consists of a sporangium (I) which is elevated above the maternal gametophyte by a seta (J). At maturity the sporogenous tissue will undergo meiosis and spore maturation.

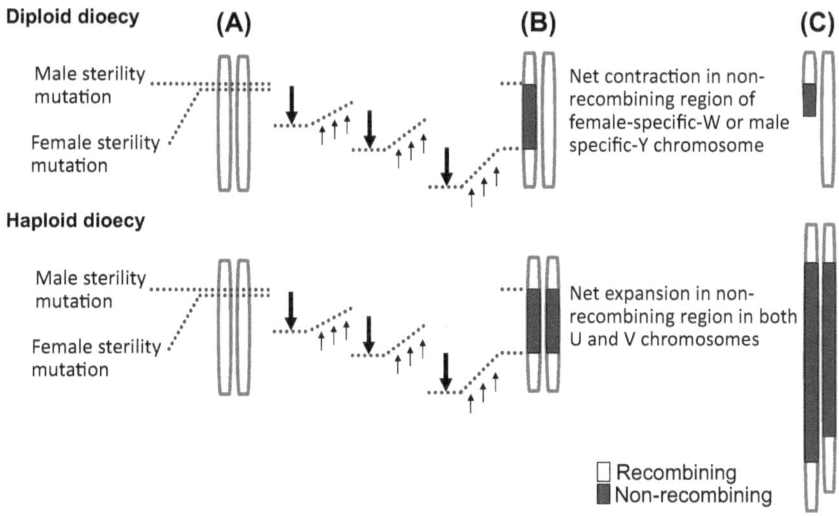

Figure 2 The evolution of the UV sex chromosome in comparison to XY sex chromosomes. The evolution of genetic sex determination involves at least two mutations (A) that are linked in repulsion (ie, on distinct haplotypes). This is typically modelled by supposing that a male sterility mutation invades a population, producing females. At this stage, the population consists of females and hermaphrodites. Next, a female-sterility mutation invades the population. This mutation arises in a distinct gene from the male sterility gene on a haplotype with the hermaphroditism allele. At this point the population consists of females, males, and hermaphrodites. This situation can lead to dioecy if the male and female-sterility mutations are on distinct haplotypes and are closely linked, so that recombination between the sterility alleles is rare. This is because any recombination between the two sterility alleles will produce chromosomes with both male and female-sterility alleles, which obviously will not persist. The population evolves complete dioecy when the hermaphroditism allele is lost from the population. This occurs most rapidly when the sterility alleles have compensatory effects on the fitness of the opposite sex (eg, a male sterility allele is favoured when it also increases female fitness). Thus, the evolution of genetic sex determination requires linked male- and female-sterility alleles with additional fitness effects on the nonsterilized sex (B). Unlike XY or ZW systems where suppressed recombination is confounded with both sex-limited transmission (to males in XY systems or females in ZW systems) and relaxed purifying selection (because Y-linked and W-linked mutations experience enforced heterozygosity with the recombining X or Z), UV systems are symmetrical. Interestingly, moss U and V sex chromosomes are often larger than the average autosome, while Y and W chromosomes are often smaller than the average autosome (C).

driving the expansion of the sex-specific region of the sex chromosome. It is this accumulation of sex-beneficial alleles and the suppression of recombination that are the hallmarks of sex chromosomes. In species with an XY system, like in mammals, it is presumably male-beneficial alleles that exert the strongest effect on the evolution of suppressed recombination (although recessive female-beneficial mutations on the X chromosome can also facilitate the evolution of suppressed recombination). Similarly, in ZW systems, like in birds, presumably mostly female-beneficial mutation drive the evolution of suppressed recombination on the W chromosome. Because both males and females in dioecious mosses possess a sex-limited genomic region, both male- and female-beneficial mutations can promote the evolution of suppressed recombination. Thus, the accumulation of sex-beneficial mutations in bryophytes may be accelerated relative to organisms in which only one sex has a sex-limited chromosome (ie, XY or ZW systems; Bachtrog et al., 2011).

In *C. purpureus*, meiotic chromosome staining and genetic mapping indicate that the UV chromosome pair has a large region over which crossing-over is suppressed (Heitz, 1932; McDaniel et al. 2007; Ramsay & Berrie, 1982). McDaniel, Neubig, et al. (2013) used a genetic map to determine that the nonrecombining regions of the sex chromosomes in the moss *C. purpureus* recently have captured at least eight genes (Fig. 3). For genes that have homologues on both the U and V chromosomes, the divergence times of the female-specific U—male-specific V pairs should reflect the times the loci stopped recombining (barring subsequent interchromosomal gene conversion; Perrin, 2009). McDaniel et al. (2013c) inferred that the nonrecombining region expanded stepwise into the adjacent and previously recombining pseudoautosomal regions (PARs) in at least two events (0.6—1.3 MYA and 2.8—3.5 MYA; the previously studied *des6* gene may represent another event at 6 MYA, McDaniel et al., 2007). The fact that they found homologues on both the U and V chromosomes suggests that the expansion was not caused by translocation events to only the U or V. Nevertheless, more complex scenarios involving a gradual expansion of heterochromatin, chromosomal inversions, or heterogeneous patterns of gene conversion between the sex chromosomes (or combinations of these processes) to explain the pattern of divergence times among the sex-linked loci could not be excluded.

The rapid gain of sex-linked genes suggests that many mutations may have sex-specific effects. Indeed, bryophytes exhibit sex-specific patterns of parental care, morphology, and life history trade-offs (Horsley, Stark, & McLetchie, 2011; McDaniel, 2005; McDaniel et al., 2007; Norrell, Jones, Payton, & McDaniel, 2014; Stark, McLetchie & Eppley, 2010; Stark, McLetchie, & Mishler, 2001). Fertilization in bryophytes requires sperm to swim through the terrestrial environment. Even in early developmental

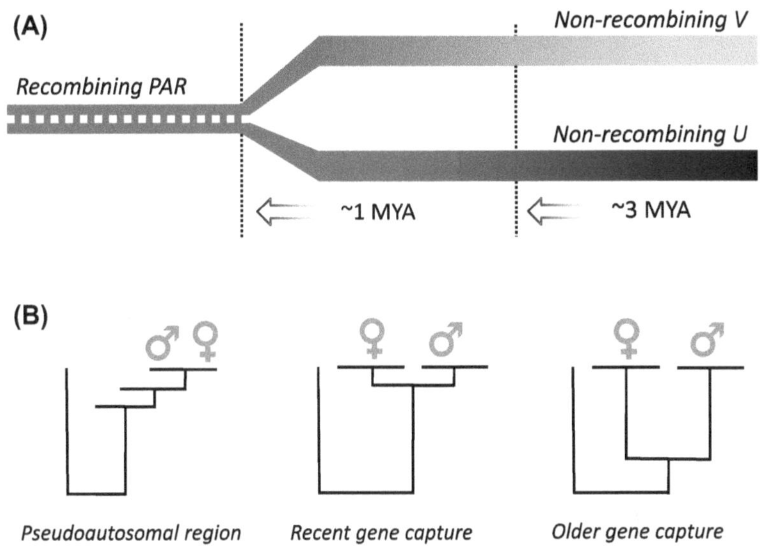

Figure 3 The expansion of the sex-specific, nonrecombining portion of the sex chromosome (A) leaves a signature in the patterns of divergence between male- and female-linked copies of homologous genes (B). Genes that were recently captured by the nonrecombining region should be close to the PAR boundary, and have shallower coalescence times between male and female homologues. Genes that were captured in the more distant past should be located further from the PAR boundary and have older coalescence times. Of course these patterns may become obscured by subsequent structural changes or gene loss on the U or V chromosome, and may not apply in the first place to genes that were acquired by translocations to only one chromosome or insertions within one of the regions of suppressed recombination.

tissue such as protonema, with little morphological complexity and only a modest number of cell types, growth is subtly sexually dimorphic (McDaniel, Willis, & Shaw, 2008), suggesting that males and females of *C. purpureus* are physiologically dimorphic almost right from spore germination, if not before.

Finding heterogeneous female-specific U—male-specific V divergence values is strikingly similar to the evolutionary strata on the mammalian XY pair (Lahn & Page, 1999; Sandstedt & Tucker, 2004), the avian ZW (Handley et al., 2004; Nam & Ellegren, 2008), the XY chromosomes in the plants *S. latifolia* and *Carica papaya* (Bergero, Forrest, Kamau, & Charlesworth, 2007; Gschwend et al., 2012; Nicolas et al., 2005; Wang et al., 2012), and the mating type loci of the fungus *Microbotryum violaceum* (Votintseva & Filatov, 2009). In the other species where strata have been discovered, the genes with lowest divergence (ie, most recent loss of recombination) cluster close to the PAR boundary, based on genetic map distances inferred from the recombining sex chromosome. In *C. purpureus*, it is not yet clear whether this is true, because genes in neither U nor V regions can be genetically mapped due to recombination suppression on both homologous chromosomes. A physical map is currently being constructed to provide a linear order for the loci on the female sex chromosome. This resource will allow investigators to determine if loci that have very similar U–V divergence values are physically close to each other on the chromosome, suggesting that they experienced recombination suppression at the same time.

It is noteworthy that in animal sex chromosomes, where the possession of separate sexes is ancient, the evolutionary strata involved many genes and formed 10 to more than 100 MYA (Nam & Ellegren, 2008; Sandstedt & Tucker, 2004), whereas in the angiosperms *C. papaya* and *S. latifolia* the strata are relatively young (Desfeux, Maurice, Henry, Lejeune, & Gouyon, 1996; Liu et al., 2004; Wang et al., 2012). The origin of the *C. purpureus*, UV sex chromosome certainly predates the gene-capture events reported by McDaniel, Neubig, et al. (2013), because the divergence times between *C. purpureus* and its dioecious sister species (*Cheilothela chloropus* and *Trichodon cylindricus*) preceded the coalescence times of the U- and V-linked copies that were studied. In fact, the screen employed by McDaniel, Neubig, et al. (2013) was probably biased towards identifying young sex-linked genes. They sampled only loci that were expressed in a modestly sexually dimorphic tissue (protonema), which had clearly identifiable homologues between *C. purpureus* and the distantly related *P. patens*, and which were found in both a male and a female lab strain (McDaniel, Neubig, et al., 2013). They reasoned that older sex-linked genes, in contrast, were likely to be expressed in more mature, sexually dimorphic tissues, and may be present in one sex and absent in the other. Nevertheless, in spite of these caveats, more than the 10% of genes tested mapped to the approximately one quarter of the *C. purpureus* genome (~80 Mbps) that is occupied by the sex chromosome (McDaniel et al., 2007). This finding suggests that the *C. purpureus* sex chromosomes may be relatively rich in new sex-linked genes, much like *S. latifolia* (Chibalina & Filatov, 2011). Data on genetic differentiation between the male- and female-specific portions of the sex chromosomes of other bryophyte species are currently very limited (Korpelainen, Bisang, Hedenas, & Kolehmainen, 2008). In the most thorough study, Yamato et al. (2007) reported 64 genes on the male chromosome of the liverwort *Marchantia polymorpha*, fewer with identifiable homologues on the female chromosome. Whether the differences between the female-specific U and male-specific V chromosomes resulted from gene addition or gene loss is still unknown.

The final stage in the evolution of heteromorphic sex chromosomes involves degeneration or the loss of previously functional genes. Degeneration can occur for three main reasons. First, in diploid dioecious species, like mammals with an XY system, the Y chromosome is always paired with an X; thus purifying selection is relaxed on the Y, because recessive deleterious alleles are never homozygous, and therefore are never directly exposed to selection. This may be true for sporophyte (ie, diploid) expressed genes in dioecious mosses, but this process should apply equally to both the U and the V chromosome. Second, purifying selection is also relaxed for genes that are necessary for sexual function in only one sex, and therefore may degenerate on a sex chromosome in the opposite sex. For example, genes necessary for sperm function may accumulate deleterious mutations on a female U chromosome. The number of genes fitting into this category is unknown. Third, regions of suppressed recombination have low effective population sizes (see Section 4). The ultimate result of this reduction in diversity is that selection becomes less effective at removing deleterious alleles and increasing the frequency of beneficial alleles. The evidence of recent gene capture suggests that there have been multiple additions on the nonrecombining parts of the U and V chromosomes in the recent past. Together relaxed selection and stronger genetic drift may drive the degeneration of the nonrecombining portions of sex chromosomes.

The balance between the expansion and degeneration of the sex chromosomes of haploid dioecious species, such as bryophytes, should differ from that on the sex chromosomes of diploid dioecious animals and angiosperms (Bachtrog et al., 2011; Bull, 1978, Fig. 2). In a haploid dioecious system, purifying selection is likely to be strong for many genes on both nonrecombining sex chromosomes (Nishiyama et al., 2003; Szövényi, Rensing, Lang, Wray, & Shaw, 2011). Thus, size differences between the U and V chromosomes are unlikely to result from asymmetric gene loss, although genes expressed only in one sex can be lost from the sex chromosome of the other sex. Indeed, although cytological studies indicate that many bryophyte sex chromosomes are heteromorphic (suggesting differential gene loss or differential accumulation of repetitive sequences between the sex-specific portions of the sex chromosomes), they are not generally smaller than the autosomes and are sometimes larger (Allen, 1935; Nakayama, Fujishita, Sone, & Ohyama, 2001). In *C. purpureus*, for example, both the male and female chromosomes appear approximately five times larger than the average autosome (Heitz, 1932; McDaniel et al., 2007).

Whether genes on the *C. purpureus* U and V chromosomes are beginning to accumulate deleterious mutations is currently uncertain. Loci on these chromosomes may be prevented from degeneration as a result of purifying selection in the haploid gametophyte stage (Bergero & Charlesworth, 2011; Chibalina & Filatov, 2011). However, McDaniel, Neubig, et al. (2013) found that the male copy of the rpS15A-1 locus showed signs of degeneration (a deletion and a stop codon). Genetic degeneration of sex-linked loci that is unchecked by purifying selection clearly can operate on shorter timescales than those relevant for the *C. purpureus* genes that were evaluated in this study. Genes on the Y chromosome in the flowering plant *S. latifolia* show some signs of degeneration (Papadopulos, Chester, Ridout, & Filatov, 2015), as do nonrecombining mating-type loci in the fungus *Neurospora tetrasperma*, including transposable element accumulation, decreased codon bias and decreased expression of Y-linked loci (Bergero & Charlesworth, 2011; Chibalina & Filatov, 2011; Marais et al., 2008; Whittle, Sun, & Johannesson, 2011).

One of the most significant mysteries concerns what happens to the sex chromosomes when hermaphroditism evolves from an ancestor with separate sexes, which presumably possessed some form of sex chromosomes (McDaniel & Perroud, 2012). Hermaphroditic mosses do not possess sex chromosomes — the genes responsible for controlling sex expression may be distributed throughout the genome. However, the newly evolved hermaphroditic moss species clearly needs genes that, had they previously resided on a sex chromosome may have degenerated in one sex or the other in the progenitor species. This could have been because they were previously necessary only in one sex, or they had evolved a sex-specific function that was not complemented by the copy on the alternate sex chromosome. In these cases, critical parts of both the U and V must contribute to the genome of the new hermaphrodite. Polyploidy is a candidate mechanism for generating gametophytes with both sex chromosomes (through unreduced meiotic products, for example) and indeed some cases clearly show this pattern (eg, *Atrichum* and possibly *Physcomitrella*; Perley & Jesson, 2015). However, phylogenetic reconstructions based on our current data indicate that transitions from separate sexes to hermaphroditism are not associated with polyploidy (Crawford, Jesson, & Garnock-Jones, 2009; McDaniel, Atwood, et al., 2013). Comparative genetic mapping and gene family reconstruction in multiple species, like those that have been initiated in *C. purpureus*, provides a means of evaluating the genomic changes that accompany transitions between sexual systems. The existing data clearly illustrate the dynamism of the evolutionary processes shaping the sex chromosomes of *C. purpureus*.

5. PATTERNS OF POLYMORPHISM IN *CERATODON PURPUREUS*

Natural variation is widely recognized as a valuable complement to laboratory mutants for understanding developmental or physiological processes (Weigel, 2012). *Ceratodon purpureus* is widespread and abundant in temperate regions of the globe. As a consequence, the species is expected to harbour a considerable amount of genetic variation. Indeed, in nature *C. purpureus* is highly variable in most measured traits (Burley & Pritchard, 1990; McDaniel, 2005; Shaw & Beer, 1999; Shaw & Gaughan, 1993), including traits like metal tolerance (Jules & Shaw 1994) and meiotic segregation ratios (Norrell et al., 2014). This phenotypic variation is underlain by comparable levels of DNA polymorphism, which is usually measured with a statistic called θ (in units of mutations/site). This number is roughly proportional to the mutation rate (which is known to vary across genomes) multiplied by the effective population size (ie, the number of individuals that can mutate). Comprehensive genome-wide surveys have not yet been published but McDaniel et al. (2013c) surveyed approximately 200 loci in a small sample of individuals from across the distribution of the species. The mean of this sample was $\theta = 0.01$ (roughly one polymorphic site in 100 bp), with a range of 0.0–0.12 (Fig. 4). This range is broad but reflects the diversity of evolutionary processes acting on different parts of the *C. purpureus* genome and is similar to that found in other model system species. Here we report on preliminary analyses of evolutionary processes that interact with various genomic features to influence the patterns of nucleotide diversity in *C. purpureus*.

We first consider variation in genetic diversity among populations. The most complete data available come from McDaniel, Neubig, et al. (2013) who reported among-population nucleotide variation in eastern North American populations of *C. purpureus*. It is possible that some among-population variation was generated by population bottlenecks or the recent colonization of one of the study sites by relatively few individuals. Both processes could reduce genetic variation in such populations. Elevated among-population variation, in contrast, could have been caused by restricted migration among populations. However, given that *C. purpureus*

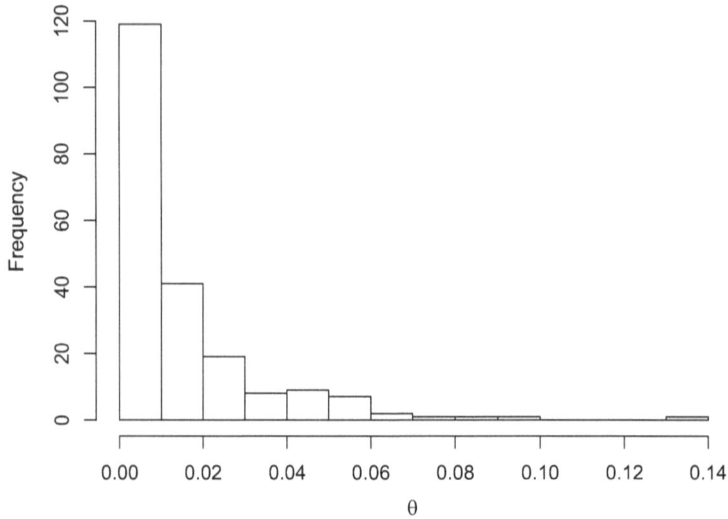

Figure 4 Variation in levels of polymorphism across the *C. purpureus* genome is likely to be driven by natural selection. Consider the fixation of a single adaptive mutation: As this mutation increases in frequency, all of the other variants that are linked to that particular haplotype also increase in frequency, at the expense of other alleles. That is, selection acting on any mutation influences the frequency of adjacent mutations, until they are uncoupled by meiotic recombination. This leads to a reduction in diversity around the adaptive mutation, much in the way that a population bottleneck reduced diversity, but its effects are specific to a small genomic region (ie, loci on the left side of the histogram). This phenomenon is called hitchhiking (Maynard-Smith & Haigh, 1974), and the loss of genetic diversity following the fixation of an adaptive mutation is known as a selective sweep. Alternatively, selection may favour different alleles in different populations, or different alleles in different sexes, or favour heterozygous sporophytes. These processes fall under the rubric of balancing selection, and generally cause elevated levels of polymorphism relative to expectations based on mutation rate and population size (ie, the right side of the histogram).

appears to be highly dispersible (McDaniel, Neubig, et al., 2013; McDaniel & Shaw, 2005), most populations are unlikely to remain isolated for long. Frequent migration among even distant populations means that the effects of these kinds of demographic processes on any one population are likely to be limited or transient. Overall, the variation in nucleotide diversity among loci in the McDaniel, Neubig, et al. (2013) study was much larger than the variation among populations. Some of the interlocus variation was demonstrably due to variation in mutation rate, based on locus-specific divergence times between *C. purpureus* and the sister species *T. cylindricus* or *C. chloropus*. However, the heterogeneity among loci in genetic variation in this species is better explained by differential patterns of natural selection.

Chromosomal inversions may be particularly likely to experience strong selection. Because meiotic recombination is often suppressed in inversions, largely due to the fact that single recombinants often produce spores (or gametes in other taxa) with lethal deletions or duplications, the net selective effects of many polymorphisms may be aggregated. Both genetic mapping and polymorphism data from *C. purpureus* suggest that inversions segregate within populations of the species (McDaniel, van Baren, et al., 2013, McDaniel et al., 2007, 2008). Linking genomic features, like chromosomal inversions, to phenotypic variation is an important next step in developing *C. purpureus* as a model system in ecological genomics and development.

Variation in nucleotide diversity on the sex chromosomes of *C. purpureus* is also very likely to contribute to the distribution of values seen in Fig. 4. Because a moss sporophyte contains two copies of each autosome for every U or V chromosome, we expect that the effective population size and therefore genetic diversity of sex-linked loci should be half that of an autosomal locus. Further reductions in the polymorphism on the nonrecombining portions of the sex chromosomes, relative to autosomes, may reflect hitchhiking effects. This is because the levels of polymorphism in regions of suppressed recombination are strongly influenced by hitchhiking effects caused by directional or purifying selection acting on the nonrecombining chromosome (Bachtrog, 2006). Indeed, McDaniel, Neubig, et al. (2013) found low relative nucleotide variation on the *C. purpureus* sex chromosomes; the U and V diversity were both significantly lower than half the autosomal estimate, similar to data from *S. latifolia* (Qiu, Bergero, Forrest, Kaiser, & Charlesworth, 2010). Interestingly the diversity in males was lower than that in females. Because both sex chromosomes experience the same meiotic process (ie, suppressed recombination in an UV sporophyte), and similar mutation rates, this difference was unexpected. McDaniel, Neubig, et al. (2013) proposed that males must have a greater variance in reproductive success than females (that is, some males fertilize many females, while many males fertilize no female). This effectively causes a male-specific population bottleneck each generation. Whether this is generally true requires sampling of several sex-linked loci from other dioecious species.

6. CONCLUSIONS

The pregenomic era of *C. purpureus* genetics has provided critical foundational knowledge to guide future studies using the genome of this species. The capstone of this effort is the transcriptome two isolates, a male and a female, that now adds to the powerful laboratory tools available for *C. purpureus* extending the utility of this emerging model system. Comparative analyses of these transcriptomes showed that *C. purpureus* and *P. patens* had highly overlapping gene sets, and that most of the genes shared between these two species evolve under strong purifying selection. However, the differences between the *C. purpureus* and *P. patens* genomes further refine our understanding of the timing of gene family gain and loss across the land plants, and that genes involved in multiple basic biological processes show an accelerated rate of evolution. This suggests considerable physiological and developmental divergence between the two species. The slightly greater codon usage bias in *C. purpureus* compared to *P. patens* may be explained by the contrasting mating system of the two species, although this requires additional confirmation. The genome doubling event ~32–38 MYA that was independent of the contemporaneous polyploidy event inferred for *P. patens*. The nonrecombining UV chromosomes of *C. purpureus* are actively capturing new genes, illustrating that at least this part of the genome is highly dynamic. Moreover, patterns of polymorphism are highly variable across the genome, suggesting that sexual recombination in other parts of the genome decouples even genes on the same chromosome, and they experience different patterns of natural selection. All of these molecular evolutionary data exist in parallel with abundant phenotypic variation. The forthcoming *C. purpureus* genome will build on these existing resources and enable us to answer definitively many questions regarding the evolution of land plant gene families, genome structure, and the genetic basis of adaptive variation.

ACKNOWLEDGEMENTS

SFM thanks Sarah Carey, Leslie Kollar, Adam Payton and Stefan Rensing for helpful suggestions that improved the clarity of this chapter. SFM was supported by NSF (DEB 1541005).

REFERENCES

Allen, C. (1935). Genetics of bryophytes. I. *Botanical Review (Lancaster), 1*, 269–291.

Bachtrog, D. (2006). A dynamic view of sex chromosome evolution. *Current Opinion In Genetics & Development, 16*, 578–585.

Bachtrog, D., Kirkpatrick, M., Mank, J., McDaniel, S. F., Pires, J. C., Rice, W., & Valenzuela, N. (2011). Are all sex chromosomes created equal? *Trends in Genetics, 27*, 350–357.

Beletskii, A., & Bhagwat, A. S. (2001). Transcription-induced cytosine-to-thymine mutations are not dependent on sequence context of the target cytosine. *Journal of Bacteriology, 183*, 6491–6493.

Bergero, R., & Charlesworth, D. (2011). Preservation of the Y transcriptome in a 10-million-year-old plant sex chromosome system. *Current Biology, 21*, 1470–1474.

Bergero, R., Forrest, A., Kamau, E., & Charlesworth, D. (2007). Evolutionary strata on the X chromosomes of the dioecious plant *Silene latifolia*: evidence from new sex-linked genes. *Genetics, 175*, 1945–1954.

Brücker, G., Mittmann, F., Hartmann, E., & Lamparter, T. (2005). Targeted site-directed mutagenesis of a heme oxygenase locus by gene replacement in the moss *Ceratodon purpureus*. *Planta, 220*, 864–874.

Bull, J. J. (1978). Sex chromosomes in haploid dioecy: a unique contrast to Muller's theory for diploid dioecy. *American Naturalist, 112*, 245–250.

Burley, J. S., & Pritchard, N. M. (1990). Revision of the genus *Ceratodon* (Bryophyta). *Harvard Papers in Botany, 2*, 1–76.

Chibalina, M. V., & Filatov, D. A. (2011). Plant Y chromosome degeneration is retarded by haploid purifying selection. *Current Biology, 21*, 1475–1479.

Cove, D., Perroud, P.-F., Charron, A., McDaniel, S., Khandelwal, A., & Quatrano, R. (2009). *The moss* Physcomitrella patens*: A novel model system for plant development and genomic studies. Emerging model organisms.* New York, NY: Cold Spring Harbor Laboratory Press.

Cove, D. J., & Quatrano, R. S. (2006). Agravitropic mutants of the moss *Ceratodon purpureus* do not complement mutants having a reversed gravitropic response. *Plant Cell and Environment, 29*, 1379–1387.

Crawford, M., Jesson, L. K., & Garnock-Jones, P. J. (2009). Correlated evolution of sexual system and life-history traits in mosses. *Evolution, 63*, 1129–1142.

Crum, H. A., & Anderson, L. E. (1981). *The mosses of eastern North America.* New York, NY: Columbia University Press.

Cutter, A. D., Wasmuth, J. D., & Blaxter, M. L. (2006). The evolution of biased codon and amino acid usage in nematode genomes. *Molecular Biology and Evolution, 23*, 2303–2315.

Desfeux, C., Maurice, S., Henry, J.-P., Lejeune, B., & Gouyon, P.-H. (1996). Evolution of reproductive systems in the genus *Silene*. *Proceedings of the Royal Society of London, Series B: Biological Sciences, 263*, 409–414.

Eppley, S. M., Taylor, P. J., & Jesson, L. K. (2006). Self-fertilization in mosses: a comparison of heterozygote deficiency between species with combined versus separate sexes. *Heredity, 98*, 38–44.

Ferris, P. J., & Goodenough, U. W. (1994). The mating-type locus of *Chlamydomonas reinhardtii* contains highly rearranged DNA-sequences. *Cell, 76*, 1135–1145.

Gschwend, A. R., Yu, Q. Y., Tong, E. J., Zeng, F. C., Han, J., VanBuren, R.… Paterson, A. H. (2012). Rapid divergence and expansion of the X chromosome in papaya. *Proceedings of the National Academy of Sciences of the United States of America, 109*, 13716–13721.

Handley, L.-J. L., Ceplitis, H., & Ellegren, H. (2004). Evolutionary strata on the chicken Z chromosome: implications for sex chromosome evolution. *Genetics, 167*, 367–376.

Heitz, E. (1928). Das heterochromatin der moose. *Jahrbücher für Wissenschaftliche Botanik, 69*, 762–818.

Heitz, E. (1932). Geschlectschromosomen bei einem Laubmoos. *Berichte der Deutschen Botanischen Gesellschaft, 30*, 204–206.

Hood, M. E. (2002). Dimorphic mating-type chromosomes in the fungus *Microbotryum violaceum*. *Genetics, 160*, 457–461.

Hood, M. E., Antonovics, J., & Koskella, B. (2004). Shared forces of sex chromosome evolution in haploid-mating and diploid-mating organisms: *Microbotryum violaceum* and other model organisms. *Genetics, 168*, 141–146.

Horsley, K., Stark, L. R., & McLetchie, D. N. (2011). Does the silver moss *Bryum argenteum* exhibit sex-specific patterns in vegetative growth rate, asexual fitness or prezygotic reproductive investment? *Annals of Botany, 107*, 897–907.

Hurst, L. (2002). The Ka/Ks ratio: diagnosing the form of sequence evolution. *Trends in Genetics, 18*, 486–489.

Jules, E. S., & Shaw, A. J. (1994). Adaptation to metal—contaminated soils in populations of the moss, Ceratodon purpureus: Vegetative growth and reproductive expression. *American Journal of Botany, 81*(6), 791–797.

Knight, C. D., Cove, D. J., Cuming, A. C., & Quatrano, R. S. (2002). Moss gene technology. In P. M. Gilmartin, & C. Bowler (Eds.), *Molecular plant biology* (vol. 2, pp. 285–299). Oxford, UK: Oxford University Press.

Korpelainen, H., Bisang, I., Hedenas, L., & Kolehmainen, J. (2008). The first sex-specific molecular marker discovered in the moss *Pseudocalliergon trifarium. Journal of Heredity, 99*, 581–587.

Lahn, B. T., & Page, D. C. (1999). Four evolutionary strata on the human X chromosome. *Science, 286*, 964–967.

Liu, Z., Moore, P. H., Ma, H., Ackerman, C. M., Ragiba, M., & Pearl, H. M. (2004). A primitive Y chromosome in papaya marks the beginning of sex chromosome evolution. *Nature, 427*, 348–352.

Marais, G. A. B., Nicolas, M., Bergero, R., Chambrier, P., Kejnovsky, E., Moneger, F. … Charlesworth, D. (2008). Evidence for degeneration of the Y chromosome in the dioecious plant *Silene latifolia. Current Biology, 18*, 545–549.

Maynard-Smith, J., & Haigh, J. (1974). The hitch-hiking effect of a favorable gene. *Genetical Research, 23*, 23–35.

McDaniel, S. F. (2005). Genetic correlations do not constrain the evolution of sexual dimorphism in the moss *Ceratodon purpureus. Evolution, 59*, 2353–2361.

McDaniel, S. F. (2009). The genetic basis of natural variation in bryophyte model systems. In C. Knight, P.-F. Perroud, & D. J. Cove (Eds.), *The moss* Physcomitrella (pp. 16–41). London: Blackwell Publishing.

McDaniel, S. F., Atwood, J., & Burleigh, J. G. (2013). Recurrent evolution of dioecy in bryophytes. *Evolution, 67*, 567–572.

McDaniel, S. F., van Baren, M. J., Jones, K. S., Payton, A. C., & Quatrano, R. S. (2013). Estimating the nucleotide diversity in *Ceratodon purpureus* (Hedw.) Brid. from 218 conserved intron-spanning nuclear loci. *Applications in Plant Sciences, 1*.

McDaniel, S. F., Neubig, K. M., Payton, A. C., Quatrano, R. S., & Cove, D. J. (2013). Recent gene capture on the UV sex chromosomes of the moss *Ceratodon purpureus. Evolution, 67*, 2811–2822.

McDaniel, S. F., & Perroud, P.-F. (2012). Invited perspective: bryophytes as models for understanding the evolution of sexual systems. *Bryologist, 115*, 1–11.

McDaniel, S. F., & Shaw, A. J. (2005). Selective sweeps and intercontinental migration in the cosmopolitan moss *Ceratodon purpureus*, (Hedw.) Brid. *Molecular Ecology, 14*, 1121–1132.

McDaniel, S. F., Willis, J. H., & Shaw, A. J. (2007). A linkage map reveals a complex basis for segregation distortion in an interpopulation cross in the moss *Ceratodon purpureus. Genetics, 176*, 2489–2500.

McDaniel, S. F., Willis, J. H., & Shaw, A. J. (2008). The genetic basis of developmental abnormalities in interpopulation hybrids of the moss *Ceratodon purpureus. Genetics, 179*, 1425–1435.

Menkis, A., Jacobson, D. J., Gustafsson, T., & Johannesson, H. (2008). The mating-type chromosome in the filamentous ascomycete *Neurospora tetrasperma* represents a model for early evolution of sex chromosomes. *PLoS Genetics, 4*, e1000030.

Mittmann, F., Dienstbach, S., Weisert, A., & Forreiter, C. (2009). Analysis of the phytochrome gene family in *Ceratodon purpureus* by gene targeting reveals the primary phytochrome responsible for photo- and polarotropism. *Planta, 230*, 27–37.

Nakayama, S., Fujishita, M., Sone, T., & Ohyama, K. (2001). Additional locus of rDNA sequence specific to the X chromosome of the liverwort, *Marchantia polymorpha. Chromosome Research, 9*, 469–473.

Nam, K., & Ellegren, H. (2008). The chicken (*Gallus gallus*) Z chromosome contains at least three nonlinear evolutionary strata. *Genetics, 180*, 1131–1136.

Nicolas, M., Marais, G., Hykelova, V., Janousek, B., Laporte, V., Vyskot, B. … Moneger, F. (2005). A gradual process of recombination restriction in the evolutionary history of the sex chromosomes in dioecious plants. *PLoS Biology, 3*, e4.

Nishiyama, T., Fujita, T., Shin-I, T., Seki, M., Nishide, H., Uchiyama, I., et al. (2003). Comparative genomics of *Physcomitrella patens* gametophytic transcriptome and *Arabidopsis thaliana*: Implication for land plant evolution. *Proceedings of the National Academy of Sciences of the United States of America, 100*, 8007–8012.

Norrell, T. E., Jones, K. S., Payton, A. C., & McDaniel, S. F. (2014). Meiotic sex ratio variation in natural populations of *Ceratodon purpureus* (Hedw.) Brid. *American Journal of Botany, 101*, 1572–1576.

Papadopulos, A. S. T., Chester, M., Ridout, K., & Filatov, D. A. (2015). Rapid Y degeneration and dosage compensation in plant sex chromosomes. *PNAS, 112*(42), 13021–13026.

Perley, D. S., & Jesson, L. K. (2015). Hybridization is associated with changes in sexual system in the bryophyte genus *Atrichum*. *American Journal of Botany, 102*, 555–565.

Perrin, N. (2009). Sex reversal: a fountain of youth for sex chromosomes? *Evolution, 63*, 3043–3049.

Perroud, P. F., Cove, D. J., Quatrano, R. S., & McDaniel, S. F. (2011). An experimental method to facilitate the identification of hybrid sporophytes in the moss *Physcomitrella patens* using fluorescent tagged lines. *New Phytologist, 191*, 301–306.

Qiu, S., Bergero, R., Forrest, A., Kaiser, V. B., & Charlesworth, D. (2010). Nucleotide diversity in *Silene latifolia* autosomal and sex-linked genes. *Proceedings of the Royal Society of London, Series B: Biological Sciences, 277*, 3283–3290.

Qiu, S., Bergero, R., Zeng, K., & Charlesworth, D. (2011). Patterns of codon usage bias in *Silene latifolia*. *Molecular Biology and Evolution, 28*, 771–780.

Ramsay, H., & Berrie, G. (1982). Sex determination in bryophytes. *Journal of the Hattori Botanical Laboratory, 52*, 255–274.

Rensing, S. A., Ick, J., Fawcett, J. A., Lang, D., Zimmer, A., Van de Peer, Y., & Reski, R. (2007). An ancient genome duplication contributed to the abundance of metabolic genes in the moss *Physcomitrella patens*. *BMC Evolutionary Biology, 7*, 130.

Rice, W. R. (1987). The accumulation of sexually antagonistic genes as a selective agent promoting the evolution of reduced recombination between primitive sex-chromosomes. *Evolution, 41*, 911–914.

Rosenstiel, T. N., Shortlidge, E. E., Melnychenko, A. N., Pankow, J. F., & Eppley, S. M. (2012). Sex-specific volatile compounds influence microarthropod-mediated fertilization of moss. *Nature, 481*, 431–433.

Sandstedt, S. A., & Tucker, P. K. (2004). Evolutionary strata on the mouse X chromosome correspond to strata on the human X chromosome. *Genome Research, 14*, 267–272.

Shaw, A. J., & Beer, S. C. (1999). Life history variation in gametophyte populations of the moss *Ceratodon purpureus* (Ditrichaceae). *American Journal of Botany, 86*, 512–521.

Shaw, A. J., & Gaughan, J. F. (1993). Control of sex-ratios in haploid populations of the moss, *Ceratodon purpureus*. *American Journal of Botany, 80*, 584–591.

Stark, L. R., McLetchie, D. N., & Eppley, S. M. (2010). Sex ratios and the shy male hypothesis in *Bryum argenteum* (Bryaceae). *Bryologist, 113*, 788–797.

Stark, L., McLetchie, N., & Mishler, B. (2001). Sex expression and sex dimorphism in sporophytic populations of the desert moss *Syntrichia caninervis*. *Plant Ecology, 157*, 181–194.

Szövényi, P., Cuming, A. C., Quatrano, R. S., Perroud, P.-F., Rensing, S. A., Stevenson, S. … McDaniel, S. F. (2015). De novo assembly and comparative analysis of the *Ceratodon purpureus* transcriptome. *Molecular Ecology Resources*. http://dx.doi.org/10.1111/1755-0998.12284.

Szövényi, P., Devos, N., Weston, D. J., Yang, X., Hock, Z., Shaw, A. J. … Wagner, A. (2014). Efficient purging of deleterious mutations in plants with haploid selfing. *Genome Biology and Evolution, 6*, 1238–1252.

Szövényi, P. T., Rensing, S. A., Lang, D., Wray, G. A., & Shaw, A. J. (2011). Generation-biased gene expression in a bryophyte model system. *Molecular Biology and Evolution, 28*, 803–812.

Trouiller, B., Charlot, F., Choinard, S., Schaefer, D., & Nogue, F. (2007). Comparison of gene targeting efficiencies in two mosses suggests that it is a conserved feature of bryophyte transformation. *Biotechnology Letters, 29*, 1591–1598.

Van Bel, M., Proost, S., Wischnitzki, E., Movahedi, S., Scheerlinck, C., Van de Peer, Y., & Vandepoele, K. (2012). Dissecting plant genomes with the PLAZA comparative genomics platform. *Plant Physiology, 158*, 590–600.

Voglmayr, H. (2000). Nuclear DNA amounts in mosses (Musci). *Annals of Botany, 85*, 531–546.

Votintseva, A. A., & Filatov, D. A. (2009). Evolutionary strata in a small mating-type-specific region of the smut fungus *Microbotryum violaceum*. *Genetics, 182*, 1391–1396.

Wang, J. P., Na, J. K., Yu, Q. Y., Gschwend, A. R., Han, J., Zeng, F. C. … Ming, R. (2012). Sequencing papaya X and Y-h chromosomes reveals molecular basis of incipient sex chromosome evolution. *Proceedings of the National Academy of Sciences of the United States of America, 109*, 13710–13715.

Weigel, D. (2012). Natural variation in arabidopsis: From molecular genetics to ecological genomics. *Plant Physiology, 158*, 2–22.

Whittle, C. A., Sun, Y., & Johannesson, H. (2011). Degeneration in codon usage within the region of suppressed recombination in the mating-type chromosomes of *Neurospora tetrasperma*. *Eukaryotic Cell, 10*, 594–603.

Wright, S. I., Kalisz, S., & Slotte, T. (2013). Evolutionary consequences of self-fertilization in plants. *Proceedings of the Royal Society of London, Series B: Biological Sciences, 280*, 20130133.

Wright, S. I., Yau, C. B. K., Looseley, M., & Meyers, B. C. (2004). Effects of gene expression on molecular evolution in *Arabidopsis thaliana* and *Arabidopsis lyrata*. *Molecular Biology and Evolution, 21*, 1719–1726.

Yamato, K. T., Ishizaki, K., Fujisawa, M., Okada, S., Nakayama, S., Fujishita, M. … Ohyama, K. (2007). Gene organization of the liverwort Y chromosome reveals distinct sex chromosome evolution in a haploid system. *Proceedings of the National Academy of Sciences of the United States of America, 104*, 6472–6477.

Yang, L., & Gaut, B. S. (2011). Factors that contribute to variation in evolutionary rate among *Arabidopsis* genes. *Molecular Biology and Evolution, 28*, 2359–2369.

Zimmer, A. D., Lang, D., Buchta, K., Rombauts, S., Nishiyama, T., Hasebe, M. … Reski, R. (2013). Reannotation and extended community resources for the genome of the non-seed plant *Physcomitrella patens* provide insights into the evolution of plant gene structures and functions. *BMC Genomics, 14*, 498.

CHAPTER FIVE

The *Sphagnum* Genome Project: A New Model for Ecological and Evolutionary Genomics

A.J. Shaw*[,1], J. Schmutz[§], N. Devos*, S. Shu[¶], A.A. Carrell*[,†], D.J. Weston[†]

*Duke University, Durham, NC, United States
[§]HudsonAlpha Institute of Biotechnology, Huntsville, AL, United States
[¶]Department of Energy Joint Genome Institute, Walnut Creek, CA, United States
[†]Oak Ridge National Laboratory, Oak Ridge, TN, United States
[1]Corresponding author: E-mail: shaw@duke.edu

Contents

Abstract

The inception of the *Sphagnum* (peat moss) genome project marks the first plant-based sequencing project aimed specifically at carbon cycling genomics in a plant system relevant to ecological and evolutionary genomics. *Sphagnum* provides considerable intra- and interspecific variation at the nucleotide level, and in physiology, morphology, net production, decomposition and carbon accumulation (peat formation). Because of the large number of peat moss species, their diversity in mating systems, and clear patterns of niche differentiation, *Sphagnum* provides an exceptionally valuable complement to *Physcomitrella patens* and *Ceratodon purpureus* as moss models for genomic research. Here we review the organismal biology of *Sphagnum* including phylogeny,

Advances in Botanical Research, Volume 78
ISSN 0065-2296
http://dx.doi.org/10.1016/bs.abr.2016.01.003

life cycle, mating systems, ecology and niche differentiation. We include the current state of *Sphagnum* genomic resources, in vitro methods and germplasm. A use-case is provided to address questions concerning epigenetics and reproduction.

1. INTRODUCTION

Sphagnum, or more broadly, phylum Bryophyta (mosses), class Sphagnopsida (peat moss), represents a valuable complement to *Physcomitrella patens* and *Ceratodon purpureus* as moss models for genomic research. The special value of peat moss derives from (1) their early diverging phylogenetic position within the mosses, (2) the speciose nature of the peat moss clade, (3) well-documented patterns of niche differentiation among *Sphagnum* species and (4) the ecological importance of peat moss for global carbon cycle dynamics. The *Sphagnum* genome project was formulated to take advantage of these unique features, which are briefly discussed below.

2. THE SPHAGNOPSIDA COMPRISES AN EARLY DIVERGING LINEAGE WITHIN THE EARLY DIVERGING MOSS LINEAGE

Sometime in the early Paleozoic Era, perhaps during the Ordovician Period (Rubinstein, Gerrienne, de la Puente, Astini, Steemans, 2010; Wellman, Osterloff, & Mohiuddin, 2003), green algal ancestors of the embryophyte land plant clade colonized terrestrial environments and diversified rapidly (Kendrick & Crane, 1997; Steemens et al., 2009). Life on land carried with it new abiotic challenges, and also new opportunities, including novel relationships with other terrestrial organisms such as fungi, eukaryotic protists and prokaryotic microbes (Knack et al., 2015; Redecker, Kodner, & Graham, 2000). Terrestrial environments presented unprecedented stresses, especially in terms of water economy, nutrient acquisition, maintaining structure in high-gravity ecosystems and elevated levels of UV radiation (Graham et al., 2003; Waters, 2003). Responses to these novel environmental pressures led to monumental changes in plant phenotypes, including morphological, developmental and biochemical/metabolic features, and to the underlying genomic bases for traits necessary to survive on land. The nonvascular embryophyte clades (ie, those plants lacking phloem and xylem), comprising the mosses (Bryophyta), liverworts (Marchantiophyta) and hornworts (Anthocerophyta), have long been considered the earliest groups of embryophyte land plants (Haeckel, 1868; Bower, 1890) based on their haploid-dominant life cycles and morphological features, and more recently this hypothesis has been strongly supported by numerous molecular phylogenetic studies (Qiu et al., 2006; Wickett et al., 2014; among others).

Comparative genomics has elucidated that many components of the molecular 'toolkit' that facilitated the diversification of land plants evolved early in embryophyte evolution and can be detected in bryophytes (eg, Floyd & Bowman, 2007; Viaene et al., 2014). These observations indicate that homologues of many of the genes critical for plant development were present in the earliest embryophyte land plants; ie, the bryophytes. Although these genes, or similar ancestral versions of modern genes, evolved early in the history of land plants, in many cases their functions have likely changed. Many good examples exist of genes present in early land plants being coopted for new functions in more recently derived lineages (Rensing et al., 2008). In contrast to *P. patens* and *C. purpureus*, which belong to the more derived group of peristomate mosses, *Sphagnum* evolved earlier in moss evolution (Chiang & Graham, 2011) and as such, provides inferences about early genomic traits in the Bryophyta. This is true notwithstanding the fact that some phenotypic traits and genomic features characterizing *Sphagnum* evolved within the Sphagnopsida lineage subsequent to their divergence from other mosses.

3. THE SPHAGNOPSIDA: A SPECIOSE CLADE OF MOSSES

The Sphagnopsida comprises a class of the phylum Bryophyta, and include four genera in three families (estimated numbers of species in each genus in parentheses): Sphagnaceae: *Sphagnum* (250–450); Flatbergiaceae: *Flatbergium* (2); Ambuchananiaceae: *Ambuchanania* (1), *Eosphagnum* (1) (Shaw, Cox, et al., 2010). *Ambuchanania, Eosphagnum* and *Flatbergium* occur in the Southern Hemisphere (although *Flatbergium sericeum* reaches southern China) and none of them occur in peatlands but rather on moist to wet rocks and soils. *Sphagnum*, by far the largest genus of Sphagnopsida, dominates mire habitats (bogs and fens) in the Northern Hemisphere but also includes tropical and Southern Hemisphere species. It is difficult to accurately estimate the number of species in *Sphagnum* because most taxa outside the boreal zone are poorly known and even within the boreal peat moss flora, species delimitation is a work in progress. Recent work has demonstrated that some common species include multiple, phylogenetically distinct lineages that may warrant individual species status (eg, the common peat-forming species, *Sphagnum fuscum*, includes two colour morphs (Andrus, 2006) that are highly distinct genetically/phylogenetically (Kyrkjeeide et al., 2016)). Allopolyploid speciation has contributed to confusing systematic patterns in *Sphagnum*; most species are gametophytically haploid but both diploid and triploid taxa are also known (Karlin et al., 2009; Ricca & Shaw, 2010). Systematic problems are directly relevant to work on peat moss genomics and ecology. For example, *Sphagnum magellanicum* is generally considered to be one of the most prominent peat moss of wetlands throughout the Northern Hemisphere, including Alaska. Recent work (Kyrkjeeide, unpublished), however, has shown that *S. magellanicum* is in fact known from only a couple sites in Alaska, and the dominant peat-forming species that has been included in numerous ecological studies as *S. magellanicum* is in fact the related allopolyploid, *Sphagnum alaskense*, which has *S. magellanicum* as one of its parents. Complexes of closely related species, most of which include one or more allopolyploids, occur in all four of the large subgenera of *Sphagnum* and clarifying these systematic problems are critical for any future genomic research on this group.

The Sphagnopsida differ from all other mosses in many developmental and morphological features. Their unique features include especially: (1) spores give rise to thallose (rather than filamentous) protonemata (shared with a few other mosses); (2) mature gametophytes typically lack rhizoids or other structures for substrate attachment; (3) gametophores have fasciculate branching with spreading and pendent branches in each fascicle; (4) dimorphic leaf cells with large, hyaline cells, dead and empty at maturity, enclosed in a network of smaller, narrow, photosynthetic chlorophyllose cells; (5) sporophytic seta lacking, capsules (sporangia) borne on pseudopodia of gametophytic origin; (6) sporogenous tissue within sporangia derived from embryonic amphithecial layers rather than endothecial tissue as in other mosses.

4. ECOLOGY AND NICHE DIFFERENTIATION WITHIN SPHAGNUM

Species of *Sphagnum* not only dominate northern peatland habitats in terms of biomass and species richness, they play an active roll in the formation of their own microenvironments (van Breemen, 1995). Boreal peatlands have long served as a model for research in community ecology because they are relatively simple systems in terms of numbers of species and community structure (compared to many tropical biomes, for example).

Nevertheless, as many as 20 or more *Sphagnum* species commonly co-occur within peatland communities of Scandinavia and northern North America. There may be no other genus of plants in which so many species occur sympatrically within a single community.

A huge body of literature exists documenting niche differentiation among northern *Sphagnum* species relative to abiotic microhabitat gradients (see Rydin & Jeglum, 2006). Species sort along several gradients including cation concentrations and pH, and especially height above the water table (Fig. 1). Most well-developed peatlands exhibit a pronounced 'hummock-hollow' structure and *Sphagnum* species occupy characteristic positions relative to these microtopographic features. Interspecific variation along the hummock-hollow gradient (ie, height above the water table) is related to phylogeny (Johnson et al., 2015) and the rate of evolution relative to this gradient is heterogeneous; interspecific differentiation among hummock species has generally evolved more quickly than among hollow species. More complex evolutionary patterns occur relative to ionic and pH gradients. The evolution of niches relative to pH may generally be constrained in *Sphagnum* by stabilizing selection, but exceptional rate

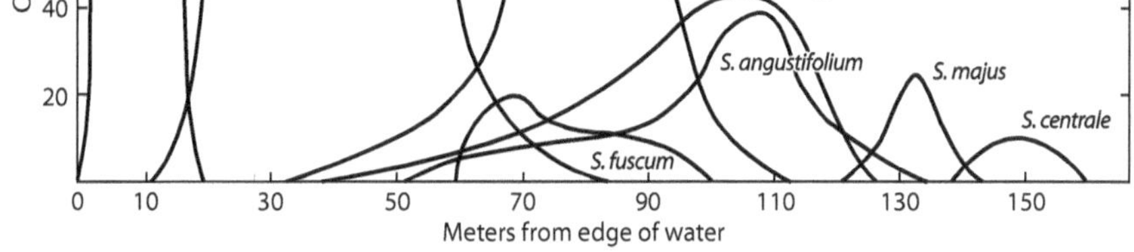

Figure 1 Niche differentiation among *Sphagnum* species in terms of distance from the mire margin. A similar sequence can be observed within the mire relative to the hummock–hummock gradient. *Sphagnum teres* occupies relatively mineral-rich microhabitats near the mire margin. *Modified from Vitt, D. H., & Slack, N. G. (1975). An analysis of the vegetation of Sphagnum-dominated kettle-hole bogs in relation to environmental gradients. Canadian Journal of Botany, 53, 332–359.* In the original figure, *Sphagnum capillifolium* was labelled *Sphagnum capillacium*, a later name sometimes used for this species; *Sphagnum angustifolium* was labelled *Sphagnum recurvum*, a name sometimes applied to a group of closely related species including *S. angustifolium*, to which the plants included in that study likely belonged.

changes appear to characterize some species in the subgenus *Sphagnum* (Johnson et al., 2015). Other examples of rapid niche changes, for example, relative to substrate electrical conductivity, have occurred at or near the tips of the *Sphagnum* phylogenetic tree; ie, relatively recently in peat moss evolution.

Mating systems also vary across *Sphagnum* (Johnson & Shaw, 2015). Of the 91 *Sphagnum* species currently recognized in North America, gametophytes are bisexual in 14 and unisexual in 58; the remaining species have unknown sexual conditions (McQueen & Andrus, 2007). Bisexual species, as expected, are characterized by high levels of selfing but some outcrossing does occur (Johnson & Shaw, 2015). Moreover, mating patterns are related to microhabitat. Among bisexual taxa, hummock species tend to have higher inbreeding coefficients that hollow species, whereas in unisexual species, taxa growing in hummocks high above the water table tend to have lower inbreeding coefficients. Individual female gametophytes generally carry multiple sporophytes (each the result of sexual fusion) and in all species studied to-date, these cohorts of sporophytic offspring are sired by multiple fathers (male gametophytes). Inbreeding depression appears to be rare in *Sphagnum*, even among species with unisexual gametophytes (Johnson & Shaw, 2015).

The large number of *Sphagnum* species, their diversity in mating patterns and their clear patterns of niche differentiation make the peat moss clade exceptionally valuable for research in ecological and evolutionary genomics. The *Sphagnum* genome project holds the potential to elucidate underlying genetic factors that determine ecological differences among closely related species and their evolution. The observation that sporophytes attached to individual maternal gametophytes generally have different fathers opens the possibility for research on genetic imprinting and other parent-of-origin effects on gene expression.

5. PEAT MOSS AND GLOBAL BIOGEOCHEMISTRY: LINKING GENES TO ECOSYSTEM FUNCTION

Clymo and Hayward (1982) famously commented that there's probably more biomass in *Sphagnum* than in any other genus of plants on earth. *Sphagnum* species dominate many wetlands and produce huge deposits of peat, partially decomposed plant material made up largely of the lower portions of *Sphagnum* plants. Peat accumulates where net production exceeds decomposition over millennia. As a result, northern peatlands are estimated to harbour more than 400 Gt of carbon and are important to global fluxes of CO_2 and CH_4 (Gorham, 1991; Yu, 2012). Boreal species of *Sphagnum* vary in rates of both production and decomposition and at least some of the physiological and morphological traits that correlate with ecologically important species differences have been identified (Cornelissen, Lang, Soudzilovskaia, & During, 2007; Lindo, Nilsson, & Gundale, 2013; Rydin & Jeglum, 2013; Turetsky, Crow, Evans, Vitt, & Wieder, 2008). Peat moss species also differ in uptake rates for nitrogen, phosphorus and other nutrients that impact photosynthetic rates, as well as interactions with other plants and animals (Hajek & Beckett, 2008; Granath, Wiedermann, & Strengbom, 2009).

Sphagnum provides an opportunity to identify plant traits that scale up to impact biogeochemical processes at a global scale (Weston et al., 2015). *Sphagnum*-dominated peatlands are natural laboratories with intra- and interspecific variations at the nucleotide level, and in physiology, morphology, net production, decomposition and carbon accumulation (peat formation). Interspecific variation in a phylogenetic context shows how traits that impact ecosystem function evolved through deep time, and intraspecific variation can reveal patterns of local adaptation and the action of ongoing natural selection at multiple levels from the genome to phenotypic traits.

6. LIFE CYCLE

Like all bryophytes, the dominant life cycle stage for *Sphagnum* is the haploid gametophyte. The gametophyte generation of the life cycle starts with the development of germinating spores that transition from an early filamentous protonematal stage (Fig. 2A) to the thalloid protonematal stage after a few cell divisions (Fig. 2B). This is in contrast to most other mosses, where the protonemata stay filamentous. In *Sphagnum*, the thalloid protonemata form a leafy gametophore that is the perennial vegetative stage of the *Sphagnum* life cycle. Rhizoids are produced by the thalloid protonemata (Fig. 2B) but are generally absent from mature gametophytes. *Sphagnum* species are typically dioecious, meaning that sperm and eggs are produced on separate gametophytes (and unlike *Physcomitrella*, where individual gametophytes form both sperm and eggs, but like *C. purpureus*, which also has unisexual gametophytes). Mature male *Sphagnum* gametophytes produce antheridia (Fig. 2C) with sperm while female gametophytes produce the egg-containing archegonia (Fig. 2D). As in other mosses, the sperm and

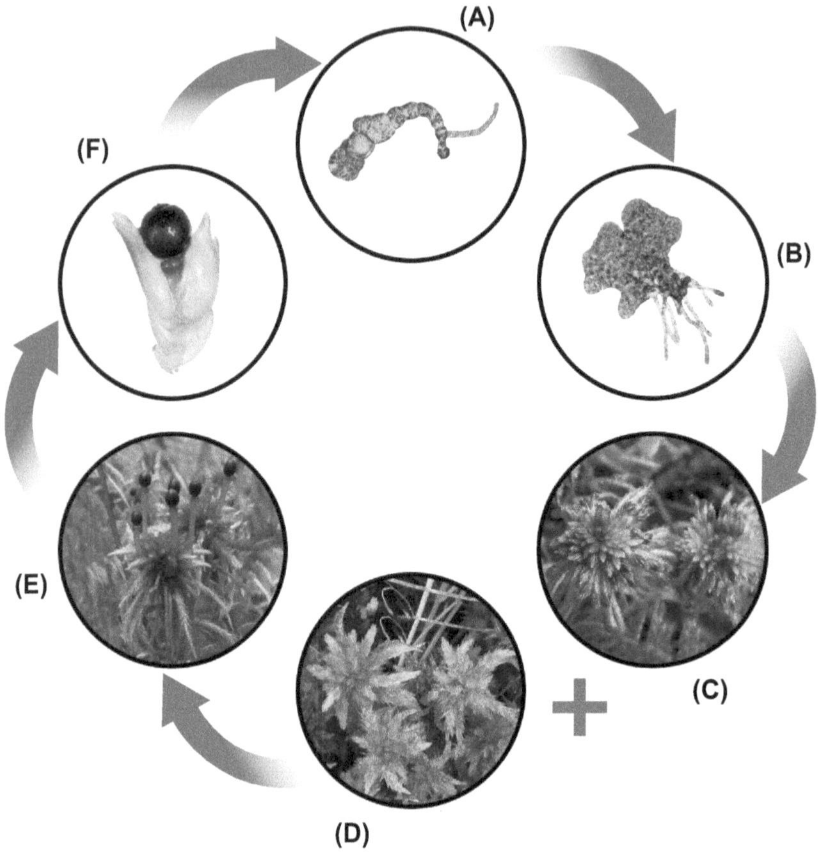

Figure 2 *Sphagnum* life cycle. (A) Young protonema with a single rhizoid. (B) Mature thalloid protonemata with rhizoids. (C) Male gametophore. (D) Female gametophore. (E) Female gametophyte bearing multiple sporophytes. (F) Mature sporophyte (round, brown capsule, bearing meiotically derived spores) borne on a short pseudopodium (mostly surrounded by perichaetial leaves). (See colour plate)

eggs are produced mitotically, so they are genetically identical. In nature, the reproductive structures are generally produced once each year, at a predictable season (like flowers). Many boreal species form antheridia and archegonia in the autumn and sporophytes mature in late spring to late summer. Reproductive seasons are species-specific, but with substantial overlap. The sperm require water to reach an archegonium, and swim down the neck of the archegonium to access the egg cell in the archegonial venter. Once the sperm penetrates the egg to form the zygote and divides to form the embryo, cell specialization occurs to form the foot, which embeds the sporophyte into the maternal gametophyte and the sporangium (capsule;

Fig. 2F) in which meiosis occurs. A seta, or capsule stalk, is not formed in *Sphagnum*. In *Physcomitrella* the seta is very short whereas in *Ceratodon*, the other developing moss model, it is well developed (>1 cm long). The formation of the zygote marks the beginning of the diploid sporophyte stage and at maturity meiosis occurs within the capsule, producing haploid spores. As in other mosses, the sporophyte is green when immature but loses its chlorophyll after meiosis as the spores mature.

7. THE GENOME PROJECT

The Department of Energy (DOE), Joint Genome Institute (JGI), has recently accepted a community supported science project aimed at sequencing two *Sphagnum* species adapted to contrasting microenvironment conditions, and the resequencing of 194 individuals from a *Sphagnum fallax* pedigree. The inception of this project provides the first plant-based genome sequencing project aimed specifically at carbon cycling genomics in an ecologically relevant system to enhance our understanding of (1) the genetic variation in natural populations of an undomesticated haploid organism and (2) genotype-to-phenotype relationships necessary for a trait-based understanding of nonvascular plants in ecosystem function. Furthermore, the resulting genomes from the *Sphagnum* species can be compared to the JGI Plant Flagship model for bryophytes, *P. patens* (Rensing et al., 2008), for which there are ample genomic resources available.

Genomic resources for *Sphagnum* are rapidly progressing. Genome size appears to be relatively uniform across *Sphagnum* species based on flow cytometric estimates, 0.47 pg, which corresponds to approximately 453 Mb. A draft genome sequence (named V0.5, phytozome.jgi.doe.gov/Sfallax_er) for *S. fallax* (subgenus *Cuspidata*) is currently in 49x coverage, assembled from short read illumina sequencing in 1228 scaffolds greater than 1 Kb, with half of the genome on 61 scaffolds that are at least 1.8 Mb in length (L50/N50: 61/1.8 Mb). Furthermore, the scaffolds are nearly complete with only 0.4% gap bases between the assembled contigs. There is a robust annotation based on 3.9 billion RNA-seq transcript reads from 16 different growth conditions, resulting in 32,298 genes (26,939 loci; 5359 alternative splice variants). The main genome is 396 Mb of sequence, the chloroplast genome is in 12 scaffolds at 179 Kb, and the mitochondrial genome is present in 33 scaffolds at 47.6 Kb. The *S. magellanicum* genome is currently in progress and appears similarly amenable to genome sequencing. The current draft genome sequence is in 6255 scaffolds, with half the genome in 464 scaffolds that are at least 294 Kb in length. The main genome, ie, total scaffold size, is 487.8 Mb. The draft genome sizes for both *Sphagnum* species compare favourably with the genome of *P. patens*, whose sequenced genome is 467 Mb in length.

A genetic map for *S. fallax* is being produced to generate a chromosome level scale genome assembly for both species. The map is being generated from a 480 member pedigree population that was developed by isolating spores from a single sporophyte, and clonally propagated through single stem descent to ensure individual genotypes. The genetic map is being constructed from SNPs and indels. A subset of a 194 individuals from the pedigree population is being used for genome resequencing and trait-based characterization.

8. CULTURE AND GERMPLASM RESOURCES

Establishing in vitro growth conditions is essential for the progression of genetic transformation efforts, culturing of organs and tissues, axenic whole plant regeneration, genotype preservation and large-scale biotechnological application (eg, bioreactors). As noted by Hohe and Reski (2005), the establishment of pure protonematal cultures of *Atrichum undulatum* and *Hypnum velutinum* marked the beginning of a long history of bryophyte in vitro culturing. The moss model, *P. patens*, benefits from well-established in vitro methods that allow for developmental manipulation of the organism, as well as axenic conditions suitable for 'omics'-based assays (protein profiling: Sarnighausen, Wurtz, Heintz, Van Dorsselaer, & Reski, 2004; metabolites: Erxleben, Gessler, Vervliet-Scheebaum, & Reski, 2012, eg, RNA-seq: Wu et al., 2014). *Sphagnum* has been grown in vitro for quite some time (eg, Simola, 1969), but it is not until relatively recently that a systematic approach was established (Beike et al., 2015). As motivated by the *Sphagnum* genome project and the use of the 'moss bag technique' for biomonitoring as part of the European FP7 'MOSSCLONE' project (www.mossclone.eu), Beike et al., (2015) developed and tested *Sphagnum*-specific axenic medium and culturing techniques for *Sphagnum fimbriatum*, *S. magellanicum*, *Sphagnum palustre* and *S. rubellum*. Large-scale bioreactor culturing techniques were also developed and demonstrated for *S. palustre* (Beike et al., 2015). Together with the genome sequences, *Sphagnum* can now benefit from the full suite of molecule 'omics' and system's biology approaches (Fig. 3A–C).

Figure 3 Representative cultivation techniques for *Sphagnum*. Gametophytes can be cultivated on solid medium in (A) liquid Knop medium in a photobioreactor, (B) magenta vessels, (C) in petri dishes or (D) populations in a series of multi-well tissue culture plates. *In photo credit for panel A: Anna Beike, Reski Lab, University of Freiburg, Germany.* (See colour plate)

In addition to the development of culturing methods, there is a growing number of *Sphagnum* clones being generated and maintained within laboratory conditions. For example, the *S. magellanicum* and *S. fallax* clonal strains used in the DOE JGI sequencing project are available in axenic in vitro culture (Weston Lab, Oak Ridge National Lab, USA). Furthermore, a 480-member pedigree derived from a single sporophyte is available in tissue culture plates and are amenable to high-throughput phenotyping for growth morphology and chlorophyll fluorescence assays.

9. CHALLENGES AND OPPORTUNITIES WITH POPULATION AND GENOMIC RESOURCES

While the development of *Sphagnum* in vitro culturing techniques, germplasm resources and draft genome assemblies mark the beginning of its use for ecological and evolutionary genomics, there are still substantial obstacles to be overcome before the utility of these resources are fully realized. For example, we are not able to manipulate sexual reproduction within laboratory conditions or perform controlled crosses that would greatly strengthen current approaches for linking traits to underlying quantitative trait loci. Furthermore, we have not yet perfected the transfer of *Sphagnum* from in vitro axenic cultures to field conditions where so many of our ecological and evolutionary questions reside. Also lacking is a stable nuclear transformation protocol that is present in most plant model systems. In *P. patens*, gene targeting through homologous recombination has been demonstrated (Kamisugi et al., 2006). In this approach, gene replacement occurs at a desired chromosomal location (locus), thereby alleviating the negative effects of undirected and multiple gene copies often observed through biolistics (gene gun technology), or delivery from *Agrobacterium*.

Although challenges in developing this emerging model system exist, there are considerable benefits in using *Sphagnum* to decipher genotype-to-phenotype relationships. For example, the ease by which large numbers of individuals can be grown, stored, assayed and phenotypically characterized is a major benefit of using bryophytes, and *Sphagnum* specifically. In most crop plants and bioenergy feedstocks, phenotype characterization including the structural, physiological and performance-related traits of genotypes in a given environment is a considerable challenge (Benfey & Mitchell-Olds, 2008; Dhondt, Wuyts, & Inzé, 2013). This may not be a major constraint for *Sphagnum* as it can be readily grown in laboratory culture with multiple media and environmental conditions (Beike et al., 2015). The 480-member *S. fallax* population discussed above is stored in a series of 12 well tissue culture plates that occupy 1.2 m^2 of illuminated shelf space (Fig. 3D). This in essence, represents the 'common garden' that requires considerably less resources in space and cultural practices relative to crop plants and bioenergy feedstocks. Furthermore, the plate-based system is amenable to phenotyping assays that include growth analyses and characterization of photosynthetically related traits such as chlorophyll-*a* fluorescence (Fig. 4) and growth responses (Fig. 5) to substrate pH. The ability to store entire populations of *Sphagnum* in axenic in vitro culture coupled with high-throughput imaging and image processing for trait characterization is a tremendous compliment to the population-based sequencing efforts that will facilitate our understanding of genotype-to-phenotype relationships.

A critical component of the *Sphagnum* research program must involve transferring experimentally grown and genomically characterized plants back into the natural environment in order to link genome attributes, phenotypic traits, and ecosystem function. This work can be done in constructed

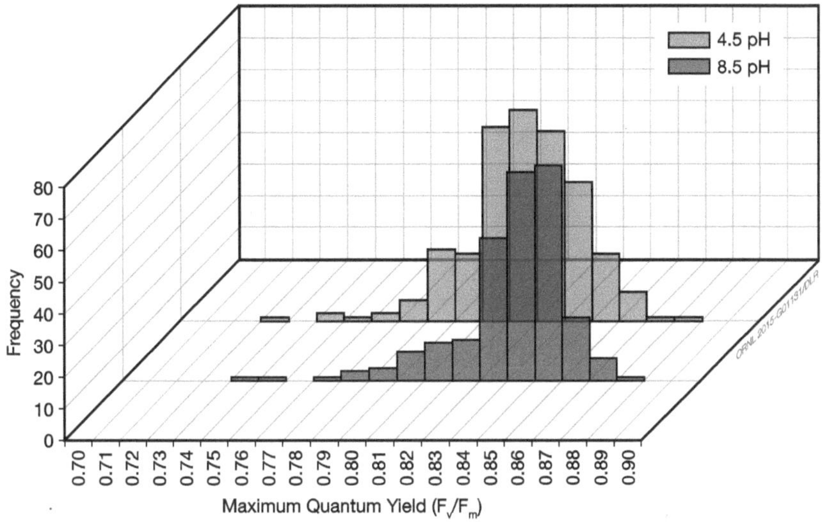

Figure 4 Distribution of photosynthetic rates (maximum quantum yield of PSII) in *Sphagnum* cultivated for 3 weeks at pH 4.5 (red (grey in print versions)) and 8.5 (blue (dark grey in print versions)).

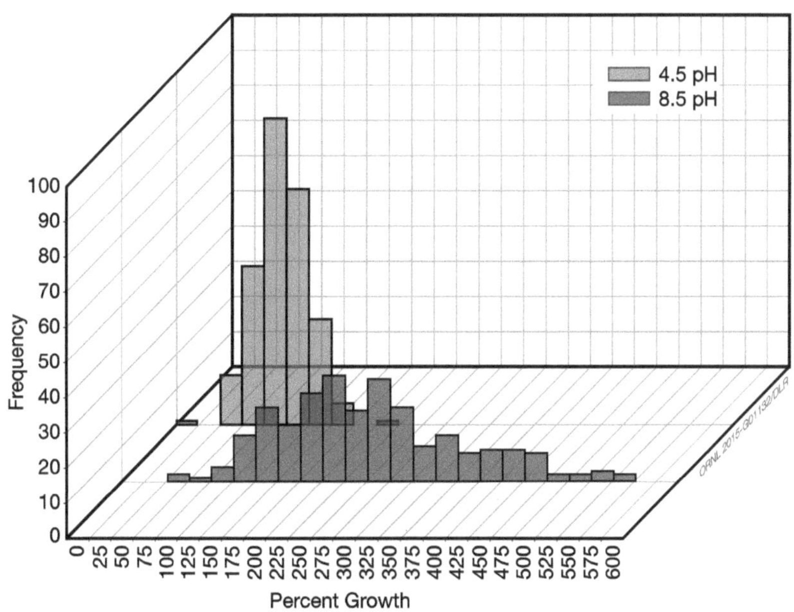

Figure 5 Percent growth (plant area after three weeks − initial plant area)/initial plant area × 100) in *Sphagnum* cultivated for 3 weeks at pH 4.5 (red (grey in print versions)) and 8.5 (blue (dark grey in print versions)).

and monitored peatland communities (for example, at the SPRUCE site; http://mnspruce.ornl.gov/), but experimental methodologies need to be refined.

10. AN EVOLUTIONARY GENOMICS EXAMPLE: EPIGENETICS AND SPHAGNUM REPRODUCTION

Several features of moss (including *Sphagnum*) mating systems make them especially valuable for studies of reproductive biology (Johnson & Shaw, 2015). First, mosses (like liverworts and hornworts) are the only land plants in which the fitness of gametophytes is a quantitative trait. This is true because moss gametophytes can parent more than one sporophytic offspring, and a single gametophyte produces many genetically identical eggs because the gametes are produced mitotically. As a result of gametophyte clonal reproduction, genetically identical eggs and sperm can participate in many independent fertilization events. In all other plants, including the spore-producing ferns, gametophytes parent either one sporophyte or none. The second feature is that multiple paternity for sporophytes attached to an individual female gametophyte appears to be the rule in *Sphagnum*, and probably in other mosses. Multiple paternity sets up the conditions in which parent—offspring conflicts can be expected, and indeed the mosses represent an ideal group for studying this phenomenon (Haig & Wilczek, 2006). Parent-of-origin effects on gene expression in offspring are well-known in angiosperms (and mammals), but it is easier to study in mosses because in angiosperms, seeds include complex combinations of tissues of maternal and paternal origin, as well as triploid endosperm with unequal contributions from the two parental genomes.

We describe here a preliminary analysis of (gametophytic) parental effects on (sporophytic) gene expression in *S. palustre*. This species is an allopolyploid (diploid gametophytes, tetraploid sporophytes), which limited our ability to unambiguously identify paternal alleles in sporophytes, but the analysis shows proof of concept for future studies.

Two female gametophytes each bearing multiple sporophytes were sampled from a natural population (West Virginia). RNA was extracted from the two maternal gametophytes, and from a total of seven attached sporophytic offspring (three from one gametophyte, four from the other) using RNAzol®RT (Molecular Research Center, Inc.). RNA-seq (Wang, Gerstein, & Snyder, 2009) paired-end libraries were constructed using the Kapa stranded mRNA library preparation kit (Kapa Biosystems). Libraries were indexed, pooled and sequenced on two lanes of an Illumina HiSeq 2500 sequencer flow cell (Illumina, San Diego, California, USA). Sequencing generated 100 pb paired-end sequences.

The sequence data were assembled for the two maternal gametophytes using Trinity_r20131110 (Grabherr et al., 2011; Haas et al., 2013). In order to remove potential contaminant transcripts (ie, from contaminating organisms) from the assemblies, the transcriptome assemblies were blasted against the Uniprot database (The UniProt Consortium 2014) using BLASTX (Altschul et al., 1997). BLASTX outputs were filtered to discard transcripts with top hits from nonland plants. For the two maternal gametophytes (M1, M2), we obtained 311,808 (N50: 1093) and 421,902 (N50: 741) contigs, respectively, for total assembly sizes of 207,293,790 and 232,315,929 bp. There were 78,040 and 87,835 hits to the *Physcomitrella* proteome and another 2992 and 4510 hits to land plants other than *Physcomitrella.* The expression level of each RNA unit was measured by the number of sequenced fragments that mapped to the transcript, which is expected to correlate directly with its abundance level. One of the gametophytic transcriptomes was used as a reference to which all the sequences from the various sporophytic RNA samples were mapped using Bowtie (Langmead, Trapnell, Pop, & Salzberg, 2009). The reads from the gametophytes were also mapped to this reference transcriptome. RSEM (Li & Dewey, 2011) was used to produce the gene count matrix for all differential expression analyses. DESeq was used on to analyse the count data and test for differential expression between gametophytes and sporophytes and between sporophytes form one maternal gametophyte versus the other (parental effect).

Gene expression profiles from the two maternal gametophytes and seven sporophytes indicate expression differences between sporophytes and gametophytes (Fig. 6); a total of 2420 genes exhibited significant expression differences between the two life cycle generations (Fig. 6B). This is also indicated by clustering the nine samples (two gametophyte, seven sporophyte) by similarities in overall expression profiles (Fig. 6A). Moreover, differential expression of genes occurred among sporophytes borne on the two different maternal gametophytes, indicating epigenetic effects associated with those gametophytes on the sporophytes they parent (Fig. 6C). This is also indicated by multigenic expression profiles, although two of the sporophytes from gametophyte M1 group with sporophytes from M2 (Fig. 6A).

The results of this preliminary experiment reveal substantial expression differences between sporophytes and gametophytes in *S. palustre.* Parent-of-origin effects on gene expression are strongly suggested although we

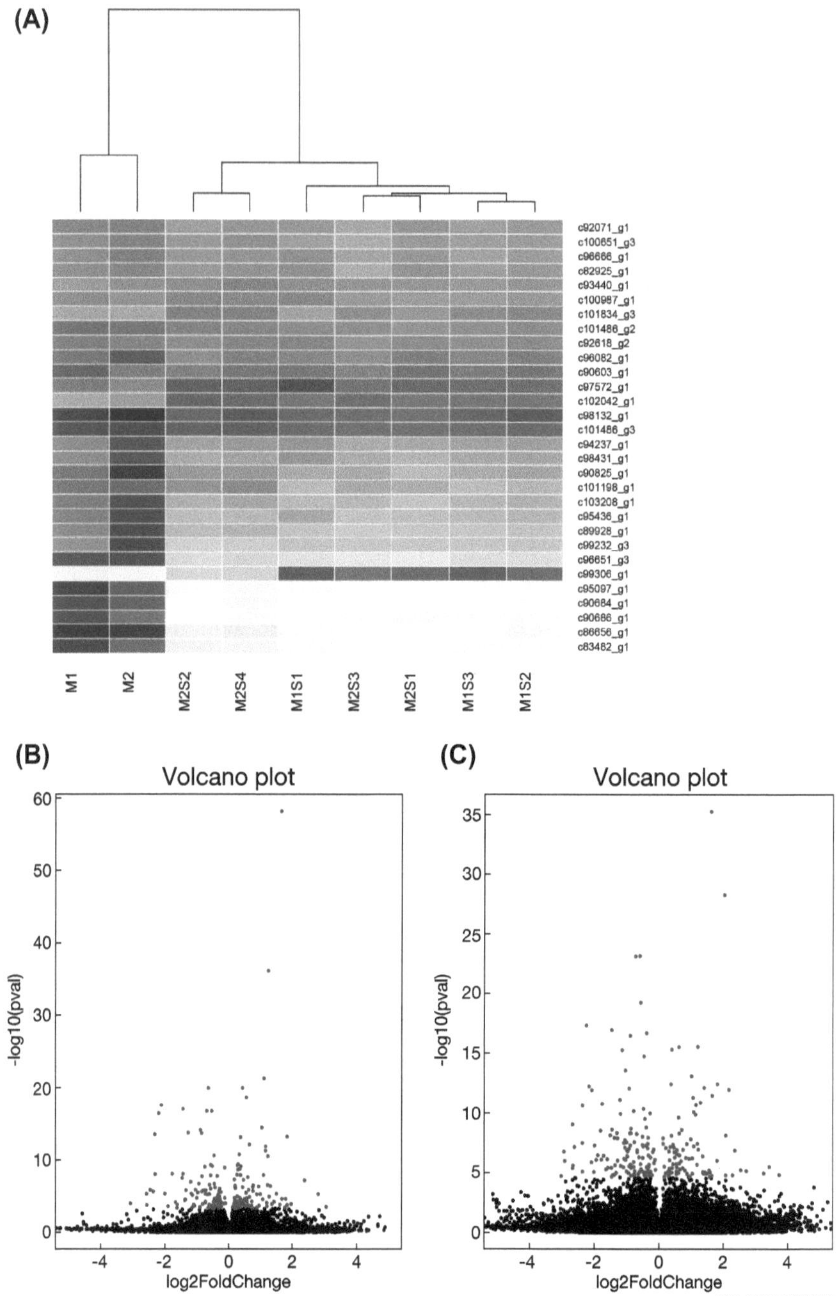

Figure 6 Gene expression in gametophytes and sporophytes of *Sphagnum palustre*. (A) Heatmap showing cluster analysis of sporophyte and gametophyte samples. M1 and M2 = maternal gametophytes; S1-S4 = sporophytes attached to M1 or M2. (B) Volcano plot showing genes (grey) that are differentially expressed (at $P < 0.0001$) in sporophytes versus gametophytes. (C) Volcano plot showing genes (grey) that are differentially expressed (at $P < 0.0001$) in sporophytes produced by maternal gametophyte M1 versus M2.

cannot determine whether these effects reflect maternal genetic effects, paternal genetic effects, or maternal environmental effects. The experiment nevertheless demonstrates the utility of *Sphagnum* for investigating epigenetic effects on gene expression under more rigorously controlled conditions and provides a valuable system for investigating mating systems at the genomic level.

11. SYNTHESIS: THE POTENTIAL FOR A GENUS LEVEL SEQUENCING PROJECT

Whereas the *Ceratodon* and *Physcomitrella* models have been developed primarily for intensive evo-devo investigations, *Sphagnum's* particular value is in ecological and evolutionary genomics. The *Sphagnum* system facilitates adding new dimensions to genomic research through comparative studies of ecologically diverse species in a well-studied clade. The genus *Physcomitrella* was traditionally thought to include three or four species (http://www.tropicos.org/NameSearch.aspx?name=Physcomitrella&commonname=) but phylogenetic analyses (Liu, Budke, & Goffinet, 2012) indicate that there are only two or three species, and they do not form a monophyletic group. *Ceratodon* is also a genus of only about four species worldwide, though *C. purpureus* is common around the Northern Hemisphere. The fact that peat moss have such profound impacts on global biogeochemistry and therefore climate provides exceptional added value for functional ecology and ecological genomics.

We envision a research program in which *Sphagnum* species representing the phylogenetic and ecological diversity of peat moss are characterized at the genomic and phenotypic levels to determine how genome structure and function translate into plant traits, and through those traits, to niche differentiation and patterns of carbon cycling. Species to be included should include taxa that inhabit nutrient-poor bogs, richer fens, those that form hummocks high above the water table and accumulate a lot of peat, and fully aquatic species that differ in relevant traits and decompose faster. Boreal, tropical, and Southern Hemisphere species open a more macro dimension to comparative analyses of ecology, and inclusion of species representing the major clades within *Sphagnum* can reveal evolutionary changes over deep time scales. Based on molecular data (Shaw, Devos, et al., 2010) the Sphagnopsida likely diverged from other mosses 100–200 mya, but *Sphagnum* s. str. appears to have diversified rapidly during the Neogene and this agrees closely with estimates of when *Sphagnum*-dominated wetlands appear in the fossil record (Greb, DiMichele, & Gastaldo, 2006). Inclusion of species from the related genera *Ambuchanania, Eosphagnum* and *Flatbergium* in future genomic work will (potentially) enable a complete picture of how peat moss became wetland dominants as global climate cooled during the tertiary and boreal peatlands came into existence.

ACKNOWLEDGEMENTS

The time and resources to produce this chapter were supported by U.S. Department of Energy, Office of Science, Biological and Environmental Research, including the SPRUCE project (http://mnspruce.ornl.gov/) and the Laboratory Directed Research and Development Program of Oak Ridge National Laboratory. Oak Ridge National Laboratory is managed by UT-Battelle, LLC, for the US Department of Energy under contract DE-AC05- 00OR22725. The work conducted by the U.S. Department of Energy Joint Genome Institute was supported by the Office of Science of the U.S. Department of Energy under Contract No. DE-AC02-05CH11231.

REFERENCES

Altschul, S. F., Madden, T. L., Schäffer, A. A., Zhang, J., Zhang, Z., Miller, W., & Lipman, D. J. (1997). Gapped BLAST and PSI-BLAST: a new generation of protein database search programs. *Nucleic Acids Research, 25*, 3389–3402.

Andrus, R. E. (2006). Six new species of *sphagnum* (Bryophyta: Sphagnaceae) from North America. *Sida, 22*, 959–972.

Beike, A. K., Spagnuolo, V., Lüth, V., Steinhart, F., Ramos-Gómez, J., Krebs, M., & Adamo, P. (2015). Clonal in vitro propagation of peat mosses (*Sphagnum* L.) as novel green resources for basic and applied research. *Plant Cell, Tissue and Organ Culture, 120*, 1037–1049.

Benfey, P. N., & Mitchell-Olds, T. (2008). From genotype to phenotype: systems biology meets natural variation. *Science, 320*, 495–497.

Bower, F. O. (1890). On antithetic as distinct from homologous alternation of generations in plants. *Annals of Botany, 4*, 347–370.

van Breemen, N. (1995). How *Sphagnum* bogs down other plants. *Trends in Ecology and Evolution, 10*, 270–275.

Chiang, Y., & Graham, S. W. (2011). Inferring the higher-order phylogeny of mosses (Bryophyta) using a large, multigene plastid dataset. *American Journal of Botany, 95*, 839–849.

Clymo, R. S., & Hayward, P. M. (1982). *The ecology of* Sphagnum. UK and New York, NY, USA. London: Chapman and Hall Ltd.

Cornelissen, J. H. C., Lang, S. I., Soudzilovskaia, N. A., & During, H. J. (2007). Comparative cryptogam ecology: a review of bryophyte and lichen traits that drive biogeochemistry. *Annals of Botany, 99*, 987–1001.

Dhondt, S., Wuyts, N., & Inzé, D. (2013). Cell to whole-plant phenotyping: the best is yet to come. *Trends in Plant Science, 18*, 428–439.

Erxleben, A., Gessler, A., Vervliet-Scheebaum, M., & Reski, R. (2012). Metabolite profiling of the moss *Physcomitrella patens* reveals evolutionary conservation of osmoprotective substances. *Plant Cell Reports, 31*, 427–436.

Floyd, S. K., & Bowman, J. L. (2007). The ancestral developmental tool kit of land plants. *International Journal of Plant Sciences, 168*, 1–35.

Gorham, E. (1991). Northern peatlands: role in the carbon cycle and probable responses to climatic warming. *Ecological Applications, 1*, 182–195.

Grabherr, M. G., Haas, B. J., Yassour, M., Levin, J. Z., Thompson, D. A., Amit, I. … Regev, A. (2011). Full-length transcriptome assembly from RNA-seq data without a reference genome. *Nature Biotechnology, 29*, 644–652.

Graham, L. E., Kodner, R. G., Fisher, M. M., Graham, J. M., Wilcox, L. W., Hackney, J. M. … Cook, M. E. (2003). Early land plant adaptation to terrestrial stress: a focus on phenolics. In A. Hemsley, & I. Pecie (Eds.), *The evolution of plant physiology* (pp. 155–171). London: Academic Press.

Granath, G., Wiedermann, M. M., & Strengbom, J. (2009). Physiological responses to nitrogen and sulphur addition and raised temperature in *Sphagnum balticum*. *Oecologia, 161*, 481–490.

Greb, S. F., DiMichele, W. A., & Gastaldo, R. A. (2006). Evolution and importance of wetlands in earth history. In S. F. Greb, & W. A. DiMichele (Eds.), *Wetlands through time: Geological Society of America Special Paper 399* (pp. 1–40).

Haas, B. J., Papanicolaou, A., Yassour, M., Grabherr, M., Blood, P. D., Bowden, J. … Regev, A. (2013). De novo transcript sequence reconstruction from RNA-seq using the Trinity platform for reference generation and analysis. *Nature Protocols, 8*, 1494–1512.

Haeckel, E. (1868). *The history of creation*. New York: Appleton and company.

Haig, D., & Wilczek, A. (2006). Sexual conflict and the alternation of haploid and diploid generations. *Philosophical Transactions: Biological Sciences, 361*, 335–343.

Hajek, T., & Beckett, R. P. (2008). Effect of water content components on desiccation and recovery in *Sphagnum* mosses. *Annals of Botany, 101*, 165–173.

Hohe, A., & Reski, R. (2005). From axenic spore germination to molecular farming – one century of bryophyte in vitro culture. *Plant Cell Reports, 23*, 513–521.

Johnson, M. J., Granath, G., Teemu, T., Pouliot, R., Stenøien, H. K., Rochefort, L. … Shaw, A. J. (2015). Evolution of niche preference in *Sphagnum* peat mosses. *Evolution, 69*, 90–103.

Johnson, M. J., & Shaw, A. J. (2015). Genetic diversity, sexual condition, and microhabitat preference determine mating patterns in *Sphagnum* (Sphagnaceae) peat-mosses. *Biological Journal of the Linnean Society, 11*, 96–113.

Kamisugi, Y., Schlink, K., Rensing, S. A., Schween, G., von Stackelberg, M., Cuming, A. C. … Cove, D. J. (2006). The mechanism of gene targeting in *Physcomitrella patens*: homologous recombination, concatenation and multiple integration. *Nucleic Acids Research, 34*(21), 6205–6214.

Karlin, E. F., Boles, S. B., Ricca, M., Temsch, E. M., Grelihuber, J., & Shaw, A. J. (2009). Three genome mosses: complex double allopolyploid origins for triploid gametophytes in *Sphagnum*. *Molecular Ecology, 18*, 1439–1454.

Kendrick, P., & Crane, P. R. (1997). The origin and early evolution of plants on land. *Nature, 389*, 33–39.

Knack, J. J., Wilcox, L. W., Delaux, P.-M., Ané, J.-M., Piotrowski, M. J., Cook, M. E. … Graham, L. E. (2015). *International Journal of Plant Science, 165*, 405–420.

Kyrkjeeide, M. O., Hassel, K., Stenøien, H. K., Prestø, T., Boström, E., Shaw, A. J., & Flatberg, K. I. (2016). The dark morph of *Sphagnum fuscum* (Schimp.) H.Klinggr. in Europe is conspecific with the North American S. beothuk. *Journal of Bryology* (in press).

Langmead, B., Trapnell, C., Pop, M., & Salzberg, S. L. (2009). Ultrafast and memory-efficient alignment of short DNA sequences to the human genome. *Genome Biology, 10*, R25.

Li, B., & Dewey, C. N. (2011). RSEM: accurate transcript quantification from RNA-Seq data with or without a reference genome. *BMC Bioinformatics, 12*, 323.

Lindo, Z., Nilsson, M. C., & Gundale, M. J. (2013). Bryophyte-cyanobacteria associations as regulators of the northern latitude carbon balance in response to global change. *Global Change Biology, 19*, 2022–2035.

Liu, Y., Budke, J. M., & Goffinet, B. (2012). Phylogenetic inference rejects sporophyte based classification of the Funariaceae (Bryophyta): Rapid radiation suggests rampant homoplasy in sporophyte evolution. *Molecular Phylogenetics and Evolution, 62*, 130–145.

McQueen, C. B., & Andrus, R. E. (2007). Sphagnaceae. In Flora of North America Editorial Committee (Ed.), *Flora of North America North of Mexico* (pp. 45–101) (New York, NY, USA and Oxford, UK).

Qiu, Y. L., Li, L., Wang, B., Chen, Z., Knoop, V., Groth-Malonek, M. … Dombrovska, O. (2006). The deepest divergences in land plants inferred from phylogenomic evidence. *Proceedings of the National Academy of Sciences of the United States of America, 103*, 15511–15516.

Redecker, D., Kodner, R., & Graham, L. E. (2000). Glomalian fungi from the Ordovician. *Science, 289*, 1920–1921.

Rensing, S. A., Lang, D., Zimmer, A. D., Terry, A., Salamov, A., Shapiro, H. … Boore, J. L. (2008). The *Physcomitrella* genome reveals evolutionary insights into the conquest of land by plants. *Science, 319*, 64–69.

Ricca, M., & Shaw, A. J. (2010). Allopolyploidy and homoploid hybridization in the *Sphagnum subsecundum* complex (Sphagnaceae: Bryophyta). *Biological Journal of the Linnaean Society, 99*, 135–151.

Rubinstein, C. V., Gerrienne, P., de la Puente, G. S., Astini, R. A., & Steemans, P. (2010). Early Middle Ordovician evidence for land plants in Argentina (eastern Gondwana). *New Phytologist, 188*, 365–369.

Rydin, H., & Jeglum, J. (2006). *The biology of peatlands*. New York, NY, USA: Oxford University Press.

Rydin, H., & Jeglum, J. (2013). *The biology of peatlands* (ed. 2). New York, NY, USA: Oxford University Press.

Sarnighausen, E., Wurtz, V., Heintz, D., Van Dorsselaer, A., & Reski, R. (2004). Mapping of the *Physcomitrella patens* proteome. *Phytochemistry, 65*, 589–1607.

Shaw, A. J., Cox, C. J., Buck, W. R., Devos, N., Buchanan, A. M., Cave, L. … Temsch, E. M. (2010). Newly resolved relationships in an early land plant lineage: Bryophyta class Sphagnopsida (peat mosses). *American Journal of Botany, 97*, 1511–1531.

Shaw, A. J., Devos, N., Cox, C. J., Boles, S. B., Shaw, B., Buchanan, A. M. … Seppelt, R. (2010). Peatmoss (*Sphagnum*) diversification associated with Miocene Northern Hemisphere climatic cooling? *Molecular Phylogenetics and Evolution, 55*, 1139–1145.

Simola, L. K. (1969). The effect of various mono- and disaccharides on the growth of *Sphagnum nemoreum* thalli in sterile cultures. *Physiologia Plantarum, 22*, 1079–1084.

Steemans, P., Hérissé, A. L., Melvin, J., Miller, M. A., Paris, F., Verniers, J. … Wellman, C. H. (2009). Origin and radiation of the earliest vascular land plants. *Science, 324*, 353.

Turetsky, M. R., Crow, S. E., Evans, R. J., Vitt, D. H., & Wieder, R. K. (2008). Tradeoffs in resource allocation among moss species control decomposition in boreal peatlands. *Journal of Ecology, 96*, 1297–1305.

Viaene, T., Landberg, K., Thelander, M., Medvecka, E., Pederson, E., Feraru … Friml, J. (2014). Directional auxin transport mechanisms in early diverging land plants. *Current Biology, 24*, 2786–2791.

Vitt, D. H., & Slack, N. G. (1975). An analysis of the vegetation of *Sphagnum*-dominated kettle-hole bogs in relation to environmental gradients. *Canadian Journal of Botany, 53*, 332–359.

Wang, Z., Gerstein, M., & Snyder, M. (2009). RNA-Seq: a revolutionary tool for transcriptomics. *Nature Reviews Genetics, 10*, 57–63.

Waters, E. R. (2003). Molecular adaptation and the origin of land plants. *Molecular Phylogenetics and Evolution, 29*, 456–463.

Wellman, C. H., Osterloff, P. L., & Mohiuddin, U. (2003). Fragments of the earliest land plants. *Nature, 425*, 282–285.

Weston, D. J., Timm, C. M., Walker, A. P., Gu, L., Muchero, W., Schmutz, J.… Wullschleger, S. D. (2015). *Sphagnum* physiology in the context of changing climate: emergent influences of genomics, modelling and host–microbiome interactions on understanding ecosystem function. *Plant, Cell and Environment, 38*, 1737–1751.

Wickett, N. J., Mirarab, S., Nguyen, N., Carpenter, E., Matasci, N., Ayyampalayam, S.… Leebens-Mack, J. (2014). Phylotranscriptomics analysis of the origin and early diversification of land plants. *Proceedings of the National Academy of Sciences of the United States of America, 11*, 4859–4868.

Wu, H. P., Su, Y. S., Chen, H. C., Chen, Y. R., Wu, C. C., Lin, W. D., & Tu, S. L. (2014). Genome-wide analysis of light-regulated alternative splicing mediated by photoreceptors in *Physcomitrella patens*. *Genome Biology, 15*, R10.

Yu, Z. C. (2012). Northern peatland carbon stocks and dynamics: a review. *Biogeosciences, 9*, 4071–4085.

CHAPTER SIX

The Genome of the Model Species *Anthoceros agrestis*

P. Szövényi
University of Zurich, Zurich, Switzerland
Swiss Institute of Bioinformatics, Lausanne, Switzerland
ELTE, Budapest, Hungary
E-mail: peter.szoevenyi@ieu.uzh.ch

Contents

Abstract

The monophyletic group of hornworts is believed to represent the immediate sister group of all vascular land plants. However, this traditional view is still debated and cannot be satisfactorily resolved owing to the lack of detailed knowledge on the general biology and genomic features of hornworts. Until now, advancement in this field was primarily hindered by the lack of genomic resources for a hornwort model species. Here we provide the first insight into the major features of the draft genome sequence

Advances in Botanical Research, Volume 78
ISSN 0065-2296
http://dx.doi.org/10.1016/bs.abr.2015.12.001

of the model hornwort, *Anthoceros agrestis*. We show that *A. agrestis* has a remarkably small genome, with few recent paralogues, which makes it appropriate for genetic analysis. Finally, we also report on the genomic features of the chloroplast, mitochondrion and nuclear genomes and compare those with algal and vascular land plant genomes. At the very end of the chapter we summarize our achievements and provide a list of issues that need to be resolved in the future.

1. INTRODUCTION

The hornwort lineage is the least specious genus among the three monophyletic groups of bryophytes. In spite of that it has the largest number of unique morphological, physiological and developmental biological features among all three lineages of bryophytes (Renzaglia, Duff, Nickrent, & Garbary, 2000; Renzaglia, Villarreal, & Duff, 2009). The hornwort lineage exhibits a unique combination of characters connecting them with both algal ancestors and other embryophyte groups (Renzaglia, 1978). For instance, many hornworts possess a single chloroplast per cell with pyrenoid that cannot be found in other extant or extinct land plants but frequently occur in green algae (Badger et al., 1998; Hanson, Renzaglia, & Villarreal, 2014; Meyer & Griffiths, 2013). On the other hand, hornwort sporophytes are fairly independent of the gametophyte and long-lived which is assumed to be a feature connecting them to vascular plants (Goffinet, 2000; Renzaglia et al., 2009). Furthermore, the hornwort sporophyte bears stomata which, depending on their functional interpretation, may be homologous with those of vascular plants (Field, Duckett, Cameron, & Pressel, 2015; Lucas & Renzaglia, 2002; Pressel, Goral, & Duckett, 2014). Nevertheless, besides the few characters shared with algae or other embryophytes biology of hornworts is rich in unique features. For instance, the presence of a basal sporophytic meristem and the asynchronous meiosis are unique features of hornworts that cannot be found in any other group of plants (Goffinet, 2000; Renzaglia et al., 2009). Their unique morphology with very few obvious homologous characters with the tracheophytes and algae, their poor fossil record and few extant species make their phylogenetic relationship within bryophytes and to the tracheophytes contentious even after 20–30 years of phylogenetic research (Renzaglia et al., 2009). Because biological features of hornworts can be only interpreted in light of their phylogenetic position within land plants, we here briefly overview the information available concerning their evolutionary origin.

1.1 Phylogenetic Position of Hornworts and Early Land Plant Evolution

In brief, early phylogenetic analyses indicated that hornworts represent the basal most lineage of bryophytes (Duff & Nickrent, 1999; Garbary & Renzaglia, 1998; Hedderson, Chapman, & Rootes, 1996; Renzaglia et al., 2000). Some other studies even proposed that bryophytes are monophyletic which was poorly accepted as a hypothesis and was largely rejected (Nishiyama et al., 2004). Lately, the accepted view supported the bryophytes as a paraphyletic lineage in which liverworts are basal and hornworts are sister to all tracheophyte land plants. This result was supported by multiple studies and has been the accepted hypothesis for many years (Chang & Graham, 2011; Karol et al., 2010; Qiu et al., 2006). Nevertheless, recent analyses suggest that when mutational biases are corrected bryophytes are monophyletic and formation of the hornwort-tracheophyte clade is just an artefact (Cox, Li, Foster, Embley, & Cicán, 2014). Finally, the difficulty in resolving the phylogenetic position of hornworts cannot be better shown by the results of two recent studies which, by applying complex analyses and using a vast amount of sequence data, were unable to provide strong support for either of the phylogenetic hypotheses mentioned above (Liu, Cox, Wang, & Goffinet, 2014; Wickett et al., 2014). Altogether, we have to accept that phylogenetic position of hornworts is still unresolved which leaves pressing evolutionary questions of land plant evolution open. For instance, determining the phylogenetic position of hornworts is crucial to understand the evolution of the land plant life cycle and to envisage how the land plant sporophyte and its complexity has evolved over the 400 million years spanned by embryophyte evolution. Assuming that bryophytes consist of a grade with three paraphyletic lineages one can argue that life cycle evolution progressed from a haploid-dominant to a diploid-dominant diplo-haplontic life cycle. However, a monophyletic bryophyte clade would suggest that haploid-dominant life cycle could be a new invention of the bryophyte clade itself, and the life cycle type of the common ancestor remains uncertain. A clear picture on the branching order of bryophyte lineages is also critical to understand how morphological complexity has evolved and to understand the homology of critical structures across land plants.

Altogether, we may have to accept that phylogenetic analysis will not be able to resolve the position of hornworts with certainty due to the issues related to saturation, biased codon usage, taxon sampling, and long-branch attraction. Nevertheless, studying the unique biology of hornworts and comparing it with those of other lineages of land plants in details may help to clarify their unique origin. To achieve this goal, establishment of a hornwort model system and determination of its genomic sequence are required. Here we provide the first preliminary analyses on the nuclear genome sequence of a hornwort and give the first insight into the genome space and genome structure of this unique group of land plants.

1.2 Genome Size and Chromosome Number of Hornworts

There is wealth of information compiled for chromosome numbers and genome size estimates (C-values) for all three monophyletic groups of bryophytes (Bainard, Forrest, Goffinet, Newmaster, 2013; Bainard & Villarreal, 2013; Temsch, Greilhuber, & Krisai, 2010; Voglmayr, 2000). Although the level of polyploidy is still debated in each lineage of bryophytes, there is a consensus that extant liverwort taxa are rarely polyploid (about 8%) while mosses have a greater albeit debated proportion of polyploids (chromosome number-based estimates are 84%, Kuta & Przywara, 2000). Extant hornworts seem to be the exceptions by having almost always four to six chromosomes (with some accessory chromosomes) and therefore assumed to have not gone through polyploidization during its evolutionary history. It is only very recently that a wide-range of hornwort C-values have been determined (Bainard & Villarreal, 2013). C-values showed a maximum of four- to fivefold difference among species investigated, and reconstruction of genome size over the hornwort phylogeny showed a gradual genome increase towards the tips of the tree. The variation detected in hornwort C-values are small compared to the variation observed in many moss and liverwort genera (a factor of 12-fold in mosses and 38-fold in liverworts). Hornworts not only show a comparably low variation in genome size but it is also a hornwort species, *Anthoceros agrestis*, that has the smallest genome size (75–85 Mb) known to date within bryophytes (Szövényi et al., 2015; Temsch et al., 2010).

1.3 Life Cycle and General Biology

To establish a hornwort model system we selected a species with a life cycle that can be easily completed under laboratory conditions and has a small genome size (Fig. 1). *Anthoceros agrestis* proved to be a good candidate for which we collected genomic information (Szövényi et al., 2015). *Anthoceros agrestis* has an estimated genome size of 85 Mb which is the smallest ever reported for bryophytes to date (1C = 0.085 pg ca. 83 Mbp; Leitch &

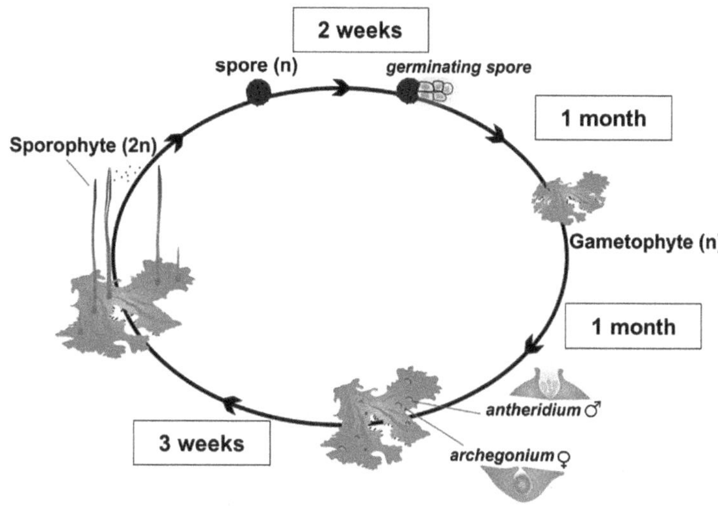

Figure 1 Simplified life cycle and approximate timing of major developmental stages of *Anthoceros agrestis* according to Szövényi et al. *Reproduced and modified from Szövényi, P., Frangedakis, E., Ricca, M., Quandt, D., Wicke, S., Langdale, J.A. (2015). Establishment of Anthoceros agrestis as a model species for studying the biology of hornworts.* BMC Plant Biology, 15*(98). http://dx.doi.org/10.1186/s12870-015-0481-x.*

Bennett, 2007). The *A. agrestis* genome is distributed across five chromosomes, one of them being smaller than the other four, and is referred to as an accessory 'm' chromosome in cytological studies (Kuta, Ochyra, & Przywara, 1990). In its natural habitat, *A. agrestis* can be found on damp soil and ditches frequently in and around of agricultural fields. As all bryophytes its life cycle alternates between a haploid gametophyte and a diploid sporophyte generation. The haploid phase is the vegetative body of the plant which forms a 1- to 1.5-cm-large thallus with an irregularly lobed surface. Each cell of the thallus contains a large solitary chloroplast, and the bottom of the thallus is covered by rhizoids. Chloroplasts have pyrenoids that cannot be found in other land plants but frequently occur in algae. Mucilage-filled cavities occur in the thallus with some containing *Nostoc* colonies. Evidence suggests that *Nostoc* colonies and the hornwort thallus are in a symbiotic relationship (Adams & Duggan, 2008, 2012; Meeks, 2003). It is a summer annual with male and female organs produced on the very same plant. Male sexual organs are red while female organs are transparent and both are embedded into cavities sitting on the surface and margin of the thallus. After fertilization the zygote develops into a cylindrical structure, the sporophyte, which is almost entirely derived from a basal meristem with stomata in its surface. The sporophyte turns black when mature and opens up with two longitudinal slits to release the spores after meiosis.

2. THE ORGANELLAR GENOMES

In the following paragraph we present general features of the hornwort organellar genomes. We note that the following discussion is not based on the sequence data available for *A. agrestis*, but it is a review of previous publications. Assembly and annotation of the *A. agrestis* organellar genomes are underway and will be reported elsewhere.

2.1 The Chloroplast Genome

The structure of the hornwort (*Anthoceros formosae*) chloroplast genome has been fully described in Kugita et al. (2003). Large-scale structure of the hornwort chloroplast resembles those of other bryophytes, but it is more similar to algal groups both in its structural features and size such as *Chaetosphaeridium*. With its 161 kb length it is one of the largest chloroplast genomes ever reported for land plants. Nevertheless, it is estimated that many gene transfers to the nuclear genome and multiple large-scale inversions were necessary to derive the hornwort plastid genome from an algal ancestor. A huge inversion of the complete *mat*K—*atp*A-I—*rpo*B-C1/2-region is shared between ferns and seed plants whereas liverworts (Ohyama et al., 1986), mosses (Oliver et al., 2010; Sugiura, Kobayashi, Aoki, Sugita, & Sugita, 2003), hornworts (Kugita et al., 2003) and lycophytes (Karol et al., 2010; Tsuji et al., 2007; Wolf et al., 2005) show a more ancestral organization similar to that of *Chaetosphaeridium* (Turmel, Otis, & Lemieux, 2002).

The chloroplast genome of *A. formosae* shows multiple peculiarities compared to the two other monophyletic lineages of bryophytes, the mosses and liverworts (Kugita et al., 2013). The structure of the genome resembles to that of most land plants, that is, a large single copy region separates two about 15 kb large inverted repeat regions. The chloroplast genome of *A. formosae* is one of the largest ever reported for land plants with a length of 161 kbp. Interestingly, this difference is not due to an expansion in the number of genes (protein-coding, tRNA and rRNA genes) and orfs because its gene number is identical to those found in other groups of bryophytes. Arrangement of its genes and the general structure of the plastid genome are most similar to that of *Marchantia polymorpha*. The major length increase is in the LIR (large inverted repeat) which increases the copies of the *ndhB* and *rps7* and the 3′ exon of the *rps12* gene. Orfs show 60–70% identity to those of *M. polymorpha*. Interestingly the overall GC content of the genome is 32.9% which is closer to *Chlorella* than to land plants or to *Chaetosphaeridium*. The *MatK* and *rps15* genes that are usually found in other plant plastid genomes are pseudogenes in the *A. formosae* chloroplast genome. Furthermore the *rrn23* gene has an intron which is lacking from plastid genomes of other land plants.

The most striking feature of the plastid genome of hornworts is that more than half of the protein-coding genes have nonsense codons which are, however, converted back to functional codons by the process of RNA editing (Kugita et al., 2003). RNA editing of plastid RNAs occurs in all major lineages of land plants but it is lacking in complex thalloid liverworts and in algae. Leafy liverworts and also mosses show plastid RNA editing. Hornworts as a group show some peculiarities concerning their plastid RNA editing which are either unique to the group or are shared with seedless plants. First of all, plastid RNA editing from C to U is a commonplace in most land plants. Nevertheless, reverse editing from U to C occurs in hornworts, ferns and lycopods but is missing from mosses, liverworts and flowering plants (Takenaka, Zehrmann, Verbitskiy, Härtel, & Brennicke, 2013). The enzymatic reactions and enzymes that are necessary for this process are unknown and have to be determined in the future. The extent of RNA editing sites in hornworts is also unique and is rather comparable to lycopods than to bryophytes or to ferns. After the lycopod *Selaginella uncinata*, the hornwort plastid exhibits the greatest number of edited plastid RNA sites among all land plants studied so far (Kugita et al., 2003; Oldenkott, Yamaguchi, Tsuji-Tsukinoki, Knie, & Knoop, 2014). Nevertheless, this editing significantly differs from the one observed in lycopods and *Isoetes*, where RNA editing frequently happens in intronic and noncoding regions, whereas in the hornwort, editing is mainly in the coding regions (Takenaka et al., 2013).

2.2 The Mitochondrial Genome

The hornwort mitochondrial genome also possesses some advanced land plant-like but also some algal features. The mitochondrial genome of *Megaceros aenigmaticus* (Li, Wang, Liu, & Qiu, 2009) was determined first which was followed by the complete mitochondrial genome sequence of *Phaeoceros laevis* (Xue, Liu, Li, Wang, & Qiu, 2010). This has approximately the size of that of *M. polymorpha* (184 kb) mitochondrial genome but smaller than of the moss *Physcomitrella patens*. It is, however, smaller than any seed plant mitochondrial genomes. In terms of spacers and intergenic regions it is similar to those of other bryophytes, but the exons are slightly longer. Furthermore this size is significantly greater than that of the charophyte and prasinophyte algae which is due to the extension of the intergenic regions that occupies only 9% of the mitochondrial genome in *Chara* but 50% in *Megaceros*. In general, gene content of the *Megaceros* mitochondrial genome is similar to that of other bryophytes and red algae and thus remained stable over the shared common ancestor of rhodophytes and Viridiplantae. Nevertheless, some groups of genes are preferentially lost in the mitochondrial genome of *Megaceros* such as cytochrome c biogenesis genes, ribosomal proteins, tRNA genes and the succinate-ubiquinone oxidoreductase gene. The loss of many genes in the hornwort mitochondrial genome represents a unique feature probably restricted to the lineage of hornworts. Interestingly and in contrast to seed plants, lost tRNA genes are not replaced by their plastid counterparts in the hornwort mitochondrial genome.

In terms of gene order, eight inversions and one translocation are necessary to make the *P. patens* and hornwort mitochondrial genomes fully collinear. This is surprising knowing the extreme dynamics involved in the mitochondrial genome evolution of seed plants. This may be explained by the lack of group II trans-spliced introns in the bryophyte lineage. Another derived feature of the mitochondrial genome is that it has gained many introns (14 hornwort-specific introns are group II introns). The *Phaeoceros* mitochondrion shows similar features to that of *Megaceros* with some more genes pseudogenized. Mitochondrial genes are also known to undergo editing in hornworts of which full extent is still questionable (Takenaka et al., 2013). Analyses suggest extensive editing in the *nad5* (Duff, 2006) gene in both the U to C and C to U directions. Editing sites seem to be also well-conserved across hornworts (Duff, 2006; Duff & Moore, 2005).

3. THE NUCLEAR GENOME

3.1 Sequencing of the Nuclear Genome

To begin revealing basic features of the *A. agrestis* genome we carried out a low-coverage sequencing of its nuclear genome. We provide preliminary information on the draft genome sequence of the BONN strain. Tissue culture, DNA isolation and sequencing were done as described in Szövényi et al. (2015).

3.2 Features of the Draft Genome of *Anthoceros agrestis*

3.2.1 Assembly

As described above, size of the *A. agrestis* genome has been estimated to be about 85 Mb using flow cytometry which was confirmed by our k-mer-based estimate of 71 Mb (Szövényi et al., 2015). This is the smallest genome size ever reported for bryophytes. After several filtering steps total length of our assembly was 95 Mb which well approached our previous size estimates. The current version 1.1 assembly is shattered into 25,571 scaffolds with an N50 length of 9500 bp and an average coverage of ca. 9×. We are currently in the process of improving the v 1.1 assembly by increasing its depth of coverage using short read and its contiguity using long read technologies. The genome is GC rich with a GC content of 48.71%. This is at the upper end of the GC content range reported for plant genomes so far (33.6% in *Juncus inflexus* and 48.9% in *Triticum monococcum*, Smarda et al., 2014). It is also considerably higher than the GC content of the model moss *P. patens* genome (38.7%), but it is similar to the GC content of diatom genomes (48.5% in *Phaeodactylum tricornutum* Bowler et al., 2008; 47% in *Thalassiosira pseudonana* Armbrust et al., 2004).

3.2.2 Repeat Content

The remarkably small size of the genome can be achieved by a reduced gene complement or with the low abundance of repetitive elements and compact genes. Therefore, we used homology-based assessment (using Repeatmasker (Smit, Hubley, & Green, 2013–2015), Repeatmodeler (http://www.repeatmasker.org/RepeatModeler.html) and repbase libraries (Jurka et al., 2005)) to investigate the repetitive element content of the draft *A. agrestis* assembly. We estimated that about 6.98% of the draft genome consists of repetitive elements (Table 1). This is well below the estimates obtained for other plant genomes (10–70% repetitive elements) and for the moss *P. patens* (48%) (Rensing et al., 2008; Kejnovsky, Hawkins, & Feschotte, 2012). Low repetitive element content of the *A. agrestis* genome resembles of the recently sequenced minute plant genome, *Utricularia gibba* (78 Mb), with a ca. 3% repetitive element content (Ibarra-Laclette et al., 2013). Therefore, small size of the *A. agrestis* genome appears to be partly due to its small repetitive element content.

In particular, we found that retroelements abundant in angiosperm genomes make up the largest set of mobile elements of the repetitive fraction of the genome (4.8%). Most of these are composed of LTR retrotransposons of the Ty1/Copia (2.01%) and Gipsy/Dirs1 (1.41%) types followed by the

Table 1 Repeat Content of the *Anthoceros agrestis* Genome

Elements			Percentage of the Genome Occupied
Retroelements			4.8
	Non-LTR elements		
		SINEs	0.19
		LINEs	0.73
		Other	0.69
	LTR elements		3.87
		Ty1/Copia	1.41
		Gypsy/DIRS1	2.01
		Retroviral	0.39
DNA transposons			0.26
Helitrons			0
Unclassified			0.03
Total interspersed repeats			5.09
Satellites			0.65
Simple repeats			1.03
Low complexity			0.15

non-LTR retrotransposons LINES (0.79%) and SINES (0.19%). Although helitrons occur in almost all sequenced plant genomes, we detected none in the *A. agrestis* genome (Kejnovsky et al., 2012). DNA transposons made up only 0.26% of the genome. Simple repeats and low-complexity sequences made up roughly 1.5% of the genome. Altogether, this implies that the *A. agrestis* genome has a low repetitive element content which is either due to the low activity of repetitive elements or could be a consequence of active and efficient deletion of repetitive elements mainly via unequal crossing over. Our preliminary analysis suggests that most LTR elements are highly degenerated which seems to support the existence of an effective deletion mechanism. This seems to contradict the hypothesis that small genome size is ancestral in the genus *Anthoceros* (Bainard & Villarreal, 2013).

3.2.3 The Gene Space of Anthoceros agrestis

In spite of its small size the *A. agrestis* genome harbours a significant number of genes (Table 2). Using ab initio training (Augustus, Stanke, Diekhans, Baertsch, & Haussler, 2008; SNAP, Korf, 2004) and evidence-based gene prediction (MAKER_P, Campbell, Holt, Moore, & Yandell, 2014)

Table 2 General Statistics (Average) of the *Anthoceros agrestis* Gene Set and of Other Plant Genomes Representing the Major Lineages of Plants Including the Chlorophytes. Median Values Are Provided in Brackets

Features		***Anthoceros agrestis***	***Physcomitrella patens* (v1.6)**	***Chlamydomonas reinhardtii* (v4.1)**	***Selaginella moellendorffii* (FM3)**	***Arabidopsis thaliana* (TAIR10)**
Gene length (bp)		1550 (1107)	2369 (1809)	5363 (4273)	1699 (1368)	2190 (1896)
Exon number		2.86 (1)	5 (3)	9 (7)	5.6 (4)	5.9 (4)
Exon length		1313 (963)	275 (155)	322 (155)	213 (128)	261 (147)
Intron length						
	All genes	485.4 (308)	278 (213)	308 (238)	110 (59)	164 (100)
	Intron-containing genes	543.7 (350)				
CDS length		710.3 (486)	1062 (813)	2043 (1425)	1145 (951)	1234 (1053)
Multi exon transcripts		17226	29378	15322	18789	29050
Single exon genes		15469	8979	613	3470	6126

approaches we predicted the genome to contain approximately 31,695 genes. The structure of genes also suggests that the *A. agrestis* genome is highly compact. The average gene length being around 1550 bp is on the smallest end of the values reported so far for bryophytes. Gene compactness is achieved by the presence of many intron-less genes which number is close to 16,000. This is about 50% of the full gene set and represents the highest value recorded so far among bryophytes and *Selaginella moellendorffii*. In fact, on average, intron-containing genes have 544-bp-long introns which are close to the average seen in nonseed plants. Nevertheless genes have on average three exons that seem to be the smallest average value reported for bryophytes and also for nonseed plants. In contrast to the small number and shortness of the introns, exons seem to be few but large compared to other bryophytes and seedless plants. Altogether, our preliminary analysis suggests that *A. agrestis* has relatively short genes, a large proportion of which is devoid of introns. Nevertheless, whenever introns are present they are larger than those seen in algae and in other land plants.

3.2.4 *The* Anthoceros agrestis *Gene Set and Its Affinity With Other Land Plant Genomes*

To describe the genome space of *A. agrestis* we blasted all predicted *A. agrestis* protein sequences against the combined proteome of all plants included in the PLAZA2.5 database (Van Bel et al., 2011). Only 55.4% of all protein translations had a valid match to the 25 fully sequenced plant proteomes of the PLAZAv2.5 database using a blastP e-value threshold of 10^{-6} (Fig. 2). This indicates that a large proportion of the *A. agrestis* gene set is either new to land plants, or it is too divergent to provide a valid match with other land plant proteomes. This finding is not unusual as about 48% of the *P. patens* proteome has no homologues in other land plant genomes (Zimmer et al., 2013).

Affinities of the *A. agrestis* proteins to various taxonomic groups of land plants were also used to gather information about the most likely phylogenetic position of *A. agrestis* (Fig. 2). The largest proportion (42%) of the proteins mapped to moss sequences (*P. patens*). Interestingly, the second largest set of proteins showed affinity with angiosperms (39%) and then with *Selaginella* genes (17.6%). As explained earlier, hornworts also show morphological and developmental features that connect them with algae, such as the presence of the pyrenoids in their plastids. In spite of that, only about 1% of the proteins had their best hit on chlorophyte genes. The proportion of the *A. agrestis* gene set that shows homology to algal

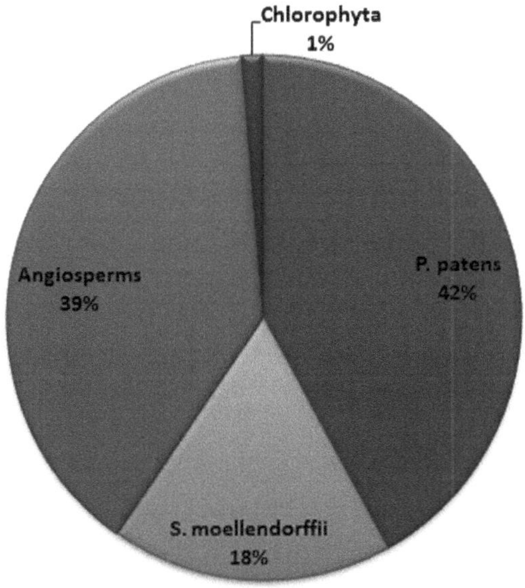

Figure 2 Proportion of *Anthoceros agrestis* gene models having their best hit on algal (Chlorophyta), moss (*Physcomitrella patens*), lycophyte (*Selaginella moellendorffii*) and angiosperm proteomes.

genes is similar to what we obtained in a parallel analyses conducted for a moss, *Funaria hygrometrica*, and for a liverwort, *M. polymorpha* (results not shown here). That is, our simple homology search does not seem to provide evidence that the gene space of *A. agrestis* would be more similar to algae than what would be expected in a moss and in a liverwort. This seems to contradict the 'hornworts basal' phylogenetic hypothesis. Importantly, we also found that about 39% of the *A. agrestis* genes had their best hit on angiosperm and 18% on *Selaginella* genes. This suggests a strong affinity of the hornwort gene set to those of tracheophytes which is more than what we would expect in other mosses and in the liverwort *M. polymorpha* (unpublished results). Therefore, we argue that this finding is in favour of the hornwort sister to land plant hypothesis, however, more careful and detailed analysis will be necessary to verify this hint. Nevertheless, the finding that about half of the hornwort proteins did not have recognizable homologues in any of the plant proteomes included in our analysis suggests that hornworts disposes over a large set of unique genes. This finding is in line with the many unique morphological, biochemical and developmental biological characteristics of the hornwort lineage.

3.2.5 *Protein Domain Abundance of the* Anthoceros agrestis *Proteome*

Searching the *A. agrestis* protein database against the PFAM database (version 28.0, Finn et al., 2014) using HMMER searches we found that 51% of the predicted proteins had detectible PFAM domains (Fig. 3). This figure is similar to what is seen in the moss *P. patens* in which 52% of the predicted gene models have no detectible PFAM domains. In total we identified 3577 PFAM domains in the predicted gene set of *A. agrestis*. We found that the most abundant domains were the pentatricopeptide (PPR) domains. A large proportion of PPR repeat-containing proteins are involved with RNA editing processes which known to be extensive in the organelles in hornwort species (see above). Another abundant PFAM domain the DYW deaminase domain is also known to be responsible for RNA editing processes in particular in the C to U editing process (Takenaka et al., 2013). This suggests that complex organellar RNA editing of hornworts is probably associated with the expansion of PPR repeat proteins. We also compared the 10 most abundant PFAM domains of the *A. agrestis* proteome with that of the predicted

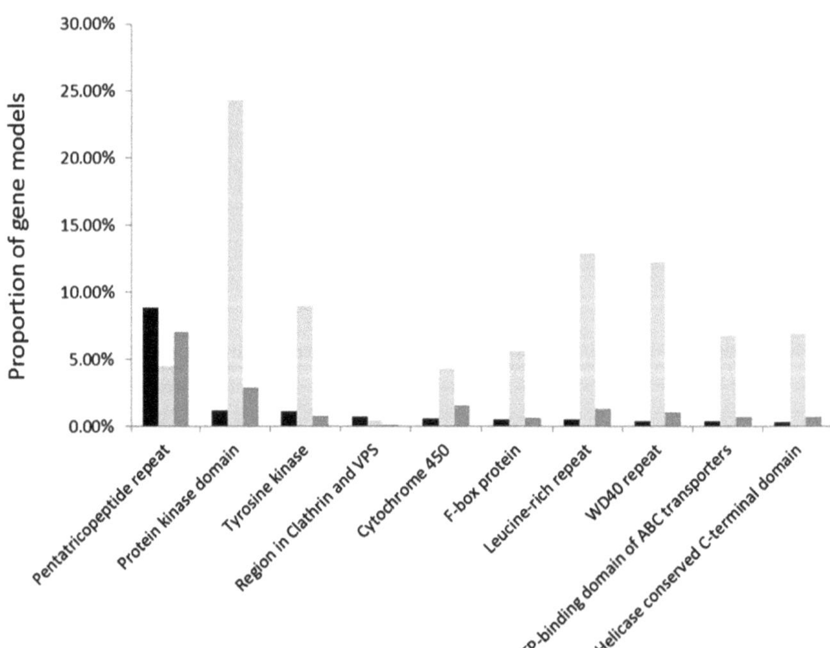

Figure 3 Proportional abundance of PFAM domains in the *Anthoceros agrestis, Physcomitrella patens* and *Selaginella moellendorffii* proteomes. The most abundant PFAM domains of *A. agrestis* are shown. Black: *A. agrestis*; light grey: *P. patens*; dark grey: *Selaginella moellendorffii*.

gene set of the moss genome (*P. patens* v 1.6). In general, the list of the 10 most abundant PFAM domains of the *A. agrestis* and *P. patens* proteomes sowed highly divergent domain composition (Fig. 4). Furthermore, almost all domains were proportionally more abundant in *P. patens* than in *A. agrestis*. Only the PPR repeat domain showed the opposite pattern and was proportionally more abundant in *A. agrestis* than in *P. patens*. In particular, the most abundant domains of *P. patens* were mainly related to signal transduction, to transport or to transcription factors. In contrast, in *A. agrestis* the most abundant domains were either represented by PPR or TPR repeats, and kinase domains occurred at the end of the list of the 10 most abundant domains. Altogether, this suggests that abundance of protein domains is considerably different in the hornwort and in the *P. patens* genome. Furthermore, this difference is primarily due to the overwhelming abundance of protein domains putatively involved in RNA editing in *A. agrestis*.

Owing to the great abundance of protein domains likely involved with organellar RNA editing we thought to repeat the previous analysis using *S. moellendorffii*. This lycophyte shows extensive organellar RNA editing and has over 600 predicted PPR proteins (Takenaka et al., 2013; Tsuji et al., 2007). We reasoned that if the abundance of particular PFAM domains in the *A. agrestis* genome is related to extensive RNA editing then these domains should be similarly abundant in the *S. moellendorffii* genome as well. Indeed, our analysis confirmed this assertion and revealed a greater extent of similarity in PFAM domain abundance between the proteomes of *A. agrestis* and *S. moellendorffii* than between *A. agrestis* and *P. patens* (Figs 3 and 4). In particular, the PPR domain turned out to be the most abundant in both the *A. agrestis* and *S. moellendorffii* proteomes which are in sharp contrast to the proteome of *P. patens*. Nevertheless, some abundant domains putatively involved with RNA editing such as the DYW deaminase and the tetratricopeptide repeat domains were less abundant in the *S. moellendorffii* genome (Fig. 4). Therefore, their functional significance in RNA editing in the *A. agrestis* genome is not yet clear. Finally, we also found a better agreement in the relative abundance of PFAM domains between the *A. agrestis* and *S. moellendorffii* than the *A. agrestis* and *P. patens* proteomes (Fig. 4). Altogether, these additional analyses suggest that highly abundant PFAM domains of the *A. agrestis* genome are best explained by their functional significance in organellar RNA editing.

3.2.6 Gene Family Sharing With Algae and Vascular Plants

In order to further assess the evolutionary relationship among land plants and the hornwort gene set we extended the above analysis and assigned

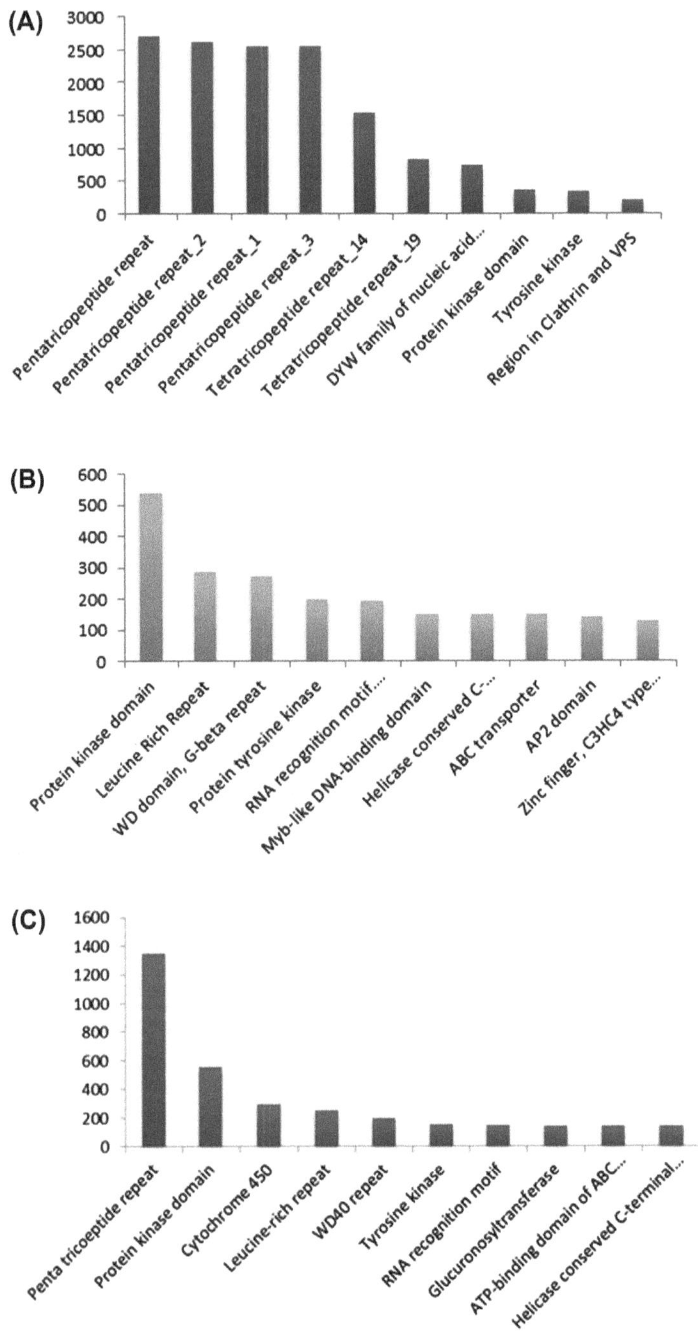

Figure 4 Abundance (absolute number) of the 10 most abundant PFAM domains in the *Anthoceros agrestis* (A), *Physcomitrella patens* (B) and *Selaginella moellendorffii* (C) proteomes.

hornwort genes to PLAZA2.5 gene families (Van Bel et al., 2011). We used the more inclusive and larger HOM families for our classification. Hornwort proteins sorted into 5333 PLAZAv2.5 gene families (Fig. 5).

We first compared gene family sharing among *A. agrestis*, the moss *P. patens* and the alga *Chlamydomonas reinhardtii* (Fig. 5). We found that the largest number of families (1500) is uniquely shared by *P. patens*. We also found that of the 5833 gene families *A. agrestis* genes were sorted in 314 were devoid of *P. patens* and *C. reinhardtii* sequences. That is, these families are uniquely shared by *A. agrestis* and tracheophytes but are lacking from other bryophytes and algae. Finally, we found 143 gene families that were uniquely shared by *C. reinhardtii* and *A. agrestis*. These may be families encoding the algal features of the chloroplasts. When we look at the three-way comparison among *A. agrestis*, *P. patens* and *S. moellendorffii* the figure is very similar. It is important to note that the number of families uniquely shared between *A. agrestis* and *S. moellendorffii* is greater than the ones uniquely shared by the moss and *S. moellendorffii* which again is in favour of a closer relationship between hornworts and tracheophytes than *P. patens* and tracheophytes. Altogether, the extent of gene family sharing supports the findings obtained by the analysis of the taxonomic distribution of the best hits and is in favour of the hornwort-tracheophyte sister relationship. It is important to note that 877 families of the Plazav2.5 database contain only *P. patens* genes, that is, they are moss- or bryophyte-specific. Nevertheless, in 90 of these families we also found *A. agrestis* gene models;

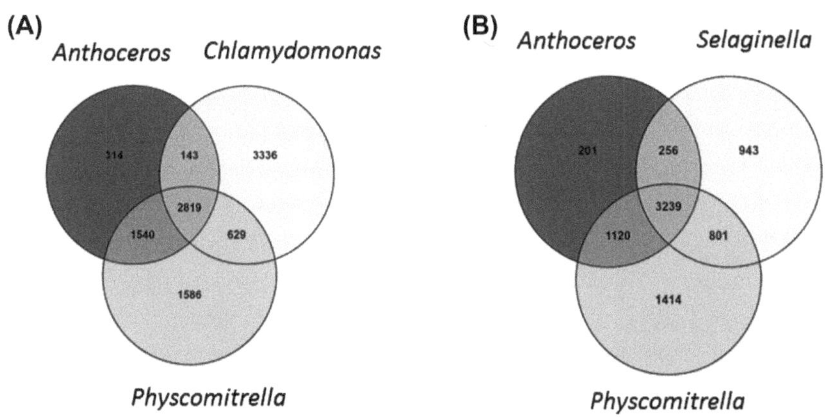

Figure 5 Gene family sharing between (A) *Anthoceros agrestis, Physcomitrella patens* and *Chlamydomonas reinhardtii* and between (B) *A. agrestis, P. patens* and *Selaginella moellendorffii.*

thus some of these moss-specific gene families seem to be shared with hornworts and may represent bryophyte- rather than moss-specific families.

3.2.7 *Gene Family Expansions Compared to* Physcomitrella patens, Selaginella moellendorffii *and* Arabidopsis thaliana

Gene family expansion provides an effective way to generate new functions, increase complexity and potentially contribute to phenotypic evolution (Rensing, 2013). Therefore, here we investigated the number and putative function of gene families that went through an expansion or depletion in gene number in *A. agrestis* compared to *P. patens*, *S. moellendorffii* and the *A. thaliana* genomes.

When comparing the number of *A. agrestis* and *P. patens* genes in shared gene families we found that in a large proportion of gene families *A. agrestis* genes were less numerous than *P. patens* genes. In particular, we found that 1550 gene families showed a twofold depletion while only 741 showed a twofold expansion in *A. agrestis* gene numbers compared to *P. patens*. This may be due to the three rounds of gene duplication *P. patens* genome went through in the past (Rensing et al., 2008). In contrast to that, 1098 gene families showed a two fold expansion while only 710 showed a contraction in *A. agrestis* gene numbers compared to the gene set of *S. moellendorffii*. We detected a twofold expansion in 839 and contractions in 1389 families of the *A. agrestis* gene set compared to *A. thaliana*. Finally, 1085 families showed an expansion and 334 showed a contraction compared to the gene set of the *C. reinhardtii* genome. Altogether, this shows that gene family expansions dominate in the gene families that are shared with *C. reinhardtii* and with *S. moellendorffii*. In contrast, gene family contractions predominate in the gene families shared with *P. patens* and *A. thaliana*.

After that we more carefully examined the families that did not have members in all chlorophyte algae and in *P. patens* but contained *A. agrestis* sequences. There are 255 such families. Interestingly, these families have many members in the tracheophytes and in *A. agrestis* which may support the genetic link between the tracheophytes and the hornworts. There are many families with various biological functions involved in these families. Among the families with the greatest expansion are some that may be of particular interest. A gene family containing Toll-like receptors show about a 100-fold increase in the number of *A. agrestis* genes compared to *P. patens*. We also found that protein-serine threonine kinase genes are also greatly expanded compared to *P. patens*. Interestingly, *A. agrestis* has 16-fold more auxin-responsive GH3 family protein sequences than *P. patens* that are known to be responsible for conducting the effect of the hormone auxin. Also gene families with putatively many functions in RNA metabolism such as RNA-processing protein, HAT helix proteins are more numerous in *A. agrestis* than in *P. patens*. The exact functional significance of these expansions still remains to be discovered.

As explained above many more families seem to have contracted in *A. agrestis* compared to *P. patens*, the greatest contraction we found in a family containing Myb-like homeodomain transcription factors. Similarly, MADS box transcription factor families went through a dramatic reduction which is also true for the families containing pathogenesis-related transcription factors. Many other families containing primarily transcription factor genes had dramatically less *A. agrestis* than *P. patens* genes such as the NAC (no apical meristem family), zinc-finger dof-type, zinc-finger c2h2 and ring type, bZIP1 and B3 transcription factor families containing ARF genes. Similarly, we found auxin-responsive SAUR protein containing families to be seriously contracted compared to *P. patens*. This suggests that dramatic contractions have taken place in gene families with primary function in development such as transcription factors and in genes involved with hormonal regulation. This latter seems to be similar to the observation made in the *M. polymorpha* genome (Flores-Sandoval, Eklund, & Bowman, 2015).

4. CONCLUSIONS AND FUTURE PROSPECTS

Preliminary analysis of the *A. agrestis* genome presented here provides the first insights into the genome of a hornwort species. We note that this analysis is preliminary, and with the further improvement of the genome sequence and its annotation our interpretation of the genome may change. In spite of its preliminary nature, some general conclusions can be drawn and summarized. Our analysis suggests that structure of the *A. agrestis* genome resembles that of the minute plant genome of *U. gibba* (Ibarra-Laclette et al., 2013). In particular, the *A. agrestis* genome has a highly reduced repeat content and a relatively large set of compact genes. Its gene set shows close affinity to mosses and vascular plants favouring the hornworts sister to land plant phylogenetic hypothesis. Nevertheless, about half of its genes seem to have no homologues in any other land plants suggesting a unique set of genes that is in line with their unique biology. Functional annotation also implies considerable differences between the moss and *A. agrestis* gene sets. Finally, the number of transcription factor genes and genes related to hormonal regulation seem to be highly reduced in the *A. agrestis* genome compared to *P. patens* which will likely aid their functional analyses. Our analysis is still lacking a wealth of information including the analysis of small RNA complement, epigenetic landscape and genome duplication history of the *A. agrestis* genome. We hope that availability of the genome of the hornwort model species, *A. agrestis*, will trigger evo-devo studies spanning all three monophyletic lineages of bryophytes and addressing large-scale questions on the evolutionary trajectory of developmental mechanisms in land plants.

REFERENCES

Adams, D. G., & Duggan, P. S. (2008). Cyanobacteria-bryophyte symbioses. *Journal of Experimental Botany, 59*, 1047–1058.

Adams, D. G., & Duggan, P. S. (2012). Signalling in cyanobacteria–plant symbioses. In S. Perotto, & F. Baluska (Eds.), *Signaling and communication in plant symbiosis, signaling and communication in plants* (Vol. 11, pp. 93–121). Berlin-Heidelberg: Springer-Verlag.

Armbrust, E. V., Berges, J. A., Bowler, C., Green, B. R., Martinez, D., Putnam, N. H. … Rokhsar, D. S. (2004). The genome of the diatom *Thalassiosira pseudonana*: ecology, evolution, and metabolism. *Science, 306*, 79–86.

Badger, M. R., Andrews, T. J., Whitney, S. M., Ludwig, M., Yellowlees, D. C., Leggat, W., & Price, G. D. (1998). The diversity and coevolution of Rubisco, plastids, pyrenoids and chloroplast-based CO_2-concentrating mechanisms in the algae. *Canadian Journal of Botany, 76*, 1052–1071.

Bainard, J. D., Forrest, L. L., Goffinet, B., & Newmaster, S. G. (September 2013). Nuclear DNA content variation and evolution in liverworts. *Molecular Phylogenetics and Evolution, 68*(3), 619–627.

Bainard, J. D., & Villarreal, J. C. (2013). Genome size increases in recently diverged hornwort clades. *Genome, 56*, 431–435. http://dx.doi.org/10.1139/gen-2013-0041.

Bowler, C., Allen, A. E., Badger, J. H., Grimwood, J., Jabbari, K., Kuo, A. … Grigoriev, I. V. (2008). The *Phaeodactylum* genome reveals the evolutionary history of diatom genomes. *Nature, 456*, 239–244.

Campbell, M. S., Holt, C., Moore, B., & Yandell, M. (2014). Genome annotation and curation using MAKER and MAKER-P. *Current Protocols in Bioinformatics, 48*, 4.11.1–4.11.39.

Chang, Y., & Graham, S. W. (2011). Inferring the higher-order phylogeny of mosses (Bryophyta) and relatives using a large, multigene plastid data set. *American Journal of Botany, 98*, 839–849. http://dx.doi.org/10.3732/ajb.0900384.

Cox, C. J., Li, B., Foster, P. G., Embley, T. M., & Civán, P. (2014). Conflicting phylogenies for early land plants are caused by composition biases among synonymous substitutions. *Systematic Biology, 63*, 272–279. http://dx.doi.org/10.1093/sysbio/syt109.

Duff, R. J. (2006). Divergent RNA editing frequencies in hornwort mitochondrial nad5 sequences. *Gene, 366*, 285–291.

Duff, R. J., & Moore, F. B. (2005). Pervasive RNA editing among hornwort *rbcL* transcripts except *Leiosporoceros*. *Journal of Molecular Evolution, 61*, 571–578.

Duff, R. J., & Nickrent, D. L. (1999). Phylogenetic relationships of land plants using mitochondrial small-subunit rDNA sequences. *American Journal of Botany, 86*, 372–386.

Field, K. J., Duckett, J. G., Cameron, D. D., & Pressel, S. (2015). Stomatal density and aperture in non-vascular land plants are non-responsive to above-ambient atmospheric CO_2 concentrations. *Annals of Botany, 115*, 915–922.

Finn, R. D., Bateman, A., Clements, J., Coggill, P., Eberhardt, R. Y., Eddy, S. R.... Punta, M. (2014). The Pfam protein families database. *Nucleic Acids Research, 42*(Database Issue), D222–D230.

Flores-Sandoval, E., Eklund, D. M., & Bowman, J. L. (2015). A simple auxin transcriptional response system regulates multiple morphogenetic processes in the liverwort *Marchantia polymorpha. PLoS Genetics, 11*, e1005207. http://dx.doi.org/10.1371/journal.pgen.1005207.

Garbary, D. J., & Renzaglia, K. S. (1998). Bryophyte phylogeny and the evolution of land plants: evidence from development and ultrastructure. In W. J. Bates, N. W. Ashton, & J. G. Duckett (Eds.), *Bryology for the twenty first century* (pp. 45–63). London: Maney Publications.

Goffinet, B. (2000). Origin and phylogenetic relationships of bryophytes. In B. Goffinet, & A. J. Shaw (Eds.), *Bryophyte biology* (1st ed., pp. 124–149). Cambridge: Cambridge University Press.

Hanson, D., Renzaglia, K. S., & Villarreal, J. C. (2014). Diffusion limitation and CO_2 concentrating mechanisms in bryophytes. In D. T. Hanson, & S. K. Rice (Eds.), *Advances in photosynthesis and respiration: Photosynthesis in early land plants* (Vol. 37, pp. 95–112). Berlin: Springer.

Hedderson, T. A., Chapman, R. L., & Rootes, W. L. (1996). Phylogenetic relationships of bryophytes inferred from nuclear-encoded rRNA gene sequences. *Plant Systematics and Evolution, 200*, 213–224.

Ibarra-Laclette, E., Lyons, E., Hernández-Guzmán, G., Pérez-Torres, C. A., Carretero-Paulet, L., Chang, T. H.... Herrera-Estrella, L. (2013). Architecture and evolution of a minute plant genome. *Nature, 498*, 94–98. http://dx.doi.org/10.1038/nature12132.

Jurka, J., Kapitonov, V. V., Pavlicek, A., Klonowski, P., Kohany, O., & Walichiewicz, J. (2005). Repbase Update, a database of eukaryotic repetitive elements. *Cytogenetic and Genome Research, 110*, 462–467.

Karol, K. G., Arumuganathan, K., Boore, J. L., Duffy, A. M., Everett, K. D., Hall, J. D.... Wolf, P. G. (2010). Complete plastome sequences of *Equisetum arvense* and *Isoetes flaccida*: implications for phylogeny and plastid genome evolution of early land plant lineages. *BMC Evolutionary Biology, 10*, 321. http://dx.doi.org/10.1186/1471-2148-10-321.

Kejnovsky, E., Hawkins, J. S., & Feschotte, C. (2012). Plant transposable elements: biology and evolution. In J. F. Wendel, J. Greilhuber, J. Doležel, & I. J. Leitch (Eds.), *Plant genome diversity* (Vol. 1). Springer-Verlag Wien 2012. http://dx.doi.org/10.1007/978-3-7091-1130-7_2.

Korf, I. (2004). Gene finding in novel genomes. *BMC Bioinformatics, 5*, 59.

Kugita, M., Kaneko, A., Yamamoto, Y., Takeya, Y., Matsumoto, T., & Yoshinaga, K. (2003). The complete nucleotide sequence of the hornwort (*Anthoceros formosae*) chloroplast genome: insight into the earliest land plants. *Nucleic Acids Research, 31*, 716–721.

Kuta, E., Ochyra, R., & Przywara, L. (1990). Chromosome studies on Polish bryophytes: V. *Polish Botanical Studies, 1*, 127–147.

Kuta, E., & Przywara, L. (2000). Cytotaxonomy of bryophytes in the Department of Plant Cytology and Embryology, Institute of Botany, Jagiellonian University, Kraców, Poland. *IOPB Newsletter, 32*, 3–8.

Leitch, I. J., & Bennett, M. D. (2007). Genome size and its uses: the impact of flow cytometry. In *Flow cytometry with plant cells: Analysis of genes chromosomes and genomes* (pp. 153–176). Weinheim: John Wiley & Sons.

Li, L., Wang, B., Liu, Y., & Qiu, Y. L. (2009). The complete mitochondrial genome sequence of the hornwort *Megaceros aenigmaticus* shows a mixed mode of conservative yet dynamic evolution in early land plant mitochondrial genomes. *Journal of Molecular Evolution, 68*, 665–678. http://dx.doi.org/10.1007/s00239-009-9240-7.

Liu, Y., Cox, C. J., Wang, W., & Goffinet, B. (2014). Mitochondrial phylogenomics of early land plants: mitigating the effects of saturation, compositional heterogeneity, and codon-usage bias. *Systematic Biology, 63*, 862–878. http://dx.doi.org/10.1093/sysbio/syu049.

Lucas, J. R., & Renzaglia, K. S. (2002). Structure and function of hornwort stomata. *Microscopy and Microanalysis, 8*(Suppl. 2), 1090–1091.

Meeks, J. C. (2003). Symbiotic interactions between *Nostoc punctiforme*, a multicellular cyanobacterium, and the hornwort *Anthoceros punctatus*. *Symbiosis, 35*, 55–71.

Meyer, M., & Griffiths, H. (2013). Origins and diversity of eukaryotic CO_2-concentrating mechanisms: lessons for the future. *Journal of Experimental Botany, 64*, 769–786.

Nishiyama, T., Wolf, P. G., Kugita, M., Sinclair, R. B., Sugita, M., Sugiura, C. … Yamaguchi, K. (2004). Chloroplast phylogeny indicates that bryophytes are monophyletic. *Molecular Biology and Evolution, 21*, 1813–1819.

Ohyama, K., Fukuzawa, H., Kohchi, T., Shirai, H., Sano, T., Sano, S. … Ozeki, H. (1986). Chloroplast gene organization deduced from complete sequence of liverwort *Marchantia polymorpha* chloroplast DNA. *Nature, 322*, 572–574.

Oldenkott, B., Yamaguchi, K., Tsuji-Tsukinoki, S., Knie, N., & Knoop, V. (2014). Chloroplast RNA editing going extreme: more than 3400 events of C-to-U editing in the chloroplast transcriptome of the lycophyte *Selaginella uncinata*. *RNA, 20*, 1499–1506. http://dx.doi.org/10.1261/rna.045575.114.

Oliver, M. J., Murdock, A. G., Mishler, B. D., Kuehl, J. V., Boore, J. L., Mandoli, D. F. … Karol, K. G. (2010). Chloroplast genome sequence of the moss *Tortula ruralis*: gene content, polymorphism, and structural arrangement relative to other green plant chloroplast genomes. *BMC Genomics, 11*, 143. http://dx.doi.org/10.1186/1471-2164-11-143.

Pressel, S., Goral, T., & Duckett, J. G. (2014). Stomatal differentiation and abnormal stomata in hornworts. *Journal of Bryology, 36*, 87–103.

Qiu, Y. L., Li, L., Wang, B., Chen, Z., Knoop, V., Groth-Malonek, M. … Davis, C. C. (2006). The deepest divergences in land plants inferred from phylogenomic evidence. *Proceedings of the National Academy of Sciences of the United States of America, 103*, 15511–15516.

Rensing, S. A. (2013). Gene duplication as a driver of plant morphogenetic evolution. *Current Opinion in Plant Biology, 17*, 43–48.

Rensing, S. A., Lang, D., Zimmer, A. D., Terry, A., Salamov, A., Shapiro, H. … Boore, J. L. (2008). The *Physcomitrella* genome reveals evolutionary insights into the conquest of land by plants. *Science, 319*, 64–69.

Renzaglia, K. S. (1978). A comparative morphology and developmental anatomy of the Anthocerotophyta. *Journal of the Hattori Botanical Laboratory, 44*, 31–90.

Renzaglia, K. S., Duff, R. J., Nickrent, D. L., & Garbary, D. J. (2000). Vegetative and reproductive innovations of early land plants: implications or a unified phylogeny. *Philosophical Transactions of the Royal Society of London B Biological Sciences, 355*, 769–793.

Renzaglia, K. S., Villarreal, J. C., & Duff, R. J. (2009). New insights into morphology, anatomy, and systematics of hornworts. In B. Goffinet, & J. Shaw (Eds.), *Bryophyte biology* (2nd ed., pp. 139–171). Cambridge: Cambridge University Press.

Šmarda, P., Bureš, P., Horová, L., Leitch, I. J., Mucina, L., Pacini, E. … Rotreklová, O. (2014). Ecological and evolutionary significance of genomic GC content diversity in monocots. *Proceedings of the National Academy of Sciences of the United States of America, 111*, E4096–E4102.

Smit, A. F. A., Hubley, R., & Green, P. (2013–2015). *RepeatMasker Open-4.0*. http://www.repeatmasker.org.

Stanke, M., Diekhans, M., Baertsch, R., & Haussler, D. (2008). Using native and syntenically mapped cDNA alignments to improve de novo gene finding. *Bioinformatics, 24*, 637–644. http://dx.doi.org/10.1093/bioinformatics/btn013.

Sugiura, C., Kobayashi, Y., Aoki, S., Sugita, C., & Sugita, M. (2003). Complete chloroplast DNA sequence of the moss *Physcomitrella patens*: evidence for the loss and relocation of rpoA from the chloroplast to the nucleus. *Nucleic Acids Research, 31*, 5324–5331.

Szövényi, P., Frangedakis, E., Ricca, M., Quandt, D., Wicke, S., & Langdale, J. A. (2015). Establishment of *Anthoceros agrestis* as a model species for studying the biology of hornworts. *BMC Plant Biology, 15*, 98. http://dx.doi.org/10.1186/s12870-015-0481-x.

Takenaka, M., Zehrmann, A., Verbitskiy, D., Härtel, B., & Brennicke, A. (2013). RNA editing in plants and its evolution. *Annual Review of Genetics, 47*, 335–352. http://dx.doi.org/10.1146/annurev-genet-111212-133519.

Temsch, E. M., Greilhuber, J., & Krisai, R. (2010). Genome size in liverworts. *PRESLIA, 82*, 63–80.

Tsuji, S., Ueda, K., Nishiyama, T., Hasebe, M., Yoshikawa, S., Konagaya, A. ... Yamaguchi, K. (2007). The chloroplast genome from a lycophyte (microphyllophyte), *Selaginella uncinata*, has a unique inversion, transpositions and many gene losses. *Journal of Plant Research, 120*, 281–290.

Turmel, M., Otis, C., & Lemieux, C. (2002). The chloroplast and mitochondrial genome sequences of the charophyte *Chaetosphaeridium globosum*: Insights into the timing of the events that restructured organelle DNAs within the green algal lineage that led to land plants. *Proceedings of the National Academy of Sciences of the United States of America, 99*, 11275–11280.

Van Bel, M., Proost, S., Wischnitzki, E., Movahedi, S., Scheerlinck, C., Van de Peer, Y., & Vandepoele, K. (2011). Dissecting plant genomes with the PLAZA comparative genomics platform. *Plant Physiology, 158*, 590–600.

Voglmayr, H. (2000). Nuclear DNA amounts in mosses (Musci). *Annals of Botany, 85*, 531–546.

Wickett, N. J., Mirarab, S., Nguyen, N., Warnow, T., Carpenter, E., Matasci, N. ... Leebens-Mack, J. (2014). Phylotranscriptomic analysis of the origin and early diversification of land plants. *Proceedings of the National Academy of Sciences of the United States of America, 111*, E4859–E4868. http://dx.doi.org/10.1073/pnas.1323926111.

Wolf, P. G., Karol, K. G., Mandoli, D. F., Kuehl, J., Arumuganathan, K., Ellis, M. W. ... Boore, J. L. (2005). The first complete chloroplast genome sequence of a lycophyte, *Huperzia lucidula* (Lycopodiaceae). *Gene, 350*, 117–128.

Xue, J. Y., Liu, Y., Li, L., Wang, B., & Qiu, Y. L. (2010). The complete mitochondrial genome sequence of the hornwort *Phaeoceros laevis*: retention of many ancient pseudogenes and conservative evolution of mitochondrial genomes in hornworts. *Current Genetics, 56*, 53–61. http://dx.doi.org/10.1007/s00294-009-0279-1.

Zimmer, A. D., Lang, D., Buchta, K., Rombauts, S., Nishiyama, T., Hasebe, M. ... Reski, R. (2013). Reannotation and extended community resources for the genome of the non-seed plant *Physcomitrella patens* provide insights into the evolution of plant gene structures and functions. *BMC Genomics, 14*, 498. http://dx.doi.org/10.1186/1471-2164-14-498.

SECTION THREE

Transcriptomes

CHAPTER SEVEN

Evolutionary Genomics of Ferns and Lycophytes

E.B. Sessa*,[1], J.P. Der[§]
*University of Florida, Gainesville, FL, United States
[§]California State University, Fullerton, CA, United States
[1]Corresponding author: E-mail: emilysessa@ufl.edu

Contents

Advances in Botanical Research, Volume 78
ISSN 0065-2296
http://dx.doi.org/10.1016/bs.abr.2016.02.001

Abstract

Ferns and lycophytes are ubiquitous and charismatic members of many terrestrial ecosystems. Members of these lineages play key ecological roles in many of Earth's biomes and have an ancient fossil record dating back to the Devonian. Modern ferns underwent significant diversification in the Cretaceous, and the clade as a whole occupies a pivotal position in land plant evolution as the sister lineage to seed plants; lycophytes, in turn, are sister to ferns plus seed plants together. Both ferns and lycophytes are spore-bearing vascular plants, and each clade includes both homosporous and heterosporous members. Many taxa have large genomes with high numbers of chromosomes, particularly among the homosporous members of both lineages. They are the only two lineages of land plants that maintain independent gametophyte and sporophyte phases in their life cycle. Ferns are known to be prone to hybridization and polyploidy, but the roles these events have played in producing their large genomes are still unclear. Genome science has advanced dramatically in recent years, but genomic research in lycophytes and ferns has lagged behind other groups of plants. While nuclear genome sequences are now available for over 100 species of seed plants, there is only one sequenced genome of a lycophyte available (heterosporous *Selaginella moellendorffii*), and none from any fern, or from any homosporous vascular plant. This chapter reviews what is known about fern and lycophyte genomes (nuclear, chloroplast, and mitochondrial) and transcriptomes and presents an outlook on the future of genome research in these groups, including outstanding challenges in plant biology that will be illuminated by incorporating information on ferns and lycophytes.

1. INTRODUCTION

1.1 Homospory Versus Heterospory, Genome Size, and Polyploidy

Ferns have long been recognized for their unique genomic features, including having remarkably large genomes and high chromosome numbers. Manton's monumental 1950 work, *Problems of Cytology and Evolution in the Pteridophyta* (Manton, 1950), drastically increased the number of fern taxa for which chromosome counts and karyotypes were available, and it clarified not only that base numbers in ferns are higher on average than in angiosperms, but that many groups of ferns include polyploid series, with diploids, tetraploids, hexaploids, and even octaploids occurring within the same genus or sometimes species. Manton's work was also pivotal in the development of fern cytological studies for her pioneering new techniques for staining and separating fern chromosomes, which are often long, slender, and highly entangled (Stebbins, 1951). She established for the first time the base numbers of many fern genera and used these numbers to infer relationships between genera and among putative members of polyploid complexes within genera. While she did not discuss at length the obvious differences in base chromosome numbers between ferns and angiosperms (Stebbins, 1951), Manton did suggest that aneuploidy had been the primary driver of the large counts seen in ferns. Another hypothesis, which gained in popularity and remained the favoured explanation for homosporous ferns' high chromosome numbers for several decades, was suggested by Stebbins (1951) and Klekowski and Baker (1966), among others, and involved extensive ancient polyploidy among ancestral ferns.

Klekowski and Baker's oft-cited 1966 publication began by providing a set of estimates for average base chromosome number in various plant lineages, including flowering plants and both homosporous and heterosporous ferns and lycophytes. The fern and lycophyte clades are unique among land plants in each having both homosporous and heterosporous members (Fig. 1). When plants transition from diploidy to haploidy at meiosis, heterospory and homospory are the two options possible. In heterosporous groups, two types of haploid spores are produced via meiosis, microspores and megaspores, that are often unequal in size and which develop into strictly male (micro) or female (mega) gametophytes (Klekowski & Lloyd, 1968). In homospory, which is thought to be ancestral in land plants (Bateman & DiMichele, 1994), spores are uniform in size and the resulting gametophytes are capable of becoming male, female, or bisexual. While seed plants are uniformly heterosporous (Fig. 1), ferns and lycophytes are largely homosporous, but each lineage includes a clade of heterosporous species (Fig. 1).

Klekowski and Baker (1966) established the average base chromosome number in heterosporous ferns and lycophytes as $n = 13.62$ and in flowering plants as $n = 15.99$. These numbers are quite similar to one another and substantially lower than the average base number in homosporous ferns and lycophytes: $n = 57.05$. Homosporous ferns also include the highest chromosome number known from eukaryotes: $n = 720$, in *Ophioglossum reticulatum* (Khandelwal, 1990). Stebbins, in his 1951 review of Manton's (1950)

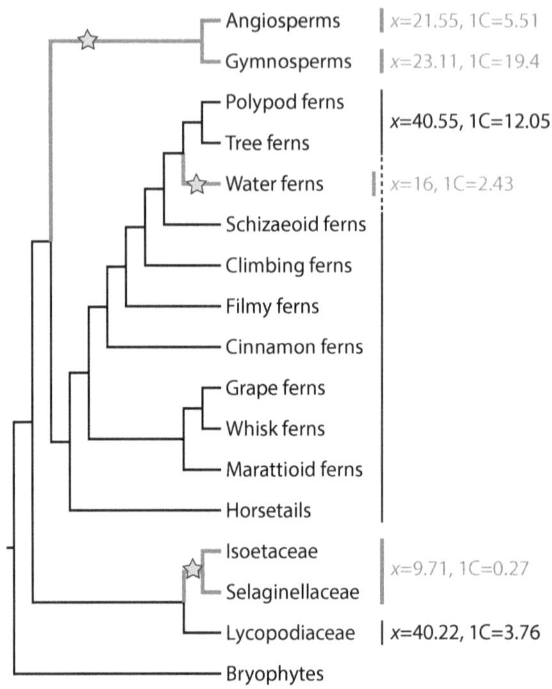

Figure 1 Overview of land plant phylogeny showing inferred transitions (*yellow stars* (light gray in print versions)) from homospory (inferred to be the ancestral state) to heterospory (*green lineages* (gray in print versions)). Transitions to heterospory could have happened at any point along the *green branches* (gray in print versions). Base chromosome numbers (*x*) and average genome sizes (1C, pg) are given.

book, suggested that 'the [homosporous fern] genera themselves are of ancient allopolyploid origin'. Klekowski and Baker (1966) formalized this hypothesis with the explanation that 'polyploidy increases gene redundancy', and the extra genomes present in polyploids allow them to experiment with new, potentially advantageous genetic functions, which, in turn, leads to selection for retention of the duplicated genes and chromosomes. This ability to retain genetic diversity would be particularly advantageous in homosporous ferns and lycophytes as it could compensate for their presumed propensity for gametophytic selfing, a breeding system only possible in homosporous groups, in which sporophytes produced via selfing of a single gametophyte are homozygous at all loci (Klekowski, 1973; Klekowski & Lloyd, 1968). Because heterosporous lineages cannot undergo this extreme form of selfing, they do not share the same reductions in heterozygosity and thus the need for extra genomes to generate non-Mendelian genetic variation (Klekowski & Baker, 1966). Several subsequent studies were perceived as supporting Klekowski and Baker's hypothesis, such as Chapman, Klekowski, & Selander's (1979) analyses of isozyme banding patterns consistent with 'homoeologous heterozygosity', the storage of genetic variation in duplicate, unlinked loci. It was thus long assumed that homosporous ferns' high chromosome numbers were unquestionably linked to their also being highly polyploid and highly selfing (Grant, 1981; Hickok, 1978; Klekowski, 1973; Love, Love, & Pichi-Sermolli, 1977; Wagner & Wagner, 1980).

The tide began to turn for this hypothesis in the late 1970s and 1980s, when additional lines of evidence emerged that repeatedly pointed to ferns being outcrossing, genetically diploid, and not ancestrally polyploid. Support for the 'ancient polyploid' hypothesis had largely eroded by the late 1990s. Evidence that challenged this theory included isozyme and genetic studies that showed homosporous ferns with the base chromosome numbers for their genera (even if those numbers were high) to have diploid expression profiles (Gastony & Darrow, 1983; Gastony & Gottlieb, 1985; Haufler, 1985, 1987; Haufler & Soltis, 1986; Soltis, 1986; Soltis & Soltis, 1986, 1987; Werth, Guttman, & Eshbaugh, 1985) and disomic inheritance patterns (Gastony & Gottlieb, 1982, 1985; Haufler & Soltis, 1984; Wolf, Haufler, & Sheffield, 1987). In addition, while studies of fern breeding systems found several homosporous species to be capable of gametophytic selfing (eg, Holbrook-Walker & Lloyd, 1973; Klekowski, 1969; Lloyd, 1973; Werth et al., 1985), most species were found either to have mixed mating systems or to be predominantly outcrossing (Holsinger, 1987; reviewed in Sessa et al., in review). In retrospect, this is somewhat unsurprising, as many ferns possess chemical sex-determining systems that promote outbreeding (Döpp, 1950; Näf, Nakanishi, & Endo, 1975). Known as antheridiogen systems, gametophytes of ferns with these systems produce gibberellin-like pheromones that control the timing of gametangia development (Schneller, Haufler, & Ranker, 1990; Tryon & Vitale, 1977), resulting in gametophyte populations with uneven ratios of primarily male and female individuals, and few bisexual gametophytes. These skewed sex ratios drastically reduce the number of gametophytic selfing events and promote sporophytic selfing and/or outcrossing (Hamilton & Lloyd, 1991).

To address these challenges to the ancestral polyploid theory, Haufler (1987) suggested an addendum. He hypothesized that repeated cycles of polyploidization, followed by extensive gene silencing without chromosome loss, could have produced the combination of diploid gene expression patterns and high chromosome counts seen in homosporous ferns. This expansion of the original hypothesis was supported by several studies that identified multiple copies of many genes, with evidence of extensive pseudogenization (McGrath & Hickok, 1999; McGrath, Hickok, & Pichersky, 1994; Pichersky, Soltis, & Soltis, 1990). The first genetic linkage map (Nakazato, Jung, Houswortth, Rieseberg, & Gastony, 2006) for a fern (*Ceratopteris richardii*) further revealed that up to 85% of genes in that species were present in multiple copies (Nakazato, Barker, Rieseberg, & Gastony, 2008). However, there was no evidence for extensive synteny among these duplicated loci, suggesting that polyploidization has likely not occurred within the last several million years (Barker, 2012). Analyses of synonymous substitution rates (Ks plots) have identified at most one ancient polyploidization event that subtends the diversification of extant ferns (Barker, 2009, 2012; Nakazato et al., 2008), and which is separate from the ancestral seed-plant duplication identified by Jiao et al. (2011; see also Z. Li et al., 2015). The exact reason, and mechanism, for ferns' high chromosome numbers, still remains a mystery.

Many additional chromosome counts have been published in the 50 years since Klekowski and Baker first reported average base numbers for homosporous ferns and lycophytes. These data are also more readily available today, thanks to publicly accessible resources like the Chromosome Counts Database (CCDB) (Rice et al., 2014). We obtained counts for 12,067 ferns and lycophytes from the CCDB. (Data were downloaded from the CCDB (http://ccdb.tau.ac.il) on 25 August 2015 using the R package chromer (https://github.com/ropensci/chromer) and cleaned using the R package CCDBcurator (https://github.com/roszenil/CCDBcurator). All analyses were performed in R (R Development Core Team, 2008).) We used CCDBcurator to identify base numbers for all genera, and we then calculated average base numbers for homosporous and heterosporous ferns and lycophytes separately. For homosporous fern and lycophyte genera together, the average $x = 40.54$ ($n = 11{,}520$), and for heterosporous fern and lycophyte genera, the average $x = 12.33$ ($n = 547$) (Table 1). The former is somewhat lower than Klekowski and Baker's 57.05 ($n = 1166$), while the latter is similar to their 13.62 ($n = 50$). We also obtained genome size estimates for ferns and lycophytes from the Kew Plant DNA C-Values database (Bennett & Leitch, 2012) (Table 1) and tested for correlations between genome size and average base chromosome number (for genera with multiple genome size estimates available; $n = 13$) across pteridophytes using standard linear regressions with

Table 1 Updated Base Chromosome Numbers for Homosporous and Heterosporous Ferns and Lycophytes

Lineage	Number of Chromosome Counts	Average *x*	Number of Genome Size Estimates	Average Genome Size (1C, pg)
Homosporous ferns	11,321	40.55	55	12.05
Homosporous lycophytes	199	40.22	12	3.76
All homosporous	11,520	40.54	67	10.56
Heterosporous ferns	105	16	2	2.43
Heterosporous lycophytes	442	9.71	10	0.27
All heterosporous	547	12.33	12	0.63

Data obtained from the Chromosome Counts Database (Rice et al., 2014).

significance assessed at $P = 0.05$. (Analyses were performed in R.) We only included species which are known to be diploids, and found that average diploid genome size is significantly correlated with base chromosome number ($R^2 = 0.6$, $DF = 12$, $P = 0.001$★★) (Fig. 2). Such a correlation does not occur in the angiosperms or gymnosperms (Bainard, Bainard, & Newmaster, 2011; Leitch & Bennett, 2004; Nakazato et al., 2008), and one explanation is that this pattern is driven by retention of genetic material following polyploidy events in ferns. Ferns may experience different patterns of genome downsizing following polyploidization compared to seed plants (Barker & Wolf, 2010; Leitch & Leitch, 2012) and may lack mechanisms for discarding redundant or excessive genomic material (Wolf et al., 2015).

A fossil discovery reported in 2014 further stimulated discussion on ancestral genome sizes in ferns and suggested that in a least one lineage, genome size and structure have been unchanged for many millions of years. Members of Osmundaceae fossilized in the early Jurassic (180 million years ago) were found with their nuclei intact and well preserved (Bomfleur, McLoughlin, & Vajda, 2014), and these were used to estimate genome size and ploidy level of the fossilized plants. The authors reported that mean nuclear size in the fossilized specimens closely matched that of extant *Osmundastrum*, and they inferred that the fossil and extant taxa are therefore likely to have roughly the same genome size and chromosome count. This represents a remarkable apparent stasis in genome size over nearly 200 million years. However, as Schneider et al. (2015) pointed out in a follow-up to this work, extensive genomic turnover can occur that does

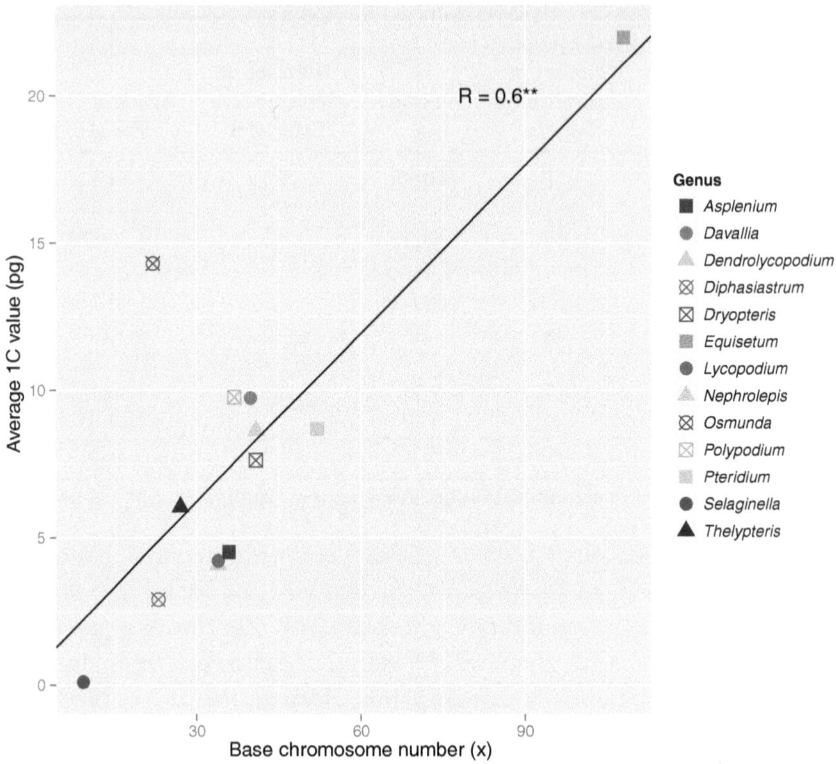

Figure 2 Average genome size (1C, pg) versus base chromosome number for 13 genera of ferns and lycophytes. There is a significant correlation ($R^2 = 0.6$, $DF = 12$, $P = 0.001^{**}$). Homosporous lycophytes include *Lycopodium*, *Dendrolycopodium*, and *Diphasiastrum*; *Selaginella* is a heterosporous lycophyte. All other taxa are homosporous ferns. (See colour plate)

not impact genome size, and this is impossible to infer from fossilized material. They reconstructed genome size evolution across Osmundaceae and found evidence for increases and decreases in genome size across the family even with the fossil taken into account.

Clearly, many aspects of the dynamics of genome size and structure evolution in ferns and lycophytes remain mysterious and deserve further study. Below we review what is currently known about genomes in ferns and lycophytes, covering the nuclear genome as well as the plastid and mitochondrial genomes, and transcriptomic work that is allowing comparative analyses across ferns and lycophytes as well as all land plants. We conclude with a discussion of current challenges and future directions for genomic studies on the pteridophyte lineages.

2. FERN GENOMES

2.1 Chloroplast and Mitochondrial Genomes

2.1.1 Chloroplast Genome Structure and Rearrangement

The first genome-scale resources for ferns were produced beginning in the early 2000s, in the form of sequenced chloroplast genomes (Table 2). Wolf, Rowe, Sinclair, and Hasebe (2003) published the first fern plastome sequence, from *Adiantum capillus-veneris* (Pteridaceae, Genbank accession

Table 2 List of Available Fern Plastid Genomes. An Updated List Is Maintained by Paul Wolf at: https://sites.google.com/site/paulwolfresearch/data-protocols/fern_plastome_list?pli=1

Species	NCBI Accession	Citation
Adiantum capillus-veneris	NC_004766	Wolf et al. (2003)
Alsophila spinulosa	NC_012818	Gao et al. (2009)
Angiopteris angustifolia	NC_026300	Zhu et al. (2015)
Angiopteris evecta	NC_008829	Roper et al. (2007)
Azolla caroliniana (partial)	KC626076	Gao et al. (2013)
Ceratopteris richardii	KM052729	Wolf et al. (2015)
Cheilanthes lindheimeri	NC_014592	Wolf et al. (2011)
Cyrtomium devexiscapulae	NC_028542	Lu et al. (2015)
Cyrtomium falcatum	NC_028705	unpublished
Cystopteris protrusa	KP136830	Wolf et al. (2015)
Dicksonia squarrosa	KJ569698	Zhong et al. (2014)
Diplopterygium glaucum	NC_024158	Kim et al. (2014)
Dipteris conjugata	KP136829	Wolf et al. (2015)
Equisetum arvense	NC_014699	Karol et al. (2010)
Equisetum hyemale	NC_020146	Grewe et al. (2013)
Lygodium japonicum	NC_022136	Gao et al. (2013) and Kim et al. (2014)
Mankyua chejuensis	NC_017006	Unpublished
Marsilea crenata	NC_022137	Gao et al. (2013)
Ophioglossum californicum	NC_020147	Grewe et al. (2013)
Osmundastrum cinnamomeum	NC_024157	Kim et al. (2014)
Plagiogyria formosana	KP136831	Wolf et al. (2015)
Polypodium glycyrrhiza	KP136832	Wolf et al. (2015)
Psilotum nudum	NC_003386	Unpublished
Pteridium aquilinum subsp. aquilinum	NC_014348	Unpublished
Salvinia molesta (partial)	KC626075	Gao et al. (2013)
Tmesipteris elongata	KJ569699	Zhong et al. (2014)
Woodwardia unigemmata	NC_028543	Lu et al. (2015)

NC_004766), and they reported several features of this plastome that shed light on phylogenetic relationships within ferns. For example, *Adiantum* was found to share a 3300 bp inversion with *Psilotum nudum* (sequenced in 2002 but not published, AP004638) that was absent from representatives of seed plants and bryophytes and which provided support for a monophyletic fern clade including leptosporangiates and Psilotaceae. Wolf et al. also confirmed the presence of an expansion of the inverted repeat (IR) region in *Adiantum* that had previously been suggested to be the result of two overlapping inversions in the ancestral fern plastome (Hasebe & Iwatsuki, 1990, 1992). Wolf et al. later showed these inversions to be 18 and 21 kb in length (Wolf, Duffy, & Roper, 2009; Wolf, Roper, & Duffy, 2010). This expansion was not present in *Osmunda*, supporting the position of Osmundaceae as sister to the remaining leptosporangiate ferns (Wolf et al., 2003). Subsequent plastome sequences for *Angiopteris evecta* (Marattiaceae, NC_008829, Roper et al., 2007) and *Equisetum arvense* (Equisetaceae, NC_014699, Karol et al., 2010) further clarified relationships within ferns, and agreed with additional sequence (Pryer et al., 2001, 2004) and morphological data (Kenrick & Crane, 1997) in supporting a monophyletic monilophyte clade that included leptosporangiates, horsetails, psilophytes, and marattioid ferns. Members of the polypod fern clade also share a second set of inversions within the large single-copy region (LSC) that is not present in the remaining ferns (Wolf et al., 2011, 2010).

Since 2013, 18 additional plastomes have been sequenced from a broad range of taxa across ferns (Table 2). In the species sequenced to date, plastid genome size ranges from 131 to 168 kb, with most of the variation due to length differences in the small single-copy region and the IRs (Grewe, Guo, Gubbels, Hansen, & Mower, 2013; Lu, Zhang, Du, Wen, & Li, 2015). Zhu, Guo, Gupta, Fan, and Mower (2015), in a study of plastid IRs, reported that multiple expansions, contractions and/or rearrangements are required to explain the diversity of IR structures across ferns. They noted several transfers of protein-coding genes from the single-copy to IR regions that must have occurred separately in the histories of *Psilotum*, *Angiopteris*, and *Adiantum* plus *Cheilanthes* (Pteridaceae). Individual studies have also identified lineage-specific gene losses, including loss of *trnK* from the leptosporangiate ferns (Gao, Yi, Yang, Su, & Wang, 2009; Wolf et al., 2011, 2003); loss of *chlB*, *chlL*, and *chlN* from *P. nudum* (Grewe et al., 2013); and loss of *rpoC1* from *Lygodium japonicum* (Kim, Chung, & Kim, 2014). Zhong, Fong, Collins, Mclenachan, and Penny (2014) used plastome data from tree ferns to explore rates of molecular evolution in this clade compared to other ferns and recovered a decrease in tree-fern rates that had previously been identified by Korall, Schuettpelz, and Pryer (2010), and which is thought to be linked to generation time effects in these long-lived, arborescent ferns. Overall, phylogenomic analyses of fern plastomes sequenced in recent years (eg, Karol et al., 2010; Lu et al., 2015) concur with phylogenetic analyses of other molecular data sets for ferns, including in uniformly supporting a monophyletic monilophyte clade.

2.1.2 Mitochondrial Genes

To date no complete mitochondrial genome has been sequenced for any fern, although several studies have generated sequence data for portions of fern mitochondrial genomes. Palmer, Soltis, and Soltis (1992) isolated restriction fragments corresponding to sections of the *E. arvense* and *Onoclea sensibilis* mitochondrial genomes and used these to detect the presence of several genes known from other plant mitochondrial sequences. Panarese, Rainaldi, De Benedetto, and Gallerani (2008) isolated a mitochondrial genome fragment from *Adiantum nidis* and found it to contain several genes apparently transferred from the plastome. Grewe (2011), in a study of a large fosmid clone from the mitochondrial genome of *Gleichenia dicarpa*, found that it contained a number of transposons, retrotransposons, and transposed introns. Finally, Wolf et al. (2015) conducted low-coverage genome skims for several species of ferns and assembled contigs containing putative mitochondrial genes for these species that ranged from approximately 23 to 415 kb. They were able to recover most, but not all, known mitochondrial genes in each of the six fern species.

2.2 Transcriptomes

2.2.1 Pteridium aquilinum *Transcriptome*

As for all other groups of organisms, transcriptomic resources for ferns (and lycophytes — see below) have increased exponentially in recent years. The first transcriptome sequence of a fern, from the gametophyte of *Pteridium aquilinum* (bracken fern), was published in 2011 (Der, Barker, Wickett, dePamphilis, & Wolf, 2011). The sequence included ~87% of the estimated complete transcriptome length of *P. aquilinum*, and the final assembly included 56,256 unigenes. 61.8% of these passed designated cutoffs in a blastx analysis against the NCBI nonredundant protein database, and the best hits for these unigenes corresponded to 22,596 unique protein accessions. The authors used a comparative genomics approach to identify genes shared with *Arabidopsis thaliana*, *Selaginella moellendorffii*, and *Physcomitrella patens*, and were able to identify 333 of 357 genes that are conserved as single-copy across all eukaryotes (ultraconserved orthologs). Of the 56,256 unigenes in the *P. aquilinum* assembly, 48.5% (27,304) had blastx hits to *A. thaliana*, *S. moellendorffii*, and *P. patens*. Sets of between 758 (1.34%) and 1627 (2.89%) genes had hits to only one or two of these species. Among the *Arabidopsis* homologues, the authors looked for evidence of genes previously identified as being associated with the gametophyte generation in *A. thaliana* and identified 1156 genes in the *P. aquilinum* gametophyte transcriptome with significant blastx hits to known gametophyte genes from *Arabidopsis*. This included over half of the genes previously characterized as being associated with the gametophyte generation in *Arabidopsis*, indicating that gametophyte-specific genes have been conserved over several hundred million years of evolutionary time and across organisms with numerous differences in life cycle and reproductive strategy. Finally, the remaining *P. aquilinum* unigenes not found to have significant blastx hits to the other species in the comparative analyses (21,425; 38.1%) may represent fern-specific genes or gene families that are not present in other lineages.

2.2.2 Lygodium japonicum *Transcriptome*

The second set of comprehensive fern transcriptome sequences was produced for *L. japonicum*, a vining member of the fern family Schizaeaceae. Aya et al. (2015) sequenced RNA from gametophytes at five developmental stages, and immature and mature sterile leaves, mature fertile leaves, and young rhizomes of *Lygodium* sporophytes. They produced a hybrid transcriptome assembly from all sequence data that consisted of 37,830 isoforms corresponding to 23,856 unigenes. Between 54.7% and 66% of these had significant hits to unique protein accessions, depending on the database used (UniProt, TAIR, or KOG). In an OrthoMCL analysis of *L. japonicum* and five other land plants (*P. patens*, *S. moellendorffii*, *Pinus taeda*, *A. thaliana*, and *Oryza sativa*), 24,280 contigs/isoforms from *Lygodium* were assigned to at least one gene cluster. Of these, 9287 (38.25%) were assigned to clusters that only contained sequences from *L. japonicum* and which may represent fern-specific genes. This percentage is similar to that found by Der et al. (2011) for *Pteridium*. An additional 9215 (37.95%) of the *Lygodium* contigs were assigned to clusters that contained genes from all six species. The remaining 5778 (23.8%) contigs belonged to clusters that included *L. japonicum* and at least one other species, but not all five.

The tissue-specific transcriptome assemblies of Aya et al. (2015) allowed analysis of organ-specific gene expression patterns, and the authors identified 17,270 unigenes that were differentially expressed across the nine sampled organs. Global expression patterns from the different gametophyte development stages were very similar to one another, while the leaf and rhizome expression patterns differed substantially from one another and from the gametophytes. Principal components analysis (PCA) first separated the young rhizome and fertilized gametophyte tissues from the other organs and developmental stages; however, this result may have reflected differences in sample handling prior to sequencing (these two tissue types underwent RNA amplification while the others did not). PC2 further separated the gametophyte and sporophyte stages from one another. Functional annotations identified characteristic sets of GO terms associated with each of the organ-specific gene clusters. For example, genes with high levels of expression in the gametophytes were enriched for functions associated with photosynthesis, translation, and ribosome construction. These and other comparative analyses have laid the groundwork for future analyses of gene function evolution across land plants.

2.2.3 1000 Plants Initiative

The largest contribution to fern and lycophyte transcriptome resources has been made by the 1000 Plants Initiative (oneKP; www.onekp.com). As of 2015 this international consortium has accomplished substantially more than its eponymous goal, and has made transcriptome assemblies available for over 1300 species of plants, from algae through angiosperms (Matasci et al., 2014). The first analyses of these data have been largely phylogenomic in nature, and those that include ferns have focused on understanding early diversifications across land plants (Wickett et al., 2014) and on producing the first broad-scale phylogeny for ferns (Rothfels et al., 2015). Wickett et al. (2014) used data from a subset of oneKP species to build a phylogeny directed at reconstructing the origin and major evolutionary transitions of land plants. Their results agree with previous studies that place lycophytes sister to ferns plus seed plants, and *Equisetum* sister to all other ferns (Pryer et al., 2001). Within the monilophyte clade, Wickett et al.'s analyses recover the topology (((ophioglossoids, marattioids) leptosporangiates) equisetoids)), with the branch subtending ((ophioglossoids, marattioids) leptosporangiates) receiving less than maximal support. These relationships, between the leptosporangiate ferns and the three major lineages of eusporangiate ferns (ophioglossoids, marattioids, and equisetoids), have longresisted resolution, and Rothfels et al. (2015) reached a different conclusion in analyses that used oneKP data for a smaller set of markers, but included a larger set of fern species than Wickett et al. Rothfels et al. found support for the topology (((leptosporangiates, marattioids) ophioglossoids) equisetoids), though again with less than maximal support at the crucial branches uniting leptosporangiates with marattioids, and that clade with ophioglossoids. This result, however, is consistent with several previous phylogenies derived from analyses of plastid data (eg, Kuo, Li, Chiou, & Wang, 2011; Rai & Graham, 2010), and for the time being seems the most reliable reconstruction of the relationships within monilophytes.

The oneKP data have also shed light on a remarkable event in the evolutionary history of ferns that involves the acquisition of light-sensing genes that are present in only a handful of plant lineages. Ferns possess a unique photoreceptor called neochrome that is a chimera of blue-light-sensing phototropin and red-light-sensing phytochrome photoreceptors (Kanegae, Hayashida, Kuramoto, & Wada, 2006; Kawai et al., 2003). Neochrome allows ferns to have enhanced photosynthetic light response, particularly in low-light environments (Kanegae et al., 2006; Kawai et al., 2003). In a series of recent studies on the evolution of plant photoreceptors using transcriptomic data from oneKP, Li et al. (2014), Li, Melkonian, et al. (2015) and Li, Rothfels, et al. (2015) found strong evidence that fern neochrome genes are most closely related to newly discovered neochrome genes found in hornworts. While there are several possible evolutionary scenarios that could account for this phylogenetic distribution of neochrome, Li et al. propose one or more instances of horizontal gene transfer (HGT) between ferns and hornworts as the most likely mechanism. This scenario begins with retrotransposition of a phototropin gene and its subsequent fusion with a phytochrome gene within the hornworts, which was followed by HGT from the hornwort lineage to ferns (Li et al., 2014).

2.3 Nuclear Genomes

2.3.1 Genome Skimming Analyses in Ferns

Despite the ecological and evolutionary importance of ferns, they are the only lineage of land plants for which a reference nuclear genome sequence still does not exist (Sessa et al., 2014). Having such a reference will facilitate numerous studies of fern biology and evolution and will fill a critical gap for comparative analyses of genome structure across land plants. In the meantime, a recent study employing genome skimming has provided the first data on protein-coding and repetitive content in the genomes of six fern species (Wolf et al., 2015). Wolf et al. found that, compared to seed plants, fern genomes contain relatively higher proportions of repeats from three classes: DNA transposons, long interspersed nuclear elements (LINEs), and simple repeats (Fig. 3). Satellite DNA made up a lower proportion of the genome in ferns than in seed plants. Proportions of the genome comprised by another seven types of repeat class did not differ significantly between ferns and seed plants (Wolf et al., 2015). Further work is required to determine whether these similarities and differences between the fern and seed plant lineages are biologically significant.

Protein-coding content was estimated as comprising from 2.85 to 6.61% of fern genomes. In seed plants, protein-coding content ranges from ~2.4% in *Picea abies*, a member of the conifers, which have notoriously large genomes (19.6 Gb; Nystedt et al., 2013), to 97% in *Utricularia gibba*, a carnivorous plant with reduced morphology and a relatively small genome

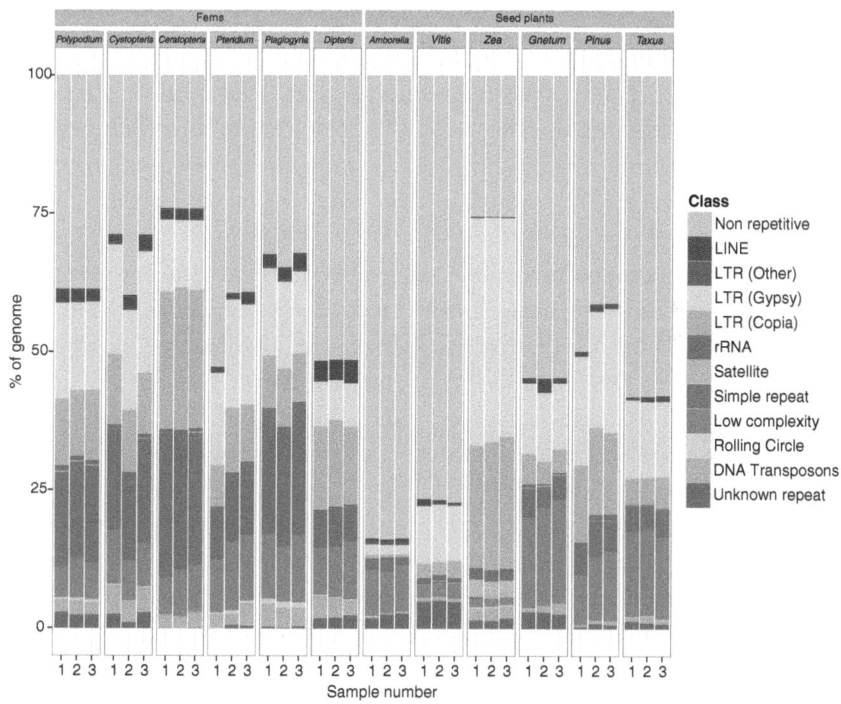

Figure 3 Genome proportions represented by ten sequence-based repeat classes, plus unknown repeats and nonrepetitive sequences, in six fern and six seed plant taxa, with three samples per taxon. *Reprinted from Wolf, P.G., Sessa, E.B., Marchant, D.B., Li, F.-W., Rothfels, C.J., Sigel, E.M. ... Der, J.P. (2015). An exploration into fern genome space.* Genome Biology and Evolution, *7(9), 2533–2544; under a Creative Commons license.* (See colour plate)

(83 Mb; Ibarra-Laclette et al., 2013). Gene number estimates for seed plants are much less variable: generally, between 20,000 and 40,000 per diploid genome (Rabinowicz et al., 2005). The single estimate for a fern, from *C. richardii*, is near this range: 42,300 genes (Rabinowicz et al., 2005), or 0.49% or 6% of the genome, depending on the calculation used (both similar to the estimates of Wolf et al., above). Ferns are therefore at the lower end of protein content estimates for land plants and tend to have relatively low gene density given their larger genome sizes (Rabinowicz et al., 2005). Wolf et al. also found that across the six fern species studied, there was no correlation between protein content and genome size. They also found no correlation between genome size and proportion of the genome comprised of repeats, though there is a positive correlation between these across land plants (Michael, 2014). This further supports the hypothesis that ferns may have different mechanisms of genome downsizing than other lineages of land plants (Barker & Wolf, 2010; Leitch & Leitch, 2012).

2.3.2 Transposon Methylation and Expression in Ferns

Wolf et al. found that ferns have higher proportions of the genome comprised of DNA transposons and LINE retrotransposons than do seed plants (Fig. 3), though within ferns these proportions were not correlated with genome size. Additional information on transposons in ferns is available from several studies. Rabinowicz et al. (2003) found that ferns are similar to other groups in having less methylation in gene-rich regions than other genomic regions. DNA methylation differences have also been found to differ by developmental stage in *C. richardii* (McGrath & Pichersky, 1997). DNA methylation and transposon activity are tightly linked in plants (Rabinowicz et al., 2005; 2003), and most methylation in plant genomes has been found to occur in transposons, which are silenced as a result (Lippman et al., 2004). Methylation linked to RNA interference also plays an important role in plant immunity, by targeting and inactivating viral DNA (Tsuboi, Sutoh, & Wada, 2014). Tsuboi et al. suggested that ferns may increase resistance to viral pathogens by integrating DNA and RNA interference systems to target viral transposon activity. There is also some evidence for expression of transposons in ferns; in a study of the gametophyte transcriptome of bracken fern (*P. aquilinum*), Der et al. (2011) found that 0.17% of the transcriptome was comprised of retrotransposons of various classes, and they identified 269 expressed DNA transposons (0.07% of the total sequence length).

2.4 *Ceratopteris richardii*, a Genetic Model Fern

2.4.1 Genetic Model System

Although ferns lack a complete reference nuclear genome sequence, genetic and genomic resources exist for the group nonetheless. Among the homosporous ferns, *C. richardii* (Pteridaceae) has emerged as a model genetic system — just as *Arabidopsis* is the model for angiosperms — for the study of biological processes best addressed in ferns (eg, stress response pathways, sex determination). *Ceratopteris* is easily cultured and has a short generation time and numerous mutant lines. Like all homosporous plants examined to date, it has a large genome: 11.26 Gb based on flow cytometry estimates (Wolf et al., 2015). The haploid chromosome number is 39 (Scott & Hickok, 1987), which is small for a homosporous plant. Analyses of repeat structure based on low-coverage genome skimming (discussed above) indicate that repetitive elements comprise ~75% of the *C. richardii* genome (Wolf et al., 2015).

The biology of sex determination in *C. richardii* has been much studied over the years and is now well characterized (Eberle, Nemacheck, Wen, Hasebe, & Banks, 1995). More than 50 sex-determining mutants have been identified and used to investigate sex determination pathways (Banks, 1994, 1997; DeYoung, Weber, Hass, & Banks, 1997; Eberle & Banks, 1986; Strain, Hass, & Banks, 2001). Additional mutants with altered responses to light (Kamachi, Matsunaga, Noguchi, & Inoue, 2004), salt tolerance (Warne, Hickok, Kinraide, & Vogelien, 1996; Warne, Hickok, Sams, & Vogelien, 1999), and herbicide resistance (Carroll, Schwarz, & Hickok, 1988) also have been identified. As in all ferns, gametophytes and sporophytes of *Ceratopteris* are independent stages of the life cycle, and so *C. richardii* and close relative *Ceratopteris thalictroides* have been used extensively to describe spore germination (Salmi, Bushart, Stout, & Roux, 2005), gamete development (Cao, Yang, & Wang, 2009; Lopez-Smith & Renzaglia, 2008; Schledbauer, Cave, & Bell, 1973), fertilization (Cao, Wang, & Bao, 2010; Cao, Wang, Yang, & Bao, 2010), and embryogenesis (Johnson & Renzaglia, 2008), and to understand asexual, apogamous pathways of sporophyte formation (Cordle, Irish, & Cheng, 2012). *Ceratopteris richardii* spores, which are unicellular, have also been used as models to study the establishment of cellular polarity in plants in response to gravity (Edwards & Roux, 1998; Murata & Sugai, 2000; Spiro, Torabi, & Cornell, 2004) and the effects of microgravity on early stages of development (Salmi, Bushart, & Roux, 2011; Salmi & Roux, 2008).

The utility of *Ceratopteris* as a genetic model should increase even further thanks to recent development of the first reliable protocols for stable transformation of gametophyte prothallus cells, with stable inheritance of transgenes demonstrated in sporophytes and subsequent generations (Plackett, Huang, Sanders, & Langdale, 2014; Plackett, Rabbinowitsch, & Langdale, 2015). Ferns have been recalcitrant to the development of efficient transformation protocols, which has been a significant impediment to detailed analyses of genetic function (Plackett et al., 2014). Previous protocols based on direct DNA uptake (Stout, Clark, Archer-Evans, & Roux, 2003), microparticle bombardment (Indriolo, Na, Ellis, Salt, & Banks, 2010; Kawai-Toyooka et al., 2004; Rutherford, Tanurdzic, Hasebe, & Banks, 2004), and *Agrobacterium*-mediated spore transformation (Muthukumar, Joyce, Elless, & Stewart, 2013) demonstrated varying degrees of stable transformation in ferns, but with uniformly low efficiency or lack of inheritance (Plackett et al., 2014). Plackett et al. (2014, 2015) developed a new technique for stable transformation via tungsten microparticle bombardment, with transformation efficiencies up to 72% in *C. richardii* and 86% in *C. thalictroides*. This should greatly increase the utility of *Ceratopteris* as a model system for studying gene function in ferns.

2.4.2 Genetic Map

Nakazato et al. (2006) produced a high-resolution genetic linkage map for *C. richardii* based on 488 doubled haploid lines that were genotyped for 368 RFLP, 358 AFLP, and 3 isozyme markers. This mapping experiment recovered 41 linkage groups and revealed that the majority of loci (over 76%) were duplicated, with most duplicates in different linkage groups. However, there was no evidence of large syntenic blocks that would support a hypothesis of extensive ancestral polyploidization, and the number of duplicated copies present varied widely among genes. Nakazato et al. therefore concluded that if *C. richardii* is polyploid, it is an ancient polyploid, and most of the duplicate loci are likely due instead to small-scale, segmental gene duplications. They did note, however, that chromosomal rearrangements as well as small-scale duplications and deletions could work to obscure any evidence that might exist of ancient polyploidy events.

2.4.3 Genome Sequencing Initiative

Plant biologists have long recognized the need for a reference genome for ferns (Pryer, Schneider, Zimmer, & Banks, 2002; Sessa et al., 2014). Fern genomes hold the keys to unlocking a number of critical questions in land plant evolution (more below). *Ceratopteris richardii* is an ideal choice to be among the first ferns sequenced because of its extensive use as a genetic model system and in education (where it is known as C-Fern and is grown in K-12 classrooms around the world). Its genome is typical in size and chromosome number for a homosporous fern, and there is currently no reference sequence available for any homosporous vascular plant; a sequence for *C. richardii* would therefore simultaneously fill two critical gaps in available land plant genomes, providing a fern sequence as well as a homosporous vascular plant reference. Sequence data from *C. richardii* will provide the broader plant community with a crucial outgroup for understanding many aspects of seed plant evolution and will provide an evolutionary reference for understanding the origins of seeds, flowers, vessel elements, wood, and other key innovations. The genome will also serve as a reference for other ferns and for exploring and improving their myriad economic uses, including biofuel development and phytoremediation. Efforts are currently underway to sequence the genome of *C. richardii*, and preliminary data from genome skimming are already available for use by the research community (Wolf et al., 2015).

2.5 *Azolla filiculoides*, a Little Fern With Big Potential

Another fern that has recently become the focus of genome-scale studies is *Azolla filiculoides* (Brouwer et al., 2014; Vries et al., 2016). *Azolla* is a small genus of five to eight species of floating aquatic plants (Metzgar, Schneider, & Pryer, 2007), sometimes called mosquito ferns, which form a symbiotic relationship with cyanobacteria capable of fixing atmospheric nitrogen (Peters & Mayne, 1974; Peters & Meeks, 1989; Plazinski et al., 1990; Wagner, 1997).

2.5.1 Agricultural and Industrial Uses

Azolla has been used for millennia in Asian agriculture in cocultivation with rice as a source of nitrogen fertilizer and in the suppression of weeds (Lumpkin & Plucknett, 1980; Manna & Singh, 1989; Pabby, Prasanna, & Singh, 2004; Wagner, 1997). A rapidly growing plant with high protein content, *Azolla* has also been used as supplementary forage for poultry (Basak, Pramanik, Rahman, & Tarafdar, 2002), fish (Abou, Fiogbé, & Micha, 2007), and livestock (Cherryl, Prasad, & Jayalaxmi, 2013; Eyini, Sujanandini, Pothiraj, Jayakumar, & Kil, 1999). Further, *Azolla* have been utilized in the phytoremediation of contaminated wastewater

(Bennicelli, Stępniewska, Banach, Szajnocha, & Ostrowski, 2004; Forni, Chen, Tancioni, & Grilli Caiola, 2001; Kooh, Lim, Lim, & Dahri, 2016; Muradov et al., 2014; Ventura, Mascarina, Furoc, & Watanabe, 1987; Vermaat & Khalid Hanif, 1998; Zazouli, Mahdavi, Bazrafshan, & Balarak, 2014; Zhao, Duncan, & Van Hille, 1999).

2.5.2 Carbon Sequestration and Paleoclimate Change

Azolla has also been implicated in playing a critical role in atmospheric carbon sequestration and the global shift from a warm to cool climate during the middle Eocene (~50 MYA) (Brinkhuis et al., 2006; Speelman et al., 2009). The so-called 'Arctic *Azolla* Event' occurred when an extensive layer of freshwater expanded over the Arctic Ocean, enabling a massive bloom of *Azolla* to grow and subsequently sink to the ocean floor where it formed a large fossil deposit under anoxic conditions. While modern cultivation of *Azolla* is unlikely to substantively affect global carbon cycling and climate change, the rapid growth of *Azolla* makes it an attractive source of alternative fuels (Brouwer et al., 2014; Brouwer, van der Werf, Schluepmann, Reichart, & Nierop, 2015; Salehzadeh, Naeemi, & Arasteh, 2014).

2.5.3 Domestication and Biotechnology Applications

To realize the full potential of *Azolla* for agricultural and industrial processes, progress still needs to be made in the breeding and improvement of genetic stocks suitable for commercialization. Toward this end, protocols for cryopreservation of plant material have been developed and transcriptome sequencing was used to investigate the molecular basis of reproduction in *A. filiculoides*, including sporocarp production and fertilization (Brouwer et al., 2014). Further investigations using transcriptome data have examined the molecular basis for the evolution of root meristems in euphyllophytes.

2.5.4 Genome Sequencing Initiative

With the reduced cost of high-throughput DNA sequencing and a recent crowd-funding effort (Li & Pryer, 2014), an international consortium has been established to sequence the genome of *A. filiculoides*. This is likely to be the first complete fern genome sequence available and will provide a key reference point for comparative analysis of genome evolution in plants (Sessa et al., 2014).

3. LYCOPHYTE GENOMES

3.1 Chloroplast and Mitochondrial Genomes

3.1.1 Lycophyte Plastid Genomes

Fewer plastome sequence resources exist for lycophytes than for ferns. The first lycophyte plastome, from *Huperzia lucidula*, was sequenced in 2005 by Wolf et al. (NC_006861). They reported that the overall structure of the *Huperzia* plastome was similar to that of other land plant plastomes, but it was more like bryophyte plastomes than those of ferns or seed plants. This supported the position of lycophytes as sister to the fern plus seed plant clade (Fig. 1). The best early evidence in support of this relationship was a 30 kb inversion in the plastid genome shared between lycophytes and bryophytes, which was discovered via chloroplast restriction site mapping by Raubeson and Jansen (1992). The remaining lycophyte plastome sequences belong to two *Selaginella* (*Selaginella uncinata*, AB197035, Tsuji et al., 2007; *S. moellendorffii*, FJ755183, Smith, 2009) and one *Isoetes* (*Isoetes flaccida*, GU191333, Karol et al., 2010) species, all members of the heterosporous lycophyte clade. These sequences further support the monophyly of lycophytes, but reveal several unique rearrangements within the group. Both *Isoetes* and *Selaginella* have undergone independent rearrangements in their plastomes relative to *Huperzia*, including inversions and translocations of the *ycf2* and *chlL/chlN* gene cluster in *Isoetes* (Karol et al., 2010), and a large translocation (~14 bp) from the LSC to the IR/SSC in both sequences from *Selaginella* (Karol et al., 2010; Tsuji et al., 2007).

3.1.2 Mitochondrial Genome Sequences From Selaginella *and* Isoetes

Sequences exist for portions of the mitochondrial genomes of *S. moellendorffii* (Smith, 2009) and *Isoetes engelmannii* (Grewe et al., 2010; Grewe, Viehoever, Weisshaar, & Knoop, 2009). Analyses of these genomes have found extremely high GC content in *Selaginella* and evidence of extensive RNA editing in both species. Smith reported that *S. moellendorffii* may, in fact, have the most GC-rich plant organellar genome studied to date. This high GC content may be linked to the high levels of RNA editing present, a relationship previously noted in analyses of other land plants (eg, Jobson & Qiu, 2008; Malek, Lättig, Hiesel, Brennicke, & Knoop, 1996). Lycophytes and ferns, as well as bryophytes, are known to have particularly high levels of RNA editing in their organellar genomes relative to seed plants (Takenaka, Zehrmann, Verbitskiy, Härtel, & Brennicke, 2013). Additional sequencing of lycophyte and fern mitochondrial genomes will be essential for further investigation of this relationship.

3.2 Transcriptomes

3.2.1 1000 Plants Initiative

As for the chloroplast and mitochondrial genomes, many fewer transcriptomic resources exist for lycophytes than for ferns. The only comprehensive transcriptome data produced to date for lycophytes have been generated as part of the 1000 Plants Initiative. As of 2015, transcriptomes were available for 22 species of lycophytes, including members of all three major clades (Selaginellaceae, Isoetaceae, and Lycopodiaceae), and some of these data have been analysed as part of the broad phylogenomic studies conducted by Wickett et al. (2014; discussed above).

3.3 Nuclear Genomes

3.3.1 Selaginella moellendorffii

Unlike the ferns, lycophytes do have a reference nuclear genome available, from *S. moellendorffii* (Banks et al., 2011). *Selaginella* is a large genus of lycophytes (c. 700 spp., Arrigo et al., 2013), and it also comprises the monogeneric family Selaginellaceae (Banks, 2009; Zhou et al., 2015). Selaginellaceae and Isoetaceae are sister to one another and share a presumptive single origin of heterospory (Banks, 2009); together they are sister to the homosporous Lycopodiaceae (Fig. 1). *Selaginella moellendorffii* is among the smallest plant genomes ever reported (~110 Mb Wang et al., 2005), which made it an attractive candidate for whole-genome sequencing. As in the ferns, homosporous and heterosporous lycophytes differ substantially in average genome size, with homosporous taxa having uniformly larger genomes: 2.6–5.5 Gb versus ~100–2000 Mb in the heterosporous lineage (Wang et al., 2005) (Fig. 1). The assembled *S. moellendorffii* genome is 212.6 Mb, consisting of two nearly identical haplotypes that include 22,285 predicted protein-coding genes. Gene density is similar to *Arabidopsis*, which has a slightly larger genome (125 Mb) (The Arabidopsis Genome Initiative, 2000). Long terminal retrotransposons make up a third of the *Selaginella* genome, and Chan et al. (2008) reported that, as in other plant lineages, high-copy repeats are the most highly methylated regions of the *Selaginella* genome, while coding regions are much less methylated (Martienssen, 1998; Rabinowicz et al., 2003).

The *S. moellendorffii* genome sequence has provided many interesting insights into the early evolution of vascular plants, particularly genetic changes involved in the shift from the gametophyte-dominant life cycle of the bryophyte lineages to the sporophyte-dominant life cycle of the tracheophytes. Many fewer new genes (519) were identified as being correlated with this shift than were found to be unique between the lycophyte and angiosperm lineages (1350) (Banks et al., 2011). Most gene families that have been found to have critical roles in angiosperm development were apparently already present in the common ancestor of land plants, with subsequent lineage-specific gene family expansions in the major groups of land plants (Banks et al., 2011). Many additional studies on gene family evolution and genome structure have been facilitated by sequencing the *Selaginella* genome (e.g., Anderberg, Kjellbom, & Johanson, 2012; Carey, Hepler, & Cosgrove, 2013; Gramzow et al., 2012; Shakirov & Shippen, 2012; Weng & Noel, 2013; Wipf, Loque, Lalonde, & Frommer, 2012; Zhou et al., 2015). Gramzow et al. (2012), in a study of MADS-domain proteins (among the most important transcription factors involved in plant development), identified 19 putative MADS-box genes in the *S. moellendorffii* genome, and determined that MADS-box genes involved in floral organ identity in angiosperms originated after the divergence of seed plants and lycophytes. Studies of two other large superfamilies of genes, expansins (Carey et al., 2013) and major intrinsic proteins (MIPS) (Anderberg et al., 2012), also found increases in gene and isoform numbers in seed plants that are independent of those in the lycophytes. The absence of a fern genome makes it impossible to know whether these expansions are truly seed plant–specific or occurred in the ancestor of all euphyllophytes (ferns plus seed plants). There is also no evidence of ancient polyploidy in the lycophytes (Banks et al., 2011); it is the only land plant lineage with no evidence of such events.

4. LOOKING TO THE FUTURE OF FERN AND LYCOPHYTE GENOMICS

4.1 Outstanding Questions in Fern and Lycophyte Genome Evolution

Although tremendous strides have been made in recent decades in generating genetic and genomic resources for numerous plant groups, the plant science research community has a pressing current need for reference genomes from homosporous and heterosporous ferns, as well as homosporous lycophytes. Sequencing fern genomes in particular will enable us to address many long-standing questions in fern biology and will facilitate innumerable studies of genome structure and evolutionary dynamics, both in ferns and for comparative analyses across all land plants. A fern genome will have utility for researchers working on the evolution and improvement of seed and wood traits, for example, and will also revolutionize the ability of the fern community to address fundamental questions about fern biology.

As described above, homosporous ferns have much larger genomes and many more chromosomes than heterosporous lineages, including angiosperms, yet similar numbers of functional genes (Nakazato et al., 2006; Rabinowicz et al., 2005). We still do not understand this most basic feature of homosporous fern genomes: why are they so large and less gene-dense than heterosporous genomes? Alternatively, why is heterospory so uniformly associated with decreases in genome size and chromosome numbers in multiple independent origins of heterospory across land plants? Is the ancestral genome size in land plants large and similar to modern homosporous lineages? What then has driven the genome downsizing seen in heterosporous groups, whose genomes are so small relative to the likely size of their ancestors' genomes? Several hypotheses have been suggested to account for the other remarkable genomic feature of homosporous ferns, the apparent correlation between chromosome number and genome size, but there is not yet strong support for any of them (see above). The question of whether ferns are ancient polyploids, which has vexed pteridologists for at least the last half century, and which has resisted resolution with the most modern forms of data available, could perhaps finally be laid to rest by analyses of complete genomic sequence data. In addition, published C-value estimates exist for only ~100 fern species (Bennett & Leitch, 2012) out of the c. 9000 known ferns (Smith et al., 2006, 2008). Additional studies are needed to establish the full range of genome sizes in ferns, which will be particularly useful for determining whether currently unsampled species have small genomes that may be suitable candidates for high-coverage sequencing and assembly. The large genome sizes and high chromosome numbers in ferns, and the challenges they represent for assembly, have contributed significantly to difficulties in obtaining a reference genome for ferns (Pryer et al., 2002; Sessa et al., 2014).

Although a homosporous genome sequence will be most representative of the majority of extant ferns, and will fill a critical gap for comparative land plant genomics, sequencing complete genomes for both heterosporous and homosporous ferns is critical. Having reference genomes from both groups will allow us to further characterize the gene and repeat content in ferns and to determine how the distribution of repetitive elements and genes differs between heterosporous and homosporous lineages, for example. A reference genome for a homosporous lycophyte is also a pressing need, for the same reason, and together, sequences from the two homosporous lineages will open entirely new avenues for exploration. Are homosporous ferns and lycophytes most similar to one another in their genomic features, or is each more similar to heterosporous relatives within its own major lineage? What patterns are correlated with the transition between heterospory and homospory, and what is unique to the separate origins of heterospory?

Ferns and lycophytes also have separate and independent phases of the life cycle, the diploid sporophyte and haploid gametophyte, and we know little about differential expression patterns between these stages in either lineage. Are genes expressed in the haploid phase of the life cycle distributed differently across the genome compared to those expressed in the diploid phase? Are there gene expression regulation mechanisms uniquely employed in the haploid and diploid phases in ferns and lycophytes? Progress is being made on these questions in bryophytes (eg, Szövényi, Rensing, Lang, Wray, & Shaw, 2011), but is hampered in ferns and lycophytes by the lack of whole genome data that would allow, in particular, investigations of regulatory mechanisms and metabolic networks. Higher-level genomic structures are also difficult to evaluate without whole genome sequences. In lycophytes, the sequencing of *Selaginella* has allowed telomere structure to be explored, and Shakirov and Shippen (2012) found that *S. moellendorffii* telomeres are composed of repeats similar to those found in telomeres of other plants and of sizes and arrangements closely resembling those of *A. thaliana*. In ferns, however, nothing is known about the structure and content of centromeres and telomeres.

Ferns and lycophytes have little perceived economic value, with the notable exception of *Azolla*, and this has contributed substantially to the difficulty in getting their genomes sequenced (Pryer et al., 2002; Sessa et al., 2014). However, many ferns and lycophytes do in fact have economic potential and various uses that can be further explored and harnessed once a reference genome becomes available. In addition to *Azolla*'s use as a biofertilizer and feed crop (Li & Pryer, 2014; Muradov et al., 2014; Pabby et al., 2004), it is also being investigated for biogas production, as are at least two other ferns (*Ceratopteris* and *Salvinia*) (Wilkie & Evans, 2010). Bracken fern, *P. aquilinum*, has long been used for a variety of purposes, including as livestock bedding, compost, mulch, and small-scale biofuel, and large stands of *Pteridium* are burnt and the ash used as a potassium-rich fertilizer (Donnelly, 2003). Numerous ferns have also been found to tolerate and even to hyperaccumulate various heavy metals and toxins. *Pteris vittata* is a hyperaccumulator of arsenic, and its sporophyte and gametophyte tissues can consist of up to 2% dry weight of inorganic arsenite (Gumaelius, Lahner, Salt, & Banks, 2004; Lessl, Guan, Sessa, Rathinasabapathi, & Ma, 2015; Ma et al., 2001; Singh & Ma, 2006). Little is known about the genes involved in arsenic sequestration and transport in *Pteris*, but identification of such genes could aid development of remediation techniques for treating arsenic-contaminated soils and groundwater (Kertulis-Tartar, Ma, Tu, & Chirenje, 2006; Salido, Hasty, Lim, & Butcher, 2003; Tu, Ma, Fayiga, & Zillioux, 2004). Other ferns that accumulate metals include *Osmundastrum cinnamomeum* (copper, cadmium, and zinc) (Francis & Petersen, 1983) and *Athyrium yokoscense* (lead) (Kamachi et al., 2005; Nishizono, Suzuki, & Ishii, 1987). In addition to being tolerant of heavy metals, many ferns are tolerant of desiccation (Pittermann, Brodersen, & Watkins, 2013). Desiccation tolerance (DT) involves more than simply being able to dry out; desiccation-tolerant plants can lose all metabolically active water from their vegetative tissues and recover from that state (Bewley, 1979; Watkins, Mack, Sinclair, & Mulkey, 2007). True DT is very rare in vascular plants, especially crops (Alpert, 2000), but gametophytes of many fern species have been found to be tolerant of desiccation to a remarkable degree (Pittermann et al., 2013; Watkins & Cardelús, 2012; Watkins et al., 2007). No study has yet been undertaken to investigate the regulatory and metabolic networks that allow ferns to be desiccation tolerant, but such studies have enormous potential for crop plant improvement, particularly as global climate patterns change and periods of drought and inconsistent precipitation become more common.

4.2 Outstanding Challenges in Land Plant Comparative Genomics

In addition to the many outstanding questions specific to fern and lycophyte biology and evolution, these lineages hold the keys to understanding many features of land plant evolution in general, including transitions between homospory and heterospory and the shift from gametophyte-dominant to sporophyte-dominant life cycles. Ferns and lycophytes are the sister groups to seed plants and euphyllophytes, respectively, and so are critical outgroups for understanding early and ongoing evolution in those lineages. In particular, the shift from homospory to heterospory, which was discussed above as an outstanding challenge for fern and lycophyte biology, has occurred independently several times during plant evolution and is critical for understanding the basis of differences in genome architecture across all land plants: what mechanisms are responsible for the universally smaller numbers of chromosomes and genome sizes in heterosporous versus homosporous lineages? And what genomic changes underlie the shift from haploid-dominant to diploid-dominant life cycles across land plants?

Homosporous fern and lycophyte genomes will allow us to reconstruct the ancestral genome structures of seed plants, euphyllophytes, and all tracheophytes and will help us to understand the numerous transitions and key innovations that have evolved during the history of these clades. Major transitions that have occurred during the evolution of tracheophytes include the independent origins of heterospory in ferns, lycophytes, and seed plants and the reduction in size of both male and female gametophytes. Gametophytes in ferns and lycophytes are multicellular, often photosynthetic, and capable of living independently. In contrast, the male gametophyte is reduced to three cells and the female gametophyte to seven cells (with eight nuclei) in seed plants, with the female gametophyte completely enclosed by and dependent on the sporophyte. These dramatic shifts were presumably accompanied by extensive alternations in gene expression, regulatory dynamics, and genome architecture, but without reference genomes from all of the tracheophyte lineages, it is impossible to understand when and by what mechanisms such changes occurred. The euphyllophyte clade is so-named for the presence of true leaves (megaphylls), once presumed to have occurred in the shared ancestor of ferns and seed plants, but recent evidence suggests that megaphylls may actually have evolved independently multiple times in the seed plant and fern lineages (Vasco, Moran, & Ambrose, 2013). Understanding the evolution of true leaves and whether they are homologous across the euphyllophytes is a crucial question in plant biology today. Finally, many key innovations evolved in the ancestors of seed plants, including a complex, multilayered apical meristem (compared to a single-cell meristem in ferns), lateral meristems, integuments and ovules, and flowers and triploid endosperm in the angiosperms. Several of these, including true leaves, wood, and seeds, have obvious implications for human use and economic applications. Genome sequences from ferns will provide the outgroup necessary to investigate the origins of these innovative features in a complete comparative framework.

In addition to understanding major evolutionary transitions and key innovations that have occurred across land plants, many elements of genome architecture remain to be studied. Examples of changes in genome architecture that could be further explored with the addition of genomes from ferns and homosporous lycophytes include distribution and sizes of introns and pseudogenes, loss and gain of various repetitive elements (e.g., retrotransposons, DNA transposons, simple repeats, satellite DNA, and others), and gene family expansion and contraction. These and other structural genomic changes are likely correlated with many long-term evolutionary processes in land plants, including the transitions and innovations described above. Whole genome sequences from ferns may also allow us to conclusively determine how many rounds of whole genome duplication have occurred in ferns, as well as in ancestral euphyllophytes and tracheophytes. Identifying where and when in the phylogeny polyploidization events have occurred unlocks further questions: how do the fates of various genomic elements compare following polyploidization in the various lineages? Are patterns of synteny detectable from these events, and how are these patterns associated with genome evolution pre- and post-polyploidization? How does genome downsizing operate in the various tracheophyte lineages, and are there differences between homosporous and heterosporous taxa that may explain their different genome sizes?

Another avenue of research in plants that would be greatly facilitated by the addition of reference genomes from ferns and homosporous lycophytes is gene expression regulation. There are many mechanisms of transcriptional and posttranscriptional regulation known in land plants, including transcription factors, DNA methylation, histone modification, RNA interference, alternative splicing, and RNA editing. Understanding how these mechanisms operate in ferns and lycophytes, and particularly in homosporous taxa, is critical for the study of individual regulatory networks and gene products and also for understanding the origins of gene regulation across plants. In particular, it is unclear whether the various extant lineages coopted ancestral regulatory mechanisms or evolved them independently, and to what extent various types of expression regulation are specialized by lineage versus sharing features common to all plants. Methylation patterns in ferns and lycophytes appear to be similar to those found in other plant groups, with most methylation occurring in gene-poor and highly repetitive regions (Chan et al., 2008; Rabinowicz et al., 2005, 2003). Studies of RNA interference (RNAi) mechanisms in the moss *P. patens* and various angiosperms have identified both similarities and differences between these lineages (reviewed by Arif, Frank, & Khraiwesh, 2013). At least one study has explored RNAi function in ferns (Tsuboi et al., 2014), but not in a comparative framework with other land plants. Alternative splicing (AS) is another important mechanism for posttranscriptional regulation in plants (Filichkin et al., 2015) and one that cannot easily be studied through gene or transcriptome sequencing alone. Very few data exist on AS in ferns, but Boothby, Zipper, van der Weele, and Wolniak (2013) identified a type of AS in the heterosporous species *Marsilea vestita* that involves retaining and subsequently removing introns from pre-mRNAs to regulate temporal patterns of translation. Whether this method of AS also occurs in other ferns is unknown. Several AS events have been found to be conserved between *Arabidopsis* and *Physcomitrella* (Iida, 2006), a tantalizing finding that would benefit immensely from expanded sampling and comparative analyses that included ferns and lycophytes. Finally, RNA editing is another posttranscriptional process that could be further investigated with data from fern and lycophyte nuclear genomes. RNA editing of organellar genomes occurs across land plants, but at much higher levels in lycophytes, ferns, and bryophytes than in seed plants (Takenaka et al., 2013). These lineages also have a unique form of RNA editing that seed plants lack, in which U-to-C editing occurs in addition to the standard C-to-U editing found in all lineages. While most research on RNA editing has focused on its role in organellar genomes, nuclear transcripts from *A. thaliana* have also been found to experience editing (Meng & Chen, 2010). It is unclear whether and to what extent this occurs in other seed plants or clades of land plants.

5. CONCLUSION

The rapid pace of whole-genome sequencing in the last decade has made abundant resources available for exploring genome evolution in plants, particularly in seed plants. Many fascinating avenues of plant evolution remain to be explored, however, for which the genomes of ferns and lycophytes are the gatekeepers. These lineages hold innumerable clues to understanding major evolutionary innovations and transitions in land plants, including the foundations of regulatory mechanisms and the ancestral structures of genomes in each of the major clades. Filling the gaps in available genomic resources by including ferns and homosporous lycophytes will allow us to address these and many other pressing questions in plant genomics and must be a fundamental goal of our community over the coming decades.

REFERENCES

Abou, Y., Fiogbé, E. D., & Micha, J.-C. (2007). A preliminary assessment of growth and production of Nile tilapia, *Oreochromis niloticus* L., fed *Azolla*-Based-Diets in earthen ponds. *Journal of Applied Aquaculture, 19*(4), 55–69.

Alpert, P. (2000). The discovery, scope, and puzzle of desiccation tolerance in plants. *Plant Ecology, 151*, 5–17.

Anderberg, H. I., Kjellbom, P., & Johanson, U. (2012). Annotation of *Selaginella moellendorffii* major intrinsic proteins and the evolution of the protein family in terrestrial plants. *Frontiers in Plant Science, 3*(33), 1–14.

Arif, M., Frank, W., & Khraiwesh, B. (2013). Role of RNA interference (RNAi) in the moss *Physcomitrella patens*. *International Journal of Molecular Sciences, 14*(1), 1516–1540.

Arrigo, N., Therrien, J., Anderson, C. L., Windham, M. D., Haufler, C. H., & Barker, M. S. (2013). A total evidence approach to understanding phylogenetic relationships and ecological diversity in *Selaginella* subg. *Tetragonostachys*. *American Journal of Botany, 100*(8), 1672–1682.

Aya, K., Kobayashi, M., Tanaka, J., Ohyanagi, H., Suzuki, T., Yano, K. ... Matsuoka, M. (2015). De Novo transcriptome assembly of a fern, *Lygodium japonicum*, and a web resource database, Ljtrans DB. *Plant and Cell Physiology, 56*(1), e5.

Bainard, J. D., Henry, T. A., Bainard, L. D., & Newmaster, S. G. (2011). DNA content variation in monilophytes and lycophytes: large genomes that are not endopolyploid. *Chromosome Research, 19*(6), 763–775.

Banks, J. A. (1994). Sex-determining genes in the homosporous fern *Ceratopteris*. *Development, 120*, 1949–1958.

Banks, J. A. (1997). The *TRANSFORMER* genes of the fern *Ceratopteris* simultaneously promote meristem and archegonia development and repress antheridia development in the developing gametophyte. *Genetics, 147*, 1885–1897.

Banks, J. A. (2009). *Selaginella* and 400 million years of separation. *Annual Review of Plant Biology, 60*(1), 223–238.

Banks, J. A., Nishiyama, T., Hasebe, M., Bowman, J. L., Gribskov, M., dePamphilis, C. W. ... Grigoriev, I. V. (2011). The *Selaginella* genome identifies genetic changes associated with the evolution of vascular plants. *Science, 332*(6032), 960–963.

Barker, M. S. (2009). Evolutionary genomic analyses of ferns reveal that high chromosome numbers are a product of high retention and fewer rounds of polyploidy relative to angiosperms. *American Fern Journal, 99*(2), 136–141.

Barker, M. S. (2012). Karyotype and genome evolution in pteridophytes. In I. J. Leitch (Ed.), *Plant genome diversity* (Vol. 2, pp. 245–253). Vienna: Springer.

Barker, M. S., & Wolf, P. G. (2010). Unfurling fern biology in the genomics age. *BioScience, 60*(3), 177–185.

Basak, B., Pramanik, M., Rahman, M. S., & Tarafdar, S. U. (2002). *Azolla (Azolla pinnata)* as a feed ingredient in broiler ration. *International Journal of Poultry Science, 1*(123), 29–34.

Bateman, R. M., & DiMichele, W. A. (1994). Heterospory: the most iterative key innovation in the evolutionary history of the plant kingdom. *Biological Reviews, 69*(3), 345–417.

Bennett, M. D., & Leitch, I. J. (2012). *Plant DNA C-values database (release 6.0, December 2012)*. http://www.kew.org/cvalues/ Accessed 15.08.15.

Bennicelli, R., Stępniewska, Z., Banach, A., Szajnocha, K., & Ostrowski, J. (2004). The ability of *Azolla caroliniana* to remove heavy metals (Hg(II), Cr(III), Cr(VI)) from municipal waste water. *Chemosphere, 55*(1), 141–146.

Bewley, J. D. (1979). Physiological aspects of desiccation tolerance. *Annual Review of Plant Physiology, 30*, 195–238.

Bomfleur, B., McLoughlin, S., & Vajda, V. (2014). Fossilized nuclei and chromosomes reveal 180 million years of genomic stasis in royal ferns. *Science, 343*(6177), 1376–1377.

Boothby, T. C., Zipper, R. S., van der Weele, C. M., & Wolniak, S. M. (2013). Removal of retained introns regulates translation in the rapidly developing gametophyte of *Marsilea vestita*. *Developmental Cell, 24*(5), 517–529.

Brinkhuis, H., Schouten, S., Collinson, M. E., Sluijs, A., Damsté, J. S. S., Dickens, G. R. ... The Expedition 302 Scientists. (2006). Episodic fresh surface waters in the Eocene Arctic Ocean. *Nature, 441*(7093), 606–609.

Brouwer, P., Bräutigam, A., Külahoglu, C., Tazelaar, A. O. E., Kurz, S., Nierop, K. G. J. … Schluepmann, H. (2014). *Azolla* domestication towards a biobased economy? *New Phytologist, 202*(3), 1069–1082.

Brouwer, P., van der Werf, A., Schluepmann, H., Reichart, G.-J., & Nierop, K. G. J. (2015). Lipid yield and composition of *Azolla filiculoides* and the implications for biodiesel production. *BioEnergy Research*, 1–9.

Cao, J.-G., Wang, Q.-X., & Bao, W.-M. (2010). Formation of the fertilization pore during oogenesis of the fern *Ceratopteris thalictroides. Journal of Integrative Plant Biology, 52*(6), 518–527.

Cao, J.-G., Wang, Q.-X., Yang, N.-Y., & Bao, W.-M. (2010). Cytological events during zygote formation of the fern *Ceratopteris thalictroides. Journal of Integrative Plant Biology, 52*(3), 254–264.

Cao, J.-G., Yang, N.-Y., & Wang, Q.-X. (2009). Ultrastructure of the mature egg and fertilization in the fern *Ceratopteris thalictroides. Journal of Integrative Plant Biology, 51*(3), 243–250.

Carey, R. E., Hepler, N. K., & Cosgrove, D. J. (2013). *Selaginella moellendorffii* has a reduced and highly conserved expansin superfamily with genes more closely related to angiosperms than to bryophytes. *BMC Plant Biology, 13*, 4.

Carroll, E. W., Schwarz, O. J., & Hickok, L. G. (1988). Biochemical studies of paraquat-tolerant mutants of the fern *Ceratopteris richardii. Plant Physiology, 98*, 651–654.

Chan, A. P., Melake-Berhan, A., O'Brien, K., Buckley, S., Quan, H., Chen, D. … Rabinowicz, P. D. (2008). The highest-copy repeats are methylated in the small genome of the early divergent vascular plant *Selaginella moellendorffii. BMC Genomics, 9*(1), 282–287.

Chapman, R. H., Klekowski, E. J., & Selander's, R. K. (1979). Homoeologous heterozygosity and recombination in the fern *Pteridium aquilinum. Science, 15*(4398), 1207–1209.

Cherryl, D. M., Prasad, R., & Jayalaxmi, P. (2013). A study on economics of inclusion of *Azolla pinnata* in swine rations. *International Journal of Agricultural Sciences and Veterinary Medicine, 1*(4), 50–56.

Cordle, A. R., Irish, E. E., & Cheng, C.-L. (2012). Gene expression associated with apogamy commitment in *Ceratopteris richardii. Sexual Plant Reproduction, 25*(4), 293–304.

Der, J. P., Barker, M. S., Wickett, N. J., dePamphilis, C. W., & Wolf, P. G. (2011). De novo characterization of the gametophyte transcriptome in bracken fern, *Pteridium aquilinum. BMC Genomics, 12*(1), 99.

DeYoung, B., Weber, T., Hass, B., & Banks, J. A. (1997). Generating autotetraploid sporophytes and their use in analyzing mutations affecting gametophyte development in the fern *Ceratopteris. Genetics, 147*, 809–814.

Donnelly, E. (2003). *Potential uses of bracken (*Pteridium aquilinum *(L.) Kuhn) in organic agriculture in Scotland.* Aberdeen, Scotland: University of Aberdeen.

Döpp, W. (1950). Eine die Antheridienbildung bei Faren forderne Substanz in den Prothallien von *Pteridium aquilinum* (L.) Kuhn. *Berichte De Deutschen Botanischen Gesellschaft, 63*, 139–147.

Eberle, J., & Banks, J. A. (1986). Genetic interactions among sex-determining genes in the fern *Ceratopteris richardii. Genetics, 142*, 973–985.

Eberle, J., Nemacheck, J., Wen, C.-K., Hasebe, M., & Banks, J. A. (1995). *Ceratopteris*: a model system for studying sex-determining mechanisms in plants. *International Journal of Plant Sciences, 156*(3), 359–366.

Edwards, E. S., & Roux, S. J. (1998). Influence of gravity and light on the developmental polarity of *Ceratopteris richardii* fern spores. *Planta, 205*, 553–560.

Eyini, M., Sujanandini, K., Pothiraj, C., Jayakumar, M., & Kil, B. S. (1999). Differential response of *Azolla microphylla* kaulf. and *Azolla filiculoides* Lam. to sodium fluoride. *Journal of Plant Biology, 42*(4), 299–301.

Filichkin, S., Priest, H. D., Megraw, M., & Mockler, T. C. (2015). Alternative splicing in plants: directing traffic at the crossroads of adaptation and environmental stress. *Current Opinion in Plant Biology, 24*, 125–135.

Forni, C., Chen, J., Tancioni, L., & Grilli Caiola, M. (2001). Evaluation of the fern Azolla for growth, nitrogen and phosphorus removal from wastewater. *Water Research, 35*(6), 1592–1598.

Francis, P. C., & Petersen, R. L. (1983). Effect of copper, cadmium, and zinc on percent spore germination of the cinnamon fern (*Osmunda cinnamomea*) and the sensitive fern (*Onoclea sensibilis*). *Bulletin of Environmental Contamination and Toxicology, 30*, 559–566.

Gao, L., Wang, B., Wang, Z. W., Zhou, Y., Su, Y. J., & Wang, T. (2013). Plastome sequences of *Lygodium japonicum* and *Marsilea crenata* reveal the genome organization transformation from basal ferns to core leptosporangiates. *Genome Biology and Evolution, 5*(7), 1403–1407.

Gao, L., Yi, X., Yang, Y.-X., Su, Y.-J., & Wang, T. (2009). Complete chloroplast genome sequence of a tree fern *Alsophila spinulosa*: insights into evolutionary changes in fern chloroplast genomes. *BMC Evolutionary Biology, 9*(1), 130.

Gastony, G. J., & Darrow, D. C. (1983). Chloroplastic and cytosolic isozymes of the homosporous fern *Athyrium filix-femina* L. *American Journal of Botany, 70*(9), 1409–1415.

Gastony, G. J., & Gottlieb, L. D. (1982). Evidence for genetic heterozygosity in a homosporous fern. *American Journal of Botany, 69*(4), 634–637.

Gastony, G. J., & Gottlieb, L. D. (1985). Genetic variation in the homosporous fern *Pellaea andromedifolia*. *American Journal of Botany, 72*(2), 257–267.

Gramzow, L., Barker, E., Schulz, C., Ambrose, B. A., Ashton, N., Theissen, G., & Litt, A. (2012). *Selaginella* genome analysis – entering the "homoplasy heaven" of the MADS world. *Frontiers in Plant Science, 3*(214), 1–14.

Grant, V. (1981). *Plant speciation* (2nd ed.). New York: Columbia University Press.

Grewe, F. (May 26, 2011). *Die mitochondriale DNA basaler Tracheophyten: Molekulare evolution komplexer Genomstrukturen*. Rheinischen Friedrich-Wilhelms-Universität Bonn.

Grewe, F., Guo, W., Gubbels, E. A., Hansen, A. K., & Mower, J. P. (2013). Complete plastid genomes from *Ophioglossum californicum, Psilotum nudum*, and *Equisetum hyemale* reveal an ancestral land plant genome structure and resolve the position of Equisetales among monilophytes. *BMC Evolutionary Biology, 13*(1), 1.

Grewe, F., Herres, S., Viehover, P., Polsakiewicz, M., Weisshaar, B., & Knoop, V. (2010). A unique transcriptome: 1782 positions of RNA editing alter 1406 codon identities in mitochondrial mRNAs of the lycophyte *Isoetes engelmannii*. *Nucleic Acids Research*, 1–13.

Grewe, F., Viehoever, P., Weisshaar, B., & Knoop, V. (2009). A trans-splicing group I intron and tRNA-hyperediting in the mitochondrial genome of the lycophyte *Isoetes engelmannii*. *Nucleic Acids Research, 37*(15), 5093–5104.

Gumaelius, L., Lahner, B., Salt, D. E., & Banks, J. A. (2004). Arsenic hyperaccumulation in gametophytes of *Pteris vittata*. A new model system for analysis of arsenic hyperaccumulation. *Plant Physiology, 136*(2), 3198–3208.

Hamilton, R. G., & Lloyd, R. M. (1991). Antheridiogen in the wild: the development of fern gametophyte communities. *Functional Ecology, 5*(6), 804–809.

Hasebe, M., & Iwatsuki, K. (1990). Chloroplast DNA from *Adian- tum capillus-veneris* L., a fern species (Adiantaceae); clone bank, physical map and unusual gene localization in comparison with angiosperm chloroplast DNA. *Current Genetics, 17*, 359–364.

Hasebe, M., & Iwatsuki, K. (1992). Gene localization on the chloroplast DNA of the maiden hair fern: *Adiantum capillus-veneris*. *Botanical Magazine Tokyo, 105*, 413–419.

Haufler, C. H. (1985). Enzyme variability and modes of evolution in *Bommeria* (Pteridaceae). *Systematic Botany, 10*(1), 92–104.

Haufler, C. H. (1987). Electrophoresis is modifying our concepts of evolution in homosporous pteridophytes. *American Journal of Botany, 74*(6), 953–966.

Haufler, C. H., & Soltis, D. E. (1984). Obligate outcrossing in a homosporous fern: field confirmation of a laboratory prediction. *American Journal of Botany, 71*(6), 878–881.

Haufler, C. H., & Soltis, D. E. (1986). Genetic evidence suggests that homosporous ferns with high chromosome numbers are diploid. *Proceedings of the National Academy of Sciences, 83*, 4389–4393.

Hickok, L. G. (1978). Homoeologous chromosome pairing and restricted segregation in the fern *Ceratopteris*. *American Journal of Botany, 65*(5), 516–521.

Holbrook-Walker, S. G., & Lloyd, R. M. (1973). Reproductive biology and gametophyte morphology of the Hawaiian fern genus *Sadleria* (Blechnaceae) relative to habitat diversity and propensity for colonization. *Botanical Journal of the Linnean Society, 67*, 157–174.

Holsinger, K. E. (1987). Gametophytic self-fertilization in homosporous plants: development, evaluation, and application of a statistical method for evaluating its importance. *American Journal of Botany, 74*(8), 1173–1183.

Ibarra-Laclette, E., Lyons, E., Hernández-Guzmán, G., Pérez-Torres, C. A., Carretero-Paulet, L., Chang, T.-H. … Herrera-Estrella, L. (2013). Architecture and evolution of a minute plant genome. *Nature, 498*, 94–98.

Iida, K. (2006). Survey of conserved alternative splicing events of mRNAs encoding SR proteins in land plants. *Molecular Biology and Evolution, 23*(5), 1085–1094.

Indriolo, E., Na, G., Ellis, D., Salt, D. E., & Banks, J. A. (2010). A vacuolar arsenite transporter necessary for arsenic tolerance in the arsenic hyperaccumulating fern *Pteris vittata* is missing in flowering plants. *The Plant Cell, 22*(6), 2045–2057.

Jiao, Y., Wickett, N. J., Ayyampalayam, S., Chanderbali, A. S., Landherr, L., Ralph, P. E. … dePamphilis, C. W. (2011). Ancestral polyploidy in seed plants and angiosperms. *Nature, 473*, 97–100.

Jobson, R. W., & Qiu, Y.-L. (2008). Did RNA editing in plant organellar genomes originate under natural selection or through genetic drift? *Biology Direct, 3*(1), 22–43.

Johnson, G. P., & Renzaglia, K. S. (2008). Embryology of *Ceratopteris richardii* (Pteridaceae, tribe Ceratopterideae), with emphasis on placental development. *Journal of Plant Research, 121*(6), 581–592.

Kamachi, H., Komori, I., Tamura, H., Sawa, Y., Karahara, I., Honma, Y. … Inoue, H. (2005). Lead tolerance and accumulation in the gametophytes of the fern *Athyrium yokoscense*. *Journal of Plant Research, 118*(2), 137–145.

Kamachi, H., Matsunaga, E., Noguchi, M., & Inoue, H. (2004). Novel mutant phenotypes of a dark-germinating mutant *dkg1* in the fern *Ceratopteris richardii*. *Journal of Plant Research, 117*(2), 163–170.

Kanegae, T., Hayashida, E., Kuramoto, C., & Wada, M. (2006). A single chromoprotein with triple chromophores acts as both a phytochrome and a phototropin. *Proceedings of the National Academy of Sciences, 103*(47), 17997–18001.

Karol, K. G., Arumuganathan, K., Boore, J. L., Duffy, A. M., Everett, K. D., Hall, J. D. … Wolf, P. G. (2010). Complete plastome sequences of *Equisetum arvense* and *Isoetes flaccida*: implications for phylogeny and plastid genome evolution of early land plant lineages. *BMC Evolutionary Biology, 10*(1), 321.

Kawai, H., Kanegae, T., Christensen, S., Kiyosue, T., Sato, Y., Imaizumi, T. … Wada, M. (2003). Responses of ferns to red light are mediated by an unconventional photoreceptor. *Nature, 421*(6920), 287–290.

Kawai-Toyooka, H., Kuramoto, C., Orui, K., Motoyama, K., Kikuchi, K., Kanegae, T., & Wada, M. (2004). DNA interference: a simple and efficient gene-silencing system for high-throughput functional analysis in the fern *Adiantum*. *Plant and Cell Physiology, 45*(11), 1648–1657.

Kenrick, P., & Crane, P. R. (1997). The origin and early evolution of plants on land. *Nature, 389*, 33–39.

Kertulis-Tartar, G. M., Ma, L. Q., Tu, C., & Chirenje, T. (2006). Phytoremediation of an arsenic-contaminated site using *Pteris vittata* L.: a two-year study. *International Journal of Phytoremediation, 8*(4), 311–322.

Khandelwal, S. (1990). Chromosome evolution in the genus *Ophioglossum* L. *Botanical Journal of the Linnean Society, 102*, 205–217.

Kim, H. T., Chung, M. G., & Kim, K.-J. (2014). Chloroplast genome evolution in early diverged leptosporangiate ferns. *Molecules and Cells, 37*(5), 372–382.

Klekowski, E. J. (1969). Reproductive biology of the Pteridophyta III. A study of the Blechnaceae. *Botanical Journal of the Linnean Society, 62*, 361–377.

Klekowski, E. J. (1973). Sexual and subsexual systems in homosporous pteridophytes: a new hypothesis. *American Journal of Botany, 60*(6), 535–544.

Klekowski, E. J., & Baker, H. G. (1966). Evolutionary significance of polyploidy in the Pteridophyta. *Science, 153*, 305–307.

Klekowski, E. J., & Lloyd, R. M. (1968). Reproductive biology of the Pteridophyta I. General considerations and a study of *Onoclea sensibilis* L. *Botanical Journal of the Linnean Society, 60*(383), 315–324.

Kooh, M. R. R., Lim, L. B. L., Lim, L. H., & Dahri, M. K. (2016). Separation of toxic rhodamine B from aqueous solution using an efficient low-cost material, *Azolla pinnata*, by adsorption method. *Environmental Monitoring and Assessment, 188*(2), 108–115.

Korall, P., Schuettpelz, E., & Pryer, K. M. (2010). Abrupt deceleration of molecular evolution linked to the origin of arborescence in ferns. *Evolution*, 1–19.

Kuo, L.-Y., Li, F.-W., Chiou, W.-L., & Wang, C.-N. (2011). First insights into fern matK phylogeny. *Molecular Phylogenetics and Evolution, 59*(3), 556–566.

Leitch, A. R., & Leitch, I. J. (2012). Ecological and genetic factors linked to contrasting genome dynamics in seed plants. *New Phytologist, 194*(3), 629–646.

Leitch, I. J., & Bennett, M. D. (2004). Genome downsizing in polyploid plants. *Biological Journal of the Linnean Society, 82*, 651–663.

Lessl, J. T., Guan, D.-X., Sessa, E. B., Rathinasabapathi, B., & Ma, L. Q. (2015). Transfer of arsenic and phosphorus from soils to the fronds and spores of arsenic hyperaccumulator *Pteris vittata* and non arsenic-hyperaccumulators. *Plant and Soil, 390*, 49–60.

Li, F.-W., Melkonian, M., Rothfels, C. J., Villarreal, J. C., Stevenson, D. W., Graham, S. W.... Mathews, S. (2015). Phytochrome diversity in green plants and the origin of canonical plant phytochromes. *Nature Communications, 6*, 7852.

Li, F.-W., & Pryer, K. M. (2014). Crowdfunding the *Azolla* fern genome project: a grassroots approach. *GigaScience, 3*, 16.

Li, F.-W., Rothfels, C. J., Melkonian, M., Villarreal, J. C., Stevenson, D. W., Graham, S. W.... Pryer, K. M. (2015). The origin and evolution of phototropins. *Frontiers in Plant Science, 6*, 1–11.

Li, F.-W., Villarreal, J. C., Kelly, S., Rothfels, C. J., Melkonian, M., Frangedakis, E.... Pryer, K. M. (2014). Horizontal transfer of an adaptive chimeric photoreceptor from bryophytes to ferns. *Proceedings of the National Academy of Sciences, 111*(18), 6672–6677.

Li, Z., Baniaga, A., Sessa, E. B., Scascitelli, M., Graham, S. W., Rieseberg, L. H., & Barker, M. S. (2015). Early genome duplications in conifers and other seed plants. *Science Advances, 1*(10), e1501084.

Lippman, Z., Gendrel, A.-V., Black, M., Vaughn, M. W., Dedhia, N., McCombie, W. R.... Martienssen, R. (2004). Role of transposable elements in heterochromatin and epigenetic control. *Nature, 430*(6998), 471–476.

Lloyd, R. M. (1973). Facultative apomixis and polyploidy in *Matteuccia orientalis*. *American Fern Journal, 63*(2), 43–48.

Lopez-Smith, R., & Renzaglia, K. (2008). Sperm cell architecture, insemination, and fertilization in the model fern, *Ceratopteris richardii*. *Sexual Plant Reproduction, 21*(3), 153–167.

Love, A., Love, D., & Pichi-Sermolli, R. E. G. (1977). *Cytotaxonomical atlas of the pteridophyta.* Vaduz, Liechtenstein: J. Cramer.
Lu, J.-M., Zhang, N., Du, X.-Y., Wen, J., & Li, D.-Z. (2015). Chloroplast phylogenomics resolves key relationships in ferns. *Journal of Systematics and Evolution, 53*(5), 448–457.
Lumpkin, T. A., & Plucknett, D. L. (1980). Azolla: botany, physiology, and use as a green manure. *Economic Botany, 34*(2), 111–153.
Ma, L. Q., Komar, K. M., Tu, C., Zhang, W., Cai, Y., & Kennelley, E. D. (2001). A fern that hyperaccumulates arsenic. *Nature, 409*, 579.
Malek, O., Lättig, K., Hiesel, R., Brennicke, A., & Knoop, V. (1996). RNA editing in bryophytes and a molecular phylogeny of land plants. *The EMBO Journal, 15*(6), 1403–1411.
Manna, A. B., & Singh, P. K. (1989). Rice yields as influenced by Azolla N_2 fixation and urea N-fertilization. *Plant and Soil, 114*(1), 63–68.
Manton, I. (1950). *Problems of cytology and evolution in the pteridophyta.* Cambridge: Cambridge University Press.
Martienssen, R. (1998). Transposons, DNA methylation and gene control. *Trends in Genetics, 14*(7), 263–264.
Matasci, N., Hung, L.-H., Yan, Z., Carpenter, E. J., Wickett, N. J., Mirarab, S. ... Wong, G. K. S. (2014). Data access for the 1000 Plants (1KP) project. *GigaScience, 3*(1), 1–10.
McGrath, J. M., & Hickok, L. G. (1999). Multiple ribosomal RNA gene loci in the genome of the homosporous fern *Ceratopteris richardii. Canadian Journal of Botany, 77*, 1199–1202.
McGrath, J. M., Hickok, L. G., & Pichersky, E. (1994). Assessment of gene copy number in the homosporous ferns *Ceratopteris thalictroides* and *C. richardii* (Parkeriaceae) by restriction fragment length polymorphisms. *Plant Systematics and Evolution, 189*, 203–210.
McGrath, J. M., & Pichersky, E. (1997). Methylation of somatic and sperm DNA in the homosporous fern *Ceratopteris richardii. Plant Molecular Biology, 35*, 1023–1027.
Meng, Y., & Chen, M. (2010). RNA editing of nuclear transcripts in *Arabidopsis thaliana. BMC Genomics, 11*(Suppl. 4), S12.
Metzgar, J. S., Schneider, H., & Pryer, K. M. (2007). Phylogeny and divergence time estimates for the fern genus Azolla (Salviniaceae). *International Journal of Plant Sciences, 168*(7), 1045–1053.
Michael, T. P. (2014). Plant genome size variation: bloating and purging DNA. *Briefings in Functional Genomics, 13*(4), 308–317.
Muradov, N., Taha, M., Miranda, A. F., Kadali, K., Gujar, A., Rochfort, S. ... Mouradov, A. (2014). Dual application of duckweed and azolla plants for wastewater treatment and renewable fuels and petrochemicals production. *Biotechnology for Biofuels, 7*, 30.
Murata, T., & Sugai, M. (2000). Photoregulation of asymmetric cell division followed by rhizoid development in the fern *Ceratopteris* prothalli. *Plant and Cell Physiology, 41*(12), 1313–1320.
Muthukumar, B., Joyce, B. L., Elless, M. P., & Stewart, C. N., Jr. (2013). Stable transformation of ferns using spores as targets: *Pteris vittata* (Chinese brake fern) and *Ceratopteris thalictroides* (C-fern 'Express'). *Plant Physiology, 163*(2), 648–658.
Näf, U., Nakanishi, K., & Endo, M. (1975). On the physiology and chemistry of fern antheridiogens. *The Botanical Review, 41*(3), 315–359.
Nakazato, T., Barker, M. S., Rieseberg, L. H., & Gastony, G. J. (2008). Evolution of the nuclear genome of ferns and lycophytes. In T. A. Ranker, & C. H. Haufler (Eds.), *Biology and evolution of ferns and lycophytes.* Cambridge, UK: Cambridge University Press.
Nakazato, T., Jung, M. K., Housworth, E. A., Rieseberg, L. H., & Gastony, G. J. (2006). Genetic map-based analysis of genome structure in the homosporous fern *Ceratopteris richardii. Genetics, 173*, 1585–1597.

Nishizono, H., Suzuki, S., & Ishii, F. (1987). Accumulation of heavy metals in the metal-tolerant fern, *Athyrium yokoscense*, growing on various environments. *Plant and Soil, 102*, 65–70.

Nystedt, B., Street, N. R., Wetterbom, A., Zuccolo, A., Lin, Y.-C., Scofield, D. G.… Jansson, S. (2013). The Norway spruce genome sequence and conifer genome evolution. *Nature, 497*(7451), 579–584.

Pabby, A., Prasanna, R., & Singh, P. K. (2004). Biological significance of *Azolla* and its utilization in agriculture. *Proceedings of the Indian National Science Academy, B70*(3), 299–333.

Palmer, J., Soltis, D. E., & Soltis, P. S. (1992). Large size and complex structure of mitochondrial DNA in two nonflowering land plants. *Current Genetics, 21*, 125–129.

Panarese, S., Rainaldi, G., De Benedetto, C., & Gallerani, R. (2008). Sequencing of a segment of a monilophyte species mitochondrial genome reveals features highly similar to those of seed plant mtDNAs. *The Open Plant Science Journal, 2*, 15–20.

Peters, G. A., & Mayne, B. C. (1974). The Azolla, anabaena azollae relationship: I. Initial characterization of the association. *Plant Physiology, 53*(6), 813–819.

Peters, G. A., & Meeks, J. C. (1989). The *Azolla-Anabaena* symbiosis: basic biology. *Annual Review of Plant Biology, 40.*

Pichersky, E., Soltis, D. E., & Soltis, P. S. (1990). Defective chlorophyll a/b-binding protein genes in the genome of a homosporous fern. *Proceedings of the National Academy of Sciences, 87*, 195–199.

Pittermann, J., Brodersen, C., & Watkins, J. E., Jr. (2013). The physiological resilience of fern sporophytes and gametophytes: advances in water relations offer new insights into an old lineage. *Frontiers in Plant Science, 4*, 1–10.

Plackett, A. R. G., Huang, L., Sanders, H. L., & Langdale, J. A. (2014). High-efficiency stable transformation of the model fern species *Ceratopteris richardii* via microparticle bombardment. *Plant Physiology, 165*(1), 3–14.

Plackett, A. R. G., Rabbinowitsch, E. H., & Langdale, J. A. (2015). Protocol: genetic transformation of the fern *Ceratopteris richardii* through microparticle bombardment. *Plant Methods, 11*(1), 1–10.

Plazinski, J., Zheng, Q., Taylor, R., Croft, L., Rolfe, B. G., & Gunning, B. E. (1990). DNA probes show genetic variation in cyanobacterial symbionts of the azolla fern and a closer relationship to free-living nostoc strains than to free-living anabaena strains. *Applied and Environmental Microbiology, 56*(5), 1263–1270.

Pryer, K. M., Schneider, H., Smith, A. R., Cranfill, R., Wolf, P. G., Hunt, J. S., & Sipes, S. D. (2001). Horsetails and ferns are a monophyletic group and the closest living relatives to seed plants. *Nature, 409*, 618–622.

Pryer, K. M., Schneider, H., Zimmer, E. A., & Banks, J. A. (2002). Deciding among green plants for whole genome studies. *Trends in Plant Science,* 7(12), 550–554.

Pryer, K. M., Schuettpelz, E., Wolf, P. G., Schneider, H., Smith, A. R., & Cranfill, R. (2004). Phylogeny and evolution of ferns (Monilophytes) with a focus on the early leptosporangiate divergences. *American Journal of Botany, 91*(10), 1582–1598.

Rabinowicz, P. D., Citek, R., Budiman, M. A., Nunberg, A., Bedell, J. A., Lakey, N.… Martienssen, R. A. (2005). Differential methylation of genes and repeats in land plants. *Genome Research, 15*, 1431–1440.

Rabinowicz, P. D., Palmer, L. E., May, B. P., Hemann, M. T., Lowe, S. W., McCombie, W. R., & Martienssen, R. A. (2003). Genes and transposons are differentially methylated in plants, but not in mammals. *Genome Research, 13*, 2658–2664.

Rai, H. S., & Graham, S. W. (2010). Utility of a large, multigene plastid data set in inferring higher-order relationships in ferns and relatives (monilophytes). *American Journal of Botany, 97*(9), 1444–1456.

Raubeson, L. A., & Jansen, R. K. (1992). Chloroplast DNA evidence on the ancient evolutionary split in vascular land plants. *Science, 255*(5052), 1697–1699.

Rice, A., Glick, L., Abadi, S., Einhorn, M., Kopelman, N., Salman-Minkov, A.... Mayrose, I. (2014). The Chromosome Counts Database (CCDB) – a community resource of plant chromosome numbers. *New Phytologist, 206*, 19–26.
Roper, J. M., Hansen, S. K., Wolf, P. G., Karol, K. G., Mandoli, D. F., Everett, K. D. E.... Boore, J. L. (2007). The complete plastid genome sequence of *Angiopteris evecta. American Fern Journal, 97*(2), 95–106.
Rothfels, C. J., Li, F.-W., Sigel, E. M., Huiet, L., Larsson, A., Burge, D. O.... Pryer, K. M. (2015). The evolutionary history of ferns inferred from 25 low-copy nuclear genes. *American Journal of Botany, 102*(7), 1089–1107.
Rutherford, G., Tanurdzic, M., Hasebe, M., & Banks, J. (2004). A systemic gene silencing method suitable for high throughput, reverse genetic analyses of gene function in fern gametophytes. *BMC Plant Biology, 4*(1), 6–10.
Salehzadeh, A., Naeemi, A. S., & Arasteh, A. (2014). Biodiesel production from *Azolla filiculoides* (Water fern). *Tropical Journal of Pharmaceutical Research, 13*(6), 957–960.
Salido, A. L., Hasty, K. L., Lim, J.-M., & Butcher, D. J. (2003). Phytoremediation of arsenic and lead in contaminated soil using Chinese brake ferns (*Pteris vittata*) and Indian mustard (*Brassica juncea*). *International Journal of Phytoremediation, 5*(2), 89–103.
Salmi, M. L., Bushart, T. J., & Roux, S. J. (2011). Autonomous gravity perception and responses of single plant cells. *Gravitational and Space Biology, 25*(1), 6–13.
Salmi, M. L., Bushart, T. J., Stout, S. C., & Roux, S. J. (2005). Profile and analysis of gene expression changes during early development in germinating spores of *Ceratopteris richardii. Plant Physiology, 138*(3), 1734–1745.
Salmi, M. L., & Roux, S. J. (2008). Gene expression changes induced by space flight in single-cells of the fern *Ceratopteris richardii. Planta, 229*(1), 151–159.
Schledbauer, M. D., Cave, C. F., & Bell, P. R. (1973). The incorporation of DL-[3-^{14}C] cysteine during spermatogenesis in *Ceratopteris thalictroides. Journal of Cell Science, 12*, 765–779.
Schneider, H., Liu, H., Clark, J., Hidalgo, O., Pellicer, J., Zhang, S.... Leitch, I. J. (2015). Are the genomes of royal ferns really frozen in time? Evidence for coinciding genome stability and limited evolvability in the royal ferns. *New Phytologist, 207*(1), 10–13.
Schneller, J., Haufler, C. H., & Ranker, T. (1990). Antheridiogen and natural gametophyte populations. *American Fern Journal, 80*(4), 143–152.
Scott, R. J., & Hickok, L. G. (1987). Genetic analysis of antheridiogen sensitivity in *Ceratopteris richardii. American Journal of Botany, 74*(12), 1872–1877.
Sessa, E. B., Banks, J. A., Barker, M. S., Der, J. P., Duffy, A. M., Graham, S. W.... Wolf, P. G. (2014). Between two fern genomes. *GigaScience, 3*, 15.
Shakirov, E. V., & Shippen, D. E. (2012). *Selaginella moellendorffii* telomeres: conserved and unique features in an ancient land plant lineage. *Frontiers in Plant Science, 3*(161), 1–7.
Singh, N., & Ma, L. Q. (2006). Arsenic speciation, and arsenic and phosphate distribution in arsenic hyperaccumulator *Pteris vittata* L. and non-hyperaccumulator *Pteris ensiformis* L. *Environmental Pollution, 141*(2), 238–246.
Smith, A. R., Pryer, K., Schuettpelz, E., Korall, P., Schneider, H., & Wolf, P. (2006). A classification for extant ferns. *Taxon, 55*(3), 705–731.
Smith, A. R., Pryer, K. M., Schuettpelz, E., Korall, P., Schneider, H., & Wolf, P. G. (2008). Fern classification. In T. A. Ranker, & C. H. Haufler (Eds.), *Biology and evolution of ferns and lycophytes* (pp. 417–462). Cambridge.
Smith, D. R. (2009). Unparalleled GC content in the plastid DNA of Selaginella. *Plant Molecular Biology, 71*(6), 627–639. http://dx.doi.org/10.1007/s11103-009-9545-3.
Soltis, D. E. (1986). Genetic evidence for diploidy in *Equisetum. American Journal of Botany, 73*(6), 908–913.
Soltis, D. E., & Soltis, P. S. (1986). Electrophoretic evidence for inbreeding in the fern *Botrychium virginianum* (Ophioglossaceae). *American Journal of Botany, 73*(4), 588–592.

Soltis, D. E., & Soltis, P. S. (1987). Breeding system of the fern *Dryopteris expansa*: evidence for mixed mating. *American Journal of Botany, 74*(4), 504–509.
Speelman, E. N., van Kempen, M. M. L., Barke, J., Brinkhuis, H., Reichart, G. J., Smolders, A. J. P. ... Sinninghe Damsté, J. S. (2009). The Eocene Arctic Azolla bloom: environmental conditions, productivity and carbon drawdown. *Geobiology, 7*(2), 155–170.
Spiro, M. D., Torabi, B., & Cornell, C. N. (2004). Cytokinins induce photomorphogenic development in dark-grown gametophytes of *Ceratopteris richardii*. *Plant and Cell Physiology, 45*(9), 1252–1260.
Stebbins, G. L. (1951). Review: problems of cytology and evolution in the pteridophyta. *Science, 113*(2940), 533–535.
Stout, S. C., Clark, G. B., Archer-Evans, S., & Roux, S. J. (2003). Rapid and efficient suppression of gene expression in a single-cell model system, *Ceratopteris richardii*. *Plant Physiology, 131*(3), 1165–1168.
Strain, E., Hass, B., & Banks, J. A. (2001). Characterization of mutations that feminize gametophytes of the fern *Ceratopteris*. *Genetics, 159*, 1271–1281.
Szövényi, P., Rensing, S. A., Lang, D., Wray, G. A., & Shaw, A. J. (2011). Generation-biased gene expression in a bryophyte model system. *Molecular Biology and Evolution, 28*(1), 803–812.
Takenaka, M., Zehrmann, A., Verbitskiy, D., Härtel, B., & Brennicke, A. (2013). RNA editing in plants and its evolution. *Annual Review of Genetics, 47*(1), 335–352.
The Arabidopsis Genome Initiative. (2000). Analysis of the genome sequence of the flowering plant *Arabidopsis thaliana*. *Nature, 408*, 796–815.
Tryon, R. M., & Vitale, G. (1977). Evidence for antheridiogen production and its mediation of a mating system in natural populations of fern gametophytes. *Botanical Journal of the Linnean Society, 74*, 243–249.
Tsuboi, H., Sutoh, K., & Wada, M. (2014). Epigenetic memory of DNAi associated with cytosine methylation and histone modification in fern. *Plant Signaling and Behavior, 7*(11), 1477–1483.
Tsuji, S., Ueda, K., Nishiyama, T., Hasebe, M., Yoshikawa, S., Konagaya, A. ... Yamaguchi, K. (2007). The chloroplast genome from a lycophyte (microphyllophyte), *Selaginella uncinata*, has a unique inversion, transpositions and many gene losses. *Journal of Plant Research, 120*(2), 281–290.
Tu, S., Ma, L. Q., Fayiga, A. O., & Zillioux, E. J. (2004). Phytoremediation of arsenic-contaminated groundwater by the arsenic hyperaccumulating fern *Pteris vittata* L. *International Journal of Phytoremediation, 6*(1), 35–47.
Vasco, A., Moran, R. C., & Ambrose, B. A. (2013). The evolution, morphology, and development of fern leaves. *Frontiers in Plant Science, 4*, 345.
Ventura, W., Mascarina, G. B., Furoc, R. E., & Watanabe, I. (1987). Azolla and Sesbania as biofertilizers for lowland rice. *Philippine Journal of Crop Science, 12*(2), 61–69.
Vermaat, J. E., & Khalid Hanif, M. (1998). Performance of common duckweed species (Lemnaceae) and the waterfern *Azolla filiculoides* on different types of waste water. *Water Research, 32*(9), 2569–2576.
Vries, J., Fischer, A. M., Roettger, M., Rommel, S., Schluepmann, H., Bräutigam, A. ... Gould, S. B. (2016). Cytokinin-induced promotion of root meristem size in the fern *Azolla* supports a shoot-like origin of euphyllophyte roots. *New Phytologist, 209*(2), 705–720.
Wagner, G. M. (1997). Azolla: a review of its biology and utilization. *The Botanical Review, 63*(1), 1–26.
Wagner, W. H., & Wagner, F. S. (1980). Polyploidy in pteridophytes. In W. H. Lewis (Ed.), *Polyploidy biological relevance* (pp. 199–214). New York.
Wang, W., Tanurdzic, M., Luo, M., Sisneros, N., Kim, H. R., Weng, J. K. ... Banks, J. A. (2005). Construction of a bacterial artificial chromosome library from the spikemoss

Selaginella moellendorffii: a new resource for plant comparative genomics. *BMC Plant Biology, 5*(10), 1–8.

Warne, T. R., Hickok, L. G., Kinraide, T. B., & Vogelien, D. L. (1996). High salinity tolerance in the *stl2* mutation of *Ceratopteris richardii* is associated with enhanced K^+ influx and loss. *Plant, Cell and Environment, 19*, 24–32.

Warne, T. R., Hickok, L. G., Sams, C. E., & Vogelien, D. L. (1999). Sodium/potassium selectivity and pleiotropy in *stl2*, a highly salt-tolerant mutation of *Ceratopteris richardii*. *Plant, Cell and Environment, 22*, 1027–1034.

Watkins, J. E., Jr., & Cardelús, C. L. (2012). Ferns in an angiosperm world: cretaceous radiation into the epiphytic niche and diversification on the forest floor. *International Journal of Plant Sciences, 173*(6), 695–710.

Watkins, J. E., Jr., Mack, M., Sinclair, T., & Mulkey, S. (2007). Ecological and evolutionary consequences of desiccation tolerance in fern gametophytes. *New Phytologist, 176*(3), 708–717.

Weng, J.-K., & Noel, J. P. (2013). Chemodiversity in *Selaginella*: a reference system for parallel and convergent metabolic evolution in terrestrial plants. *Frontiers in Plant Science, 4*(119), 1–17.

Werth, C. R., Guttman, S. I., & Eshbaugh, W. H. (1985). Electrophoretic evidence of reticulate evolution in the Appalachian *Asplenium* complex. *Systematic Botany, 10*(2), 184–192.

Wickett, N. J., Mirarab, S., Nguyen, N., Warnow, T., Carpenter, E. J., Matasci, N. ... Leebens-Mack, J. (2014). A phylotranscriptomic analysis of the origin and early diversification of land plants. *Proceedings of the National Academy of Sciences*, E4859–E4868.

Wilkie, A. C., & Evans, J. M. (2010). Aquatic plants: an opportunity feedstock in the age of bioenergy. *Biofuels, 1*(2), 311–321.

Wipf, D., Loque, D., Lalonde, S., & Frommer, W. B. (2012). Amino acid transporter inventory of the *Selaginella* genome. *Frontiers in Plant Science, 3*(36), 1–6.

Wolf, P. G., Der, J. P., Duffy, A. M., Davidson, J. B., Grusz, A. L., & Pryer, K. M. (2011). The evolution of chloroplast genes and genomes in ferns. *Plant Molecular Biology, 76*(3–5), 251–261.

Wolf, P. G., Duffy, A. M., & Roper, J. M. (2009). Phylogenetic use of inversions in fern chloroplast genomes. *American Fern Journal, 99*, 132–134.

Wolf, P. G., Haufler, C. H., & Sheffield, E. (1987). Electrophoretic evidence for genetic diploidy in the bracken fern (*Pteridium aquilinum*). *Science, 236*(4804), 947–949.

Wolf, P. G., Karol, K. G., Mandoli, D. F., Kuehl, J., Arumuganathan, K., Ellis, M. W. ... Boore, J. L. (2005). The first complete chloroplast genome sequence of a lycophyte, *Huperzia lucidula* (Lycopodiaceae). *Gene, 350*(2), 117–128.

Wolf, P. G., Roper, J. M., & Duffy, A. M. (2010). The evolution of chloroplast genome structure in ferns. *Genome, 53*(9), 731–738.

Wolf, P. G., Rowe, C. A., Sinclair, R. B., & Hasebe, M. (2003). Complete nucleotide sequence of the chloroplast genome from a leptosporangiate fern, *Adiantum capillus-veneris* L. *DNA Research, 10*, 59–65.

Wolf, P. G., Sessa, E. B., Marchant, D. B., Li, F.-W., Rothfels, C. J., Sigel, E. M. ... Der, J. P. (2015). An exploration into fern genome space. *Genome Biology and Evolution, 7*(9), 2533–2544.

Zazouli, M. A., Mahdavi, Y., Bazrafshan, E., & Balarak, D. (2014). Phytodegradation potential of bisphenolA from aqueous solution by *Azolla filiculoides*. *Journal of Environmental Health Science and Engineering, 12*(1), 66.

Zhao, M., Duncan, J. R., & Van Hille, R. P. (1999). Removal and recovery of zinc from solution and electroplating effluent using *Azolla filiculoides*. *Water Research, 33*(6), 1516–1522.

Zhong, B., Fong, R., Collins, L. J., Mclenachan, P. A., & Penny, D. (2014). Two new fern chloroplasts and decelerated evolution linked to the long generation time in tree ferns. *Genome Biology and Evolution, 6*(5), 1166–1173.

Zhou, X.-M., Rothfels, C. J., Zhang, L., He, Z.-R., Le Péchon, T., He, H. … Zhang, L. B. (2015). A large-scale phylogeny of the lycophyte genus *Selaginella* (Selaginellaceae: Lycopodiopsida) based on plastid and nuclear loci. *Cladistics*, 1–30.

Zhu, A., Guo, W., Gupta, S., Fan, W., & Mower, J. P. (2015). Evolutionary dynamics of the plastid inverted repeat: the effects of expansion, contraction, and loss on substitution rates. *New Phytologist, 209*(4), 1747–1756.

CHAPTER EIGHT

The Genomes of Charophyte Green Algae

C.F. Delwiche
University of Maryland, College Park, MD, United States
E-mail: delwiche@umd.edu

Contents

Abstract

The charophyte green algae (sometimes called basal streptophytes) belong to the lineage that gave rise to land plants (embryophytes), but are distinguished by lacking a number of features and characteristics of land plants, most notably the land plant life cycle. Individually, many charophyte algae have been used as systems to study plant biology, but the realization that these organisms belong together as an evolutionary lineage dates only to ultrastructural studies in the 1970s, and a more detailed understanding of their relationships to each other and to land plants is still emerging. Currently recognized members of the group include *Mesostigma*, *Chlorokybus*, Klebsormidiophyceae, Charophyceae *sensu stricto*, Coleochaetophyceae and Zygnematophyceae (Conjugatophyceae). Contrary to earlier molecular systematic studies, Zygnematophyceae appear as the sister to land plants in current multigene phylogenetic analyses. Early studies reported chromosome numbers and morphologies, and identified polyploid series in the Zygnematophyceae and Charophyceae *s. str.*, with less detailed chromosome counts in the other groups. Molecular studies using PCR and related methods to study individual genes and processes have been accelerated by high-throughput sequencing, particularly of transcriptomes. Both chloroplast and mitochondrial organellar genomes are available for representatives of each major subgroup, but the only near-complete nuclear genome currently available is that of *Klebsormidium*. There are several genome sequencing projects underway, and expansion of genomic data available should occur soon.

Advances in Botanical Research, Volume 78
ISSN 0065-2296
http://dx.doi.org/10.1016/bs.abr.2016.02.002

1. INTRODUCTION

The charophyte green algae (also called 'basal streptophytes') hold a unique phylogenetic position, constituting the basal branches of the lineage that includes land plants (embryophytes) (Becker & Marin, 2009). Land plants are a monophyletic group characterized by a life cycle consisting of an alternation of haploid and diploid generations (and consequently a multicellular diploid phase, hence the name embryophyte), along with complex multicellular tissue and a variety of adaptations to life on land (Delwiche & Cooper, 2015). Together with land plants, the charophyte green algae form a monophyletic group that is sister to a second major lineage of green algae, the chlorophyte green algae (chlorophytes may or may not be monophyletic) (Leliaert et al., 2012). The divergence between these two great lineages of green algae probably occurred well over a billion years ago (Fig. 1), making them extremely ancient, even compared with all land plants. Evidence from the fossil record documents green algae to about 850 Ma (Taylor, Taylor, & Krings, 2009), but molecular clock analyses consistently show a greater age for the group (Yoon, Hackett, Ciniglia, Pinto, & Bhattacharya, 2004). The green algae are thought to be the sister taxon of the red algae;

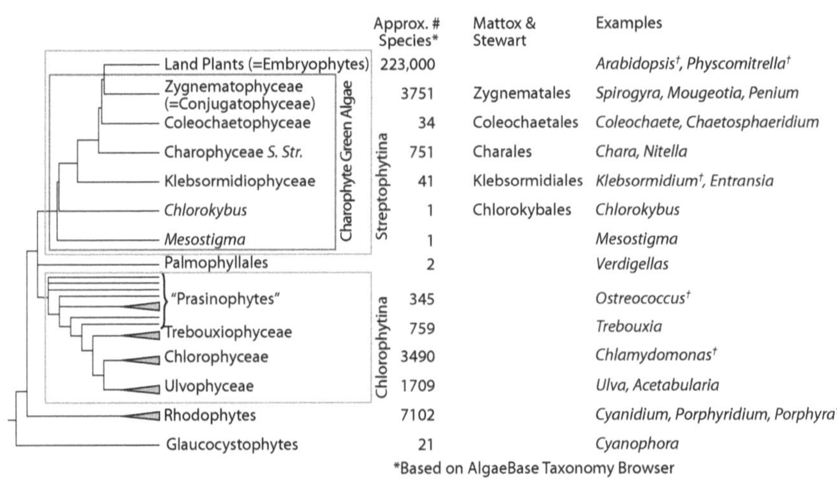

Figure 1 Phylogeny and classification of green algae, with an emphasis on the charophyte green algae. Estimates of the number of species are given based on the number of species listed in AlgaeBase (Guiry, 2012; Guiry et al., 2014). These should be viewed as very approximate, but give a sense of the relative diversity of the different lineages. The column labelled Mattox and Stewart (Mattox & Stewart, 1984) refers to the nomenclature used in that article. Species marked with a *dagger* (†) have had a complete, or nearly complete, nuclear genome sequence published.

assuming this is correct, the two groups are of the same age. *Bangiomorpha*, a superbly preserved fossil dates to ~1.2 Ga (Butterfield, 2000), and is easily recognized as a morphologically advanced red alga, implying that the group had undergone considerable diversification and specialization prior to its deposition. Consequently, the green algae as a whole are almost certainly at least 1.2 billion years old, and may well be much older.

The age of the charophyte lineage is difficult to estimate with certainty, but its phylogenetic structure, with several deep-branching, structurally simple lineages (*Mesostigma*, *Chlorokybus* and the Klebsormidiophyceae) and several phylogenetically derived and structurally complex lineages (Charophyceae *sensu stricto*, Coleochaetophyceae and the Zygnematophyceae) that all branch in phylogenetic trees at or below the land plant clade suggests that each of these lineages is at least as old as land plants. Land plants can be dated by microfossil evidence to the late Ordovician, around 460 Ma, and by the Late Silurian or Early Devonian, around 420 Ma, there was a diversifying terrestrial flora composed primarily of land plants. Molecular clock data also support the view that the charophyte lineages are ancient clades (Hedges, Dudley, & Kumar, 2006; Yoon et al., 2004). The fossil record for charophyte algae is generally sparse, with the exception being the Charophyceae *s. str.*, ie, *Chara*, *Nitella* and their relatives, which have a generous fossil record consisting primarily of gyrogonites (preserved zygote walls) and limestone formed by species that deposit calcium carbonate in their cell walls (hence the name 'stoneworts') (Feist, Liu, & Tafforeau, 2005; Tappan, 1980). A spectacular specimen of *Paleonitella* from the Rhynie Chert unambiguously dates the lineage at a minimum of 410 Ma, and they may be much older (Kelman, Feist, Trewin, & Hass, 2004). The fossil record also shows that the modern organisms in the lineage are remnants of formerly much greater biodiversity (Tappan, 1980). Fossil records of the other charophyte algal lineages are too sparse for such inferences, but the overall picture that emerges is that each of these lineages, classically treated as orders, families or even monotypic species, is a remnant lineage of great age.

There is a great deal of promise in the study of charophyte algae for improving the understanding of the origin, diversification and fundamental properties of land plants, but it is also necessary to interpret their study in the context of evolutionary history and the relationships among the organisms. There is a concept in evolutionary biology of 'heterobathmy' (the product of mosaic evolution), meaning that each organism is a mixture of ancestral ('primitive') and derived ('advanced') characters (Nelson, 2004; Schlosser, 2002). Lineages evolving independently will acquire a separate mix of derived characters, although they will share ancestral characters inherited from their common ancestor (unless, of course, those characters have undergone evolutionary change within either lineage). As a consequence, the presence of an ancestral character in an organism does not predict that other features of that organism will also be in their ancestral state. Thus, to understand the ancestral properties of a group it is necessary to study the characteristics of that group and as many of its close relatives as is practical. Taken together, the comparison of characters makes it possible to distinguish the ancestor from the derived. This is particularly important in the study of the origin of plants because of the relatively deep divergences and low diversity found in the charophyte green algae. The characteristics of any one of these organisms may not seem directly relevant to the evolution of land plants, but taken together they can provide substantial insights.

My intent here is to review the current state of knowledge of genomes and genome biology within the charophyte green algae. Genomics has come late to the charophytes, presumably because none is of great economic importance in and of itself and because they are thought to have moderately large and complex genomes. Advances in technology mean that eukaryotic genome sequencing is now feasible on relatively modest budgets and small groups of researchers, and the time for charophyte genomics is at hand.

2. OVERVIEW OF THE CHAROPHYTES

Mesostigma viride Lauterborn is a scaly, unicellular flagellate with two flagella asymmetrically inserted on one side of the cell. It is surrounded by three layers of scales, the largest of which have a spectacularly complex structure (Domozych, Wells, & Shaw, 1991) and make a layer that, when viewed with a light microscope, can be confused with a thin cell wall. The ultrastructure is reminiscent of charophyte algae and plants, which inspired Melkonian (1983, 1989) to propose that it is on the charophyte lineage. Sexual reproduction is not well described in English literature, but the Microbial Culture Collection at the National Institute for Environmental Studies in Japan (NIES) identifies strains NIES-475 and NIES-476 as heterothallic and + or − strains (respectively). Although there is some uncertainty concerning its precise placement, molecular phylogenetic studies generally support *Mesostigma* as belonging close to the base of the charophytes (Bhattacharya, Weber, An Seon, & Berning-Koch, 1998; Cocquyt, Verbruggen, Leliaert, & De Clerck, 2010). *Mesostigma* has no known close relatives and is generally classified as a monotypic taxon at or above the ordinal level (Leliaert et al., 2012), although some molecular phylogenetic analyses have placed it with *Chlorokybus*, another deep-branching charophyte, or more rarely with other taxa (Lemieux, Otis, & Turmel, 2007; Marin & Melkonian, 1999; Turmel, Otis, & Lemieux, 2007).

Chlorokybus atmophyticus Geitler, like *Mesostigma*, represents a deep-branching lineage of charophytes and has no known close relatives. It is composed of packets of cells embedded in a polysaccharide matrix and seems to primarily inhabit wet soils, although it will also grow readily in liquid culture. It seems to be rather rare in nature, but it is easy to grow and is available in culture. The zoospores are coated by a single layer of small square scales and have the lateral flagella and multilayered structure that are characteristics of charophytes. Sex seems to be completely unknown. Its placement among the charophytes is also consistently supported by molecular phylogenetic analyses (Turmel et al., 2007). Like *Mesostigma*, *Chlorokybus* is normally treated taxonomically as a monotypic, high-level taxon (Leliaert et al., 2012).

The Klebsormidiophyceae have around 40 named species (Guiry, 2012; Guiry et al., 2014), although some of these may be synonyms. An excellent monograph is available for the genus *Klebsormidium* (Lokhorst, 1996), which is widespread and common, found on wet soil, rocks and concrete around the world. A second genus, *Entransia*, which is characterized largely by its distinctive chloroplast morphology (Cook, 2004b) but is also distinct in molecular phylogenies, appears to be the sister taxon of *Klebsormidium*. Both genera are unbranched filaments. Species of *Klebsormidium* can break apart into characteristic fragments called 'hormogonia' through a process of programmed cell death, and some species are so fragmented that they appear to be almost unicellular. A third genus, *Interfilum*, appears to be a lineage within *Klebsormidium* that has largely lost its filamentous character (Rindi, Mikhailyuk, Sluiman, Friedl, & Lopez-Bautista, 2011). Many species of *Klebsormidium* can be induced to produce zoospores, which are released through apertures in the size of the cell (Lokhorst, 1996; Sanchez-Puerta & Leonard, 2001), but sex remains uncharacterized. The first (and at the time of this writing, only) extensive nuclear genome sequencing among charophytes was of *Klebsormidium flaccidum* (Kützing) P.C. Silva, K.R. Mattox and W.H. Blackwell (Hori et al., 2014).

The Charophyceae *sensu stricto*, the 'stoneworts', are the largest and the most complex charophyte green algae (excepting, of course, the land plants). There are about 750 named species (Guiry et al., 2014), but as noted above, they have a rich fossil record and had much greater diversity in the past. Important genera include *Chara, Nitella* and *Tolypella*. Although some specimens are minute, more typically individual shoots range from centimetres to metres in length, and clonal reproduction can allow a single thallus to occupy a large area. Sexual reproduction is oogamous (with eggs and sperm), but they do not produce zoospores, and the sperms are the only flagellate stage known. Charophyceae have a nodal structure that is superficially similar to that of vascular land plants, but differs dramatically in its details; internodes consist of a single cell, rather than multicellular tissue. The giant cells, which can be 10 cm or more in length, have made these organisms important in cell biology (Tazawa, 2011).

The Coleochaetophyceae are microscopic algae that rival the Charophyceae *s. str.* in structural complexity. There are about 34 named species (Guiry et al., 2014), but there is evidence that some of these are comprised of several cryptic species (Cimino & Delwiche, 2002) and the true diversity is likely substantially greater. There are two common genera, *Coleochaete* and *Chaetosphaeridium*, as well as several rare or poorly known genera, including the enigmatic *Awadhiella* (Delwiche, Karol, Cimino, & Sytsma, 2002; Nandan Prasad & Kumar Asthana, 1979). All members of the Coleochaetophyceae are comprised of branched filaments, although this structure is sometimes difficult to perceive in *Chaetosphaeridium*. Best known are the discoidal thalli of *Coleochaete scutata* and *Coleochaete orbicularis*, but most species have separate filaments and in many the structure is more three dimensional. Thalli always produce at least one sheathed hair, a complex structure, characteristic of the group, that consists of an outgrowth of the cell wall and extension of the chloroplast into the surrounding medium (McBride, 1974). Reproduction is oogamous, but badly in need of further study in *Chaetosphaeridium* (Thompson, 1969), and asexual reproduction by formation of zoospores occurs in most species (but seemingly not in *Coleochaete conchata*) (Delwiche et al., 2002; Möbius 1892).

The Zygnematophyceae (Conjugatophyceae) are the most species-rich group in the lineage and include both filamentous and unicellular species. They are abundant and diverse, and are ecologically important organisms. AlgaeBase documents 3751 named species (Guiry et al., 2014), but the literature is large and complex, and the true number of species may be quite different. Filamentous genera such as *Spirogyra, Mougeotia* and *Zygnema* can form conspicuous mats or masses of algae in lakes and streams, and the term 'pondscum' is often treated as a common name for these organisms. Unicellular Zygnematophyceae are known as 'desmids', which is a somewhat artificial terminology, but the order Desmidiales is monophyletic and is comprised either of unicells or secondary filaments (ie, chains of adherent single cells). The family Desmidiaceae itself accounts for much of the taxonomic diversity and consists of distinctive single cells that are divided into two 'semicells' with the nucleus sitting in the isthmus between them. These organisms often have elaborate cell walls and can be very visually striking. No flagellate stage is known for any member of the Zygnematophyceae, and reproduction is via 'conjugation,' a process in which the cell protoplast differentiates into a nonflagellate gamete. Although earlier studies had placed the Charales as the sister taxon to land plants, most recent, genome-scale analyses have shown the Zygnematophyceae in that position (Delwiche & Cooper, 2015; Turmel, Otis, & Lemieux, 2006; Wickett et al., 2014).

3. CYTOLOGY, CHROMOSOME COUNTS AND POLYPLOIDY

The unity of charophyte green algae is relatively recently recognized. Although land plant-like properties have long been recognized in the Charophyceae *s. str.*, Coleochaetophyceae and to a lesser extent the Zygnematophyceae, classifications through the middle of the 20th century emphasized gross morphology and did not recognize that these groups are related to each other. *The Chromosomes of Algae* (Godward, 1966) summarized the current state of knowledge of chromosome morphology among algae (including green algae) and laid the groundwork for subsequent ultrastructural studies. Subsequent study of cellular properties, pigmentation, ultrastructure, karyology and ultimately molecular systematics all helped move towards a more natural classification (Mattox & Stewart, 1984; Okelly & Floyd, 1983; Pickett-Heaps, 1975), but because these studies were themselves conducted in the context of an earlier and less natural classification, the data are sometimes confused or obscure.

Following the advent of electron microscopy, study of the patterns of cell division in green algae revealed that *Coleochaete*, *Chara* and *Nitella* undergo plant-like phragmoplastic cell division, with a somewhat different form of phragmoplastic division in filamentous members of the Zygnematophyceae including *Spirogyra* and *Zygnema* (Cook, 2004a; Pickett-Heaps, 1975). Taken together, a number of ultrastructural features, including two laterally inserted flagella with a single multilayered structure in the flagellar root and a persistent interzonal spindle following mitosis (Mattox & Stewart, 1984), emphasized the unity of charophycean green algae, which had previously been classified in disparate groups largely on the basis of their gross morphology, and also provided strong evidence for the close relationship between these algae and land plants (Waters, 2003).

Nuclear DNA content (c-values) measured by DAPI-fluorescence microscopy have been reported recently for diverse charophytes and other algae by Kapraun (2005, 2007).

Members of the Charophyceae *s. str.* have been studied fairly extensively with both classical and modern cytological methods (Casanova, 2015; Proctor, 1980). Chromosome counts vary within the group and within morphological species. Within morphological species, there are often chromosome counts that are simple multiples of a base number, implying polyploidy. Chromosome numbers are thought to correlate with a monoecious (eggs and sperms produced by a single thallus; higher numbers) or dioecious (eggs and sperms produced by two separate thalli; lower numbers) habit, but this pattern is not altogether consistent, with dioecious and monoecious strains sometimes having the same chromosome number, and strains with different chromosome numbers sometimes sharing the same breeding system (Casanova, 2015). Nonetheless, there is a broad pattern of monoecy almost exclusively occurring in strains with high chromosome numbers, presumably as the result of polyploidy. Furthermore, there is evidence of speciation occurring as a result of polyploidy, with morphologically similar strains that differ only by chromosome number being unable to interbreed (Proctor, 1980).

Chromosome numbers were reported by Godward (1966) for three species of *Coleochaete*. An unnamed species of *Coleochaete* was reported by Abbas and Godward (1964) to have 'chromosome number 42, chromosome length 0.5–3 μm (Selby (1956, unpublished))', where the Selby citation is to an unpublished dissertation. Godward cites the same authors, but provides species names and gives chromosome counts for *C. scutata* as 36 or 42, *Coleochaete nitellarum* as 42 and *C. orbicularis* as 21. *Coleochaete* (like all charophyte algae) is thought to have a haplontic life history, where the only diploid cell is the zygote, so the difference between *C. orbicularis* and the other species is unlikely to represent counts on haploid versus diploid tissue. A more likely explanation would be polyploidy. No images or voucher material are presented by Godward or Abbas, so taxon identity, cell sizes, and data cannot be verified, but it is interesting that one of the major characters distinguishing *C. orbicularis* and *C. scutata* is monoecious or dioecious habit (respectively) (Cimino, Reveal, & Delwiche, 2003). Assuming the data reported by Godward are correct, the monoecious species of *Coleochaete* has the lower chromosome count, which is inconsistent with the pattern in the Charophyceae *s. str.* Nonetheless, it would be worthwhile to look for polyploid series within *Coleochaete* analogous to those reported for Charophyceae *s. str.*

Probably more because of their abundance in inland waters and relative ease of culture than their close relationship to plants, *Spirogyra*, *Mougeotia*, and to a lesser extent other charophyte algae, played an important role in the mid-20th century for the study of nuclear organization and behaviour. These organisms can be manipulated, stained and mounted for microscopy relatively quickly and easily (Hull, Hoshaw, & Wang, 1982), and there is an extensive observational literature on their cytology (Godward, 1950, 1962; Wells & Hoshaw, 1971). A series of studies using both Fuelgen and fluorescent stains examined DNA content (c-values) in *Spirogyra* and identified variation in c-value and correlated differences in cell size that have been attributed to polyploidy (Hoshaw & McCourt, 1988; Wang, Hoshaw, & McCourt, 1986).

Clonal cultures of *Spirogyra pratensis* Transeau have been reported to undergo spontaneous polyploidization, resulting in filaments that consisted of distinct classes of wider filaments, corresponding to a polyploid series (Hoshaw & McCourt, 1988) (in turn referring to an unpublished thesis by Allen (1958)). Such polyploid series have been shown in culture in several species, but more importantly, they also seem to occur in nature (Wang et al., 1986). It is not uncommon to find several otherwise morphologically similar filaments differing only in cell width within naturally occurring mats, and these seem to correlate with nuclear DNA content (McCourt, Hoshaw, & Wang, 1986), implying that they correspond to polyploid series. This phenomenon has primarily been observed in *Spirogyra*, and the extent to which such polyploid series occur in other Zygnematophyceae is unclear.

A wide range of nuclear diameters and chromosome counts have been reported for *Klebsormidium*, possibly reflecting polyploidy (Lokhorst, 1996; Sanchez-Puerta & Leonard, 2001). Thus, there is some evidence that polyploidy may be a widespread phenomenon in the charophyte green algae, although only a handful of taxa have been well studied in this regard.

4. ORGANELLAR GENOMES; MAPPING AND MAP-BASED SEQUENCING

One of the first DNA-based data types used for phylogenetic purposes was organellar genome mapping. Before sequencing of DNA had become cheap and widespread, it was possible to map organellar genomes by isolating either total DNA (which has a higher copy-count of organellar DNA) or purified plastid and mitochondrial DNAs (which can be isolated by centrifugation of the intact organelles or the DNA itself) (Palmer, 1985). Several green algae were included in these early map-based studies, including some charophytes (Coleman & Goff, 1991; Manhart, Hoshaw, & Palmer, 1990). These studies identified some important patterns, including variation in the size and content of the inverted repeat (and its absence in some species), and a broad pattern of transfer of genes from the chloroplast to the nuclear genome, with more plant-like gene content among the charophytes, thus further reinforcing the unity of charophyte green algae and their close relationship to land plants (Baldauf, Manhart, & Palmer, 1990; Qiu, 2008). Unlike other green algae, there is no evidence of RNA editing in the plastid genomes of charophyte green algae (see chapter: Coevolution of organelle RNA editing and nuclear specificity factors in early land plants (Schallenberg-Rüdinger & Knoop, 2016)).

Related techniques also permitted the sequencing of organellar genomes; map-based sequencing would rely on the development of a large clone library, mapping these clones to the organelle map and sequencing of a selected region or the entire genome. This approach was effective, but much more labour intensive than modern methods, which rely on high-throughput sequencing technologies. Early complete chloroplast genome sequences included the chlorophyte *Chlamydomonas* and the land plant *Marchantia*, but few complete genomes of charophyte green algae were published until the mid 1990s.

Major advances were made in the late 1990s and 2000s in sequencing of both chloroplast and mitochondrial organellar genomes. The *M. viride* chloroplast (Lemieux, Otis, & Turmel, 2000) and mitochondrial (Turmel, Otis, & Lemieux, 2002b) complete sequences showed a strong affinity between *Mesostigma* and other charophytes. Similarly, the *C. atmophyticus* chloroplast (Lemieux et al., 2007) and mitochondrial (Turmel et al., 2007) organellar genomes have been sequenced and support its placement close to the base of charophyte phylogeny. From the Klebsormidiophyceae, published fully sequenced organelles genomes include the mitochondrial genomes of both *Klebsormidium* sp. (SAG 51.86) and *Entransia fimbriata* (UTEX LB 2353) (Turmel, Otis, & Lemieux, 2013), along with the chloroplast genome of *Klebsormidium flaccidum* (Kützing) Silva, Mattox, and Blackwell (1972); (SAG 121.80) (Civan, Foster, Embley, Seneca, & Cox, 2014). Complete organellar genomes have been reported from the Charophyceae *s. str.* for *Chara vulgaris* Linnaeus (collected from the wild) mitochondrion and chloroplast (Turmel, Otis, & Lemieux, 2003, 2006) and the *Nitella hookeri* (collected from the wild) chloroplast (Zhong et al., 2014). In the Coleochaetophyceae, the *Chaetosphaeridium* chloroplast and mitochondrion were reported relatively early (Turmel, Otis, & Lemieux, 2002a), but the *C. orbicularis* Pringsheim (UTEX strain unspecified) chloroplast sequence was reported only relatively recently and remains only partially analysed (Zhong et al., 2014). The Zygnematophyceae, which are the most diverse clade of charophyte green algae, have available chloroplast genomes from *Staurastrum punctulatum* de Brébisson (SAG 679-1) and *Zygnema circumcarinatum* Czurda (SAG 698-1a), *Closterium bailyanum* (SAG 50.89), *Roya obtusa* (SAG 168.80) (Turmel et al., 2013) and *Spirogyra communis* (strain unspecified) (Zhong et al., 2014).

5. PCR AND CLONE-BASED GENE SEQUENCING

Multiple studies have examined charophyte green algae with region-specific methods such as PCR and clone-based sequencing. Many of these studies were focused on phylogenetic questions and tend to make use of a relatively small number of loci, including SSU rDNA (Bhattacharya, Surek, Ruesing, Damberger, & Melkonian, 1994; Chapman & Buchheim, 1991; Devereux, Loeblich, & Fox, 1990; Gontcharov & Melkonian, 2011; Huss & Kranz, 1997; Kranz et al., 1995; Manhart, 1995; Mishler et al., 1994; Ragan, Parsons, Sawa, & Straus, 1994; Sluiman & Guihal, 1999; Surek, Beemelmanns, Melkonian, & Bhattacharya, 1994; Wilcox Lee, Fuerst Paul, & Floyd Gary, 1993). Organellar genes, particularly *rbcL*, but other organellar genes as well, also played an important role in early phylogenetic analyses (Lee & Manhart, 1997; McCourt et al., 2000; McCourt, Karol, Guerlesquin, & Feist, 1996; McCourt, Karol, Kaplan, & Hoshaw, 1995; Perez, Hall, McCourt, & Karol, 2014; Qiu & Palmer, 1999; Sakayama, Nozaki, Kasaki, & Hara, 2002). Naturally, as sequencing became more efficient, combinations of several genes became increasingly important (Gontcharov & Melkonian, 2011; Hall, Karol, Mccourt, & Delwiche, 2008; Karol, McCourt, Cimino, & Delwiche, 2001; Karol, Sanders, Kasper & McCourt, 1997; Yin-Long, 2008).

Attention has also been given to the evolution of specific properties across the charophytes. Space here will not permit discussing all of these in detail, but important studies from the single-gene era included the transfer of *tufA*, the gene encoding translation elongation factor EF-Tu, from the chloroplast to the nuclear genome (Baldauf et al., 1990; Baldauf & Palmer, 1990) and the evolution of its cytosolic homologue (Noble, Rogers, &

Keeling, 2007), the evolution of Actin (An Seon, Moepps, Weber, & Bhattacharya, 1999; Bhattacharya et al., 1998) of aldolases (Jacobshagen & Schnarrenberger, 1990) and of introns and intron splicing (Lee & Manhart, 1996; Lew & Manhart, 1993). Much less information is available from the nuclear genomes, but some tantalizing data have been reported. For example, plant-like MADS-Box genes have been reported from diverse charophyte green algae (Tanabe et al., 2005).

6. NUCLEAR GENOMES

The one charophyte for which a nearly complete nuclear genome sequence is available is *K. flaccidum* (Kützing) P.C. Silva, K.R. Mattox and W.H. Blackwell, strain NIES-2285 (Hori et al., 2014). The project also included the organellar genomes, complementing existing mitochondrial (Turmel et al., 2013) and plastid (Civan et al., 2014) genome sequences from different strains. Total DNA content was estimated at 117.1 $\pm$ 21.8 Mb on the basis of fluorescence microscopy, while the 1814 assembled scaffolds covered approximately 104 Mb, meaning approximately 89% of the nuclear genome was represented. From this, the project identified 16,215 protein-coding genes (including the organellar genomes).

Very recently, high-throughput sequencing of transcriptomes (cDNA) has opened up a window on the nuclear genome of many charophytes (Becker, Feja, & Melkonian, 2001; Becker & Marin, 2009; Simon, Glockner, Felder, Melkonian, & Becker, 2006; Timme, Bachvaroff, & Delwiche, 2012; Timme & Delwiche, 2010; Wodniok et al., 2011). There have been numerous publications taking advantage of this increased data availability that address the evolution of individual systems within the lineage, including ethylene (Ju et al., 2015), PIN auxin transporters (Viaene, Delwiche, Rensing, & Friml, 2013; Viaene et al., 2014), strigolactones (Delaux et al., 2012), defective kernel 1 (Demko et al., 2014), desiccation tolerance (Holzinger et al., 2014, 2015) and many more.

It is clear that the availability of more complete data from the nuclear genomes of these key organisms will lead to an improved understanding of the origin, early evolution and ancestral function of many critical plant processes.

ACKNOWLEDGEMENTS

The editor and publisher were far more tolerant than they should have been of an exceedingly late manuscript. Supported in part by NSF grants EF0523719 (Microbial Genome Sequencing) and DEB-1036506 (Assembling the Tree of Life) to C.F.D.

REFERENCES

Abbas, A., & Godward, M. B. E. (1964). *Journal of the Linnean Society [Botany], 58*, 499–507.
Allen, M. A. (1958). *The biology of a species complex in Spirogyra* (unpublished Ph.D. dissertation) (p. 240). Bloomington: Indiana University.
An Seon, S., Moepps, B., Weber, K., & Bhattacharya, D. (1999). *Molecular Biology and Evolution, 16*, 275–285.
Baldauf, S. L., Manhart, J. R., & Palmer, J. D. (1990). *Proceedings of the National Academy of Sciences of the United States of America, 87*, 5317–5321.
Baldauf, S. L., & Palmer, J. D. (1990). *Nature, 344*, 262–265.
Becker, B., Feja, N., & Melkonian, M. (2001). *Protist, 152*, 139–147.
Becker, B., & Marin, B. (2009). *Annals of Botany, 103*, 999–1004.
Bhattacharya, D., Surek, B., Ruesing, M., Damberger, S., & Melkonian, M. (1994). *Proceedings of the National Academy of Sciences of the United States of America, 91*, 9916–9920.
Bhattacharya, D., Weber, K., An Seon, S., & Berning-Koch, W. (1998). *Journal of Molecular Evolution, 47*, 544–550.
Butterfield, N. J. (2000). *Paleobiology, 26*, 386–404.
Casanova, M. T. (2015). *Phycologia, 54*, 149–160.
Chapman, R. L., & Buchheim, M. A. (1991). *Critical Reviews in Plant Sciences, 10*, 343–368.
Cimino, M. T., & Delwiche, C. F. (2002). *Journal of Phycology, 38*, 1213–1221.
Cimino, M. T., Reveal, J. L., & Delwiche, C. F. (2003). *Taxon, 52*, 133–134.
Civan, P., Foster, P. G., Embley, M. T., Seneca, A., & Cox, C. J. (2014). *Genome Biology and Evolution, 6*, 897–911.
Cocquyt, E., Verbruggen, H., Leliaert, F., & De Clerck, O. (2010). *Molecular Biology and Evolution, 27*, 2052–2061.
Coleman, A. W., & Goff, L. J. (1991). *Journal of Phycology, 27*, 463–473.
Cook, M. E. (2004a). *American Journal of Botany, 91*, 313–320.
Cook, M. E. (2004b). *Journal of Phycology, 40*, 424–431.
Delaux, P.-M., Xie, X., Timme, R. E., Puech-Pages, V., Dunand, C., Lecompte, E.… Sejalon-Delmas, N. (2012). *New Phytologist, 195*, 857–871.
Delwiche, C. F., & Cooper, E. D. (2015). *Current Biology, 25*, R899–R910.
Delwiche, C. F., Karol, K. G., Cimino, M. T., & Sytsma, K. J. (2002). *Journal of Phycology, 38*, 394–403.
Demko, V., Perroud, P.-F., Johansen, W., Delwiche, C. F., Cooper, E. D., Remme, P.… Olsen, O.-A. (2014). *Plant Physiology, 166*, 903–913. U684.
Devereux, R., Loeblich, A. R., III, & Fox, G. E. (1990). *Journal of Molecular Evolution, 31*, 18–24.
Domozych, D. S., Wells, B., & Shaw, P. J. (1991). *Journal of Cell Science, 100*, 397–407.
Feist, M., Liu, J. Y., & Tafforeau, P. (2005). *American Journal of Botany, 92*, 1152–1160.
Godward, M. B. E. (1950). *Nature, 165*, 653.
Godward, M. B. E. (1962). *Annals of Human Genetics, 25*, 427.
Godward, M.B.E.. (1966). Edward Arnold, London, UK.
Gontcharov, A. A., & Melkonian, M. (2011). *Protist, 162*, 253–267.
Guiry, M. D. (2012). *Journal of Phycology, 48*, 1057–1063.
Guiry, M. D., Guiry, G. M., Morrison, L., Rindi, F., Valenzuela Miranda, S., Mathieson, A. C.… Garbary, D. J. (2014). *Cryptogamie Algologie, 35*, 105–115.
Hall, J., Karol, K., Mccourt, R., & Delwiche, C. (2008). *Journal of Phycology, 44*, 467–477.
Hedges, S. B., Dudley, J., & Kumar, S. (2006). *Bioinformatics, 22*, 2971–2972.
Holzinger, A., Kaplan, F., Blaas, K., Zechmann, B., Komsic-Buchmann, K., & Becker, B. (2014). *PLoS One, 9*.
Holzinger, A., Pichrtova, M., Arc, E., Stoggl, W., Kranner, I., & Becker, B. (2015). *European Journal of Phycology, 50*, 65.

Hori, K., Maruyama, F., Fujisawa, T., Togashi, T., Yamamoto, N., Seo, M.... Ohta, H. (2014). *Nature Communications, 5.*
Hoshaw, R. W., & McCourt, R. M. (1988). *Phycologia, 27*, 511–548.
Hull, H. M., Hoshaw, R. W., & Wang, J. C. (1982). *Stain Technology, 57*, 273–282.
Huss, V. A. R., & Kranz, H. D. (1997). *Plant Systematics & Evolution*, 103–114.
Jacobshagen, S., & Schnarrenberger, C. (1990). *Journal of Phycology, 26*, 312–317.
Ju, C. L., Van de Poel, B., Cooper, E. D., Thierer, J. H., Gibbons, T. R., Delwiche, C. F., & Chang, C. R. (2015). *Nature Plants, 1.*
Kapraun, D. F. (2005). Nuclear DNA content estimates in multicellular green, red and brown algae: Phylogenetic considerations. *Annals of Botany, 95*, 7–44.
Kapraun, D. F. (2007). Nuclear DNA content estimates in green algal lineages: Chlorophyta and streptophyta. *Annals of Botany, 99*, 677–701.
Karol, K. G., McCourt, R. M., Cimino, M. T., & Delwiche, C. F. (2001). *Science, 294*, 2351–2353.
Karol, K. G., Sanders, E. R., Kasper, A., & McCourt, R. M. (1997). *Phycologia, 36*, 46.
Kelman, R., Feist, M., Trewin, N. H., & Hass, H. (2004). *Transactions of the Royal Society of Edinburgh-Earth Sciences, 94*, 445–455.
Kranz, H. D., Miks, D., Siegler, M.-L., Capesius, I., Sensen, C. W., & Huss, V. A. R. (1995). *Journal of Molecular Evolution, 41*, 74–84.
Lee, J., & Manhart, J. (1996). *American Journal of Botany, 83*, 173.
Lee, J., & Manhart, J. (1997). *American Journal of Botany, 84*, 209.
Leliaert, F., Smith, D. R., Moreau, H., Herron, M. D., Verbruggen, H., Delwiche, C. F., & De Clerck, O. (2012). *Critical Reviews in Plant Sciences, 31*, 1–46.
Lemieux, C., Otis, C., & Turmel, M. (2000). *Nature, 403*, 649–652.
Lemieux, C., Otis, C., & Turmel, M. (2007). *BMC Biology, 5.*
Lew, K. A., & Manhart, J. R. (1993). *Journal of Phycology, 29*, 500–505.
Lokhorst, G. M. (1996). *Comparative taxonomic studies on the genus Klebsormidium (Charophyceae) in Europe*. Stuttgart: Gustav Fischer.
Manhart, J. R. (1995). *American Fern Journal, 85*, 182–192.
Manhart, J. R., Hoshaw, R. W., & Palmer, J. D. (1990). *Journal of Phycology, 26*, 490–494.
Marin, B., & Melkonian, M. (1999). *Protist, 150*, 399–417.
Mattox, K. R., & Stewart, K. D. (1984). In D. E. G. Irvine, & D. M. John (Eds.), *Systematics of the green algae* (pp. 29–72). London, UK: Academic Press.
McBride, G. E. (1974). *Phycologia, 13*, 271–285.
McCourt, R. M., Hoshaw, R. W., & Wang, J. C. (1986). *Journal of Phycology, 22*, 307–313.
McCourt, R. M., Karol, K. G., Bell, J., Helm-Bychowski, K. M., Grajewska, A., Wojciechowski, M. F., & Hoshaw, R. W. (2000). *Journal of Phycology, 36*, 747–758.
McCourt, R. M., Karol, K. G., Guerlesquin, M., & Feist, M. (1996). *American Journal of Botany, 83*, 125–131.
McCourt, R. M., Karol, K. G., Kaplan, S., & Hoshaw, R. W. (1995). *Journal of Phycology, 31*, 989–995.
Melkonian, M. (1983). *British Phycological Journal, 18*, 206.
Melkonian, M. (1989). *Plant Systematics and Evolution, 164*, 93–122.
Mishler, B. D., Lewis, L. A., Buchheim, M. A., Renzaglia, K. S., Garbary, D. J., Delwiche, C. F.... Chapman, R. L. (1994). *Annals of the Missouri Botanical Garden, 81*, 451–483.
Möbius. (1892). Australische Süsswasseralgen. *Flora oder Allegeine botanische Zeitung Neue Reihe, 50*, 421–459.
Nandan Prasad, B., & Kumar Asthana, D. (1979). *Hydrobiologia, 62*, 131–135.
Nelson, G. (2004). *Milestones in Systematics*, 127–147.
Noble, G. P., Rogers, M. B., & Keeling, P. J. (2007). *BMC Evolutionary Biology, 7.*

Okelly, C. J., & Floyd, G. L. (1983). *Biosystems, 16*, 227–251.
Palmer, J. D. (1985). *Annual Review of Genetics, 19*, 325–354.
Perez, W., Hall, J. D., McCourt, R. M., & Karol, K. G. (2014). *Journal of Phycology, 50*, 776–789.
Pickett-Heaps, J. D. (1975). *Green algae: Structure, reproduction, and evolution of selected genera.* Sunderland, MA: Sinauer Associates.
Proctor, V. W. (1980). *Journal of Phycology, 16*, 218–233.
Qiu, Y.-L. (2008). Phylogeny and evolution of charophytic algae and land plants. *Journal of Systematics and Evolution, 46*, 287–306.
Qiu, Y.-L., & Palmer, J. D. (1999). Phylogeny of early land plants: insights from genes and genomes. *Trends in Plant Science, 4*, 26–30.
Ragan, M. A., Parsons, T. J., Sawa, T., & Straus, N. A. (1994). *Journal of Phycology, 30*, 490–500.
Rindi, F., Mikhailyuk, T. I., Sluiman, H. J., Friedl, T., & Lopez-Bautista, J. M. (2011). *Molecular Phylogenetics and Evolution, 58*, 218–231.
Sakayama, H., Nozaki, H., Kasaki, H., & Hara, Y. (2002). *Phycologia, 41*, 397–408.
Sanchez-Puerta, M. V., & Leonard, P. I. (2001). *Darwiniana, 39*, 223–230.
Schallenberg-Rüdinger, M., & Knoop, V. (2016). Coevolution of organelle RNA editing and nuclear specificity factors in early land plants. *Advances in Botanical Research, 78*, 37–93.
Schlosser, G. (2002). *Theory in Biosciences, 121*, 1–80.
Silva, P. C., Mattox, K. R., & Blackwell, W. H., Jr. (1972). The generic name Hormidium as applied to green algae. *Taxon, 21*, 639–645.
Simon, A., Glockner, G., Felder, M., Melkonian, M., & Becker, B. (2006). *BMC Plant Biology, 6.*
Sluiman, H. J., & Guihal, C. (1999). *Journal of Phycology, 35*, 395–402.
Surek, B., Beemelmanns, U., Melkonian, M., & Bhattacharya, D. (1994). *Plant Systematics & Evolution, 191*, 171–181.
Tanabe, Y., Hasebe, M., Sekimoto, H., Nishiyama, T., Kitani, M., Henschel, K. … Ito, M. (2005). *Proceedings of the National Academy of Sciences of the United States of America, 102*, 2436–2441.
Tappan, H. (1980). *The paleobiology of plant protists.* San Francisco, CA: W. H. Freeman and Co.
Taylor, T.N., Taylor, E.L. & Krings, M.. (2009). (2nd ed.) (pp. 1252). Burlington, MA: Academic Press.
Tazawa, M. (2011). *Progress in Botany, 72*(72), 5–34.
Thompson, R. H. (1969). *Journal of Phycology, 5*, 285–290.
Timme, R. E., Bachvaroff, T. R., & Delwiche, C. F. (2012). *PLoS One,* 7, 8.
Timme, R. E., & Delwiche, C. F. (2010). *BMC Plant Biology, 10.*
Turmel, M., Otis, C., & Lemieux, C. (2002a). *Proceedings of the National Academy of Sciences of the United States of America, 99*, 11275–11280.
Turmel, M., Otis, C., & Lemieux, C. (2002b). *Molecular Biology and Evolution, 19*, 24–38.
Turmel, M., Otis, C., & Lemieux, C. (2003). *Plant Cell, 15*, 1888–1903.
Turmel, M., Otis, C., & Lemieux, C. (2006). *Molecular Biology and Evolution, 23*, 1324–1338.
Turmel, M., Otis, C., & Lemieux, C. (2007). *BMC Genomics, 8.*
Turmel, M., Otis, C., & Lemieux, C. (2013). *Genome Biology and Evolution, 5*, 1817–1835.
Viaene, T., Delwiche, C. F., Rensing, S. A., & Friml, J. (2013). *Trends in Plant Science, 18*, 5–10.
Viaene, T., Landberg, K., Thelander, M., Medvecka, E., Pederson, E., Feraru, E. … Friml, J. (2014). *Current Biology, 24*, 2786–2791.
Wang, J. C., Hoshaw, R. W., & McCourt, R. M. (1986). *Journal of Phycology, 22*, 102–107.
Waters, E. R. (2003). *Molecular Phylogenetics And Evolution, 29*, 456–463.
Wells, C. V., & Hoshaw, R. W. (1971). *Journal of Phycology,* 7, 279.

Wickett, N. J., Mirarab, S., Nguyen, N., Warnow, T., Carpenter, E., Matasci, N.… Gitzendanner, M. A. (2014). *Proceedings of the National Academy of Sciences, 111*, E4859–E4868.
Wilcox Lee, W., Fuerst Paul, A., & Floyd Gary, L. (1993). *American Journal of Botany, 80*, 1028–1033.
Wodniok, S., Brinkmann, H., Glockner, G., Heidel, A. J., Philippe, H., Melkonian, M., & Becker, B. (2011). *BMC Evolutionary Biology, 11*.
Yin-Long, Q. (2008). *Journal of Systematics and Evolution, 46*, 287–306.
Yoon, H. S., Hackett, Y. D., Ciniglia, C., Pinto, G., & Bhattacharya, D. (2004). *Molecular Biology and Evolution, 21*, 809–818.
Zhong, B. J., Xi, Z. X., Goremykin, V. V., Fong, R., McLenachan, P. A., Novis, P. M.… Penny, D. (2014). *Molecular Biology and Evolution, 31*, 177–183.

SUBJECT INDEX

'*Note:* Page numbers followed by "f" indicate figures and "t" indicate tables.'

D

E

F

R

AUTHOR INDEX

C

D

E

F

G

H

L

N

O

P

S

T

U

V

W

X

Y

Z

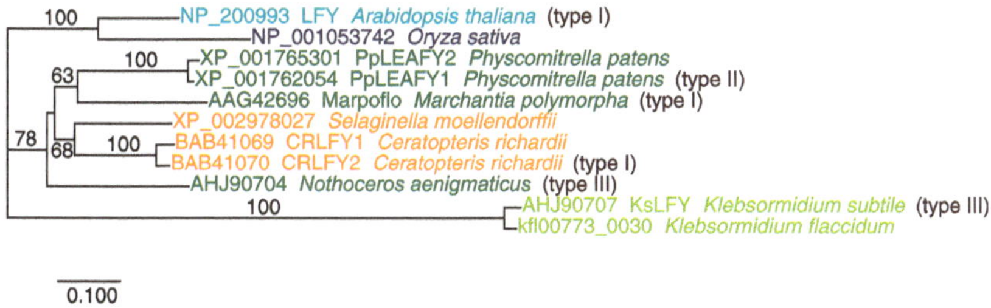

K. Sakakibara, Figure 7 A maximum likelihood phylogenetic tree of *LFY* genes constructed using 236 conserved amino acid sequences in representative green plants. Bootstrap probabilities of more than 50% are shown on the branches. The horizontal branch length is proportional to the estimated evolutionary distance. The *bar* represents 0.1 substitutions per site. The types of DNA-binding specificity are shown at the right to gene name based on Sayou et al. (2014).

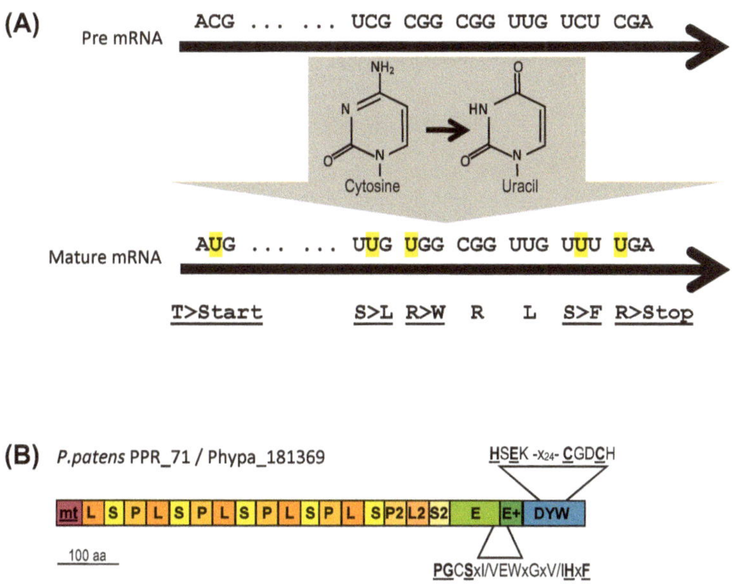

M. Schallenberg-Rüdinger and V. Knoop, Figure 1 (A) Numerous cytidines are specifically converted into uridines in plant chloroplast and mitochondrial transcripts. The C-to-U editing events mainly serve to change codon identities in the 1st and 2nd codon position to restore evolutionarily conserved amino acid identities but may also occur in silent codon positions and in introns, UTRs, rRNAs and tRNAs. (B) Pentatricopeptide Repeat (PPR) proteins are key to the specific recognition of the organelle RNA editing sites. Exemplarily shown is the structure of PPR_71 a typical "DYW-type" PPR protein crucial for editing event ccmFCeU122SF in mitochondria of the moss *Physcomitrella patens*. The DYW domain is characterized by a conserved cytidine deaminase motif. A conserved motif at the border of the E and E+ domains linking to the upstream PPR array is shown to be crucial for editing in angiosperms. The PPR arrays of editing factors are of the PLS-type featuring alternations of long (L, 35/36 aa) and short (S, 31 aa) variants of the classic P-type PPRs of 35 aa. The terminal PLS2 triplet is characterized by a deviating repeat consensus.

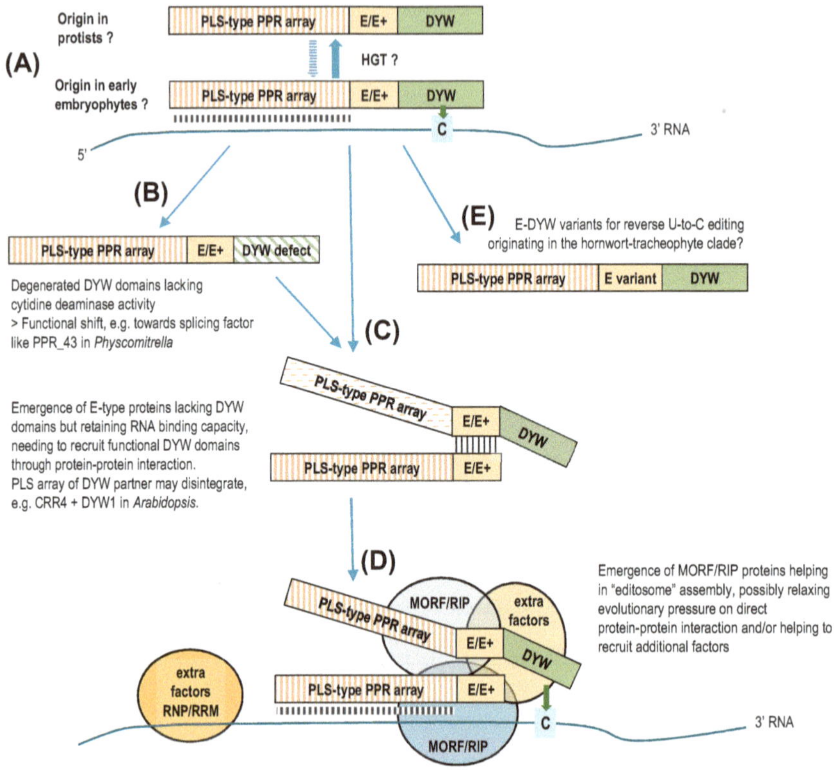

M. Schallenberg-Rüdinger and V. Knoop, Figure 2 Hypothetical scenarios for the evolution of DYW-type PPR proteins acting as editing factors as discussed in the text. (A) DYW-type PPR proteins may have originated in land plants or in protists and horizontal gene transfer (HGT) in the one or the other direction may explain their current distribution in the living world. (B) Degeneration of the cytidine deaminase functionality in the DYW domain may cause functional shifts towards intron splicing, endonucleolytic cleavage or RNA stabilization. (C) Loss of DYW domains in E-type proteins characterized as editing factors may be explained by their ability to recruit DYW domains *in trans*. (D) Other than E− or DYW-type PPR proteins, MORF/RIP proteins play a key role in angiosperm RNA editing, likely by promoting the interactions of more than one E− or DYW-type protein on their RNA target. Further protein factors (like e.g. cpRNPs, ORRMs or OZs) may contribute either by separately binding to the target transcript and making the editing site accessible or by directly interacting in a complex editosome. (E) An alteration in the terminal E/E+/DYW domains may explain the later origin of reverse U-to-C RNA editing in some land plant clades.

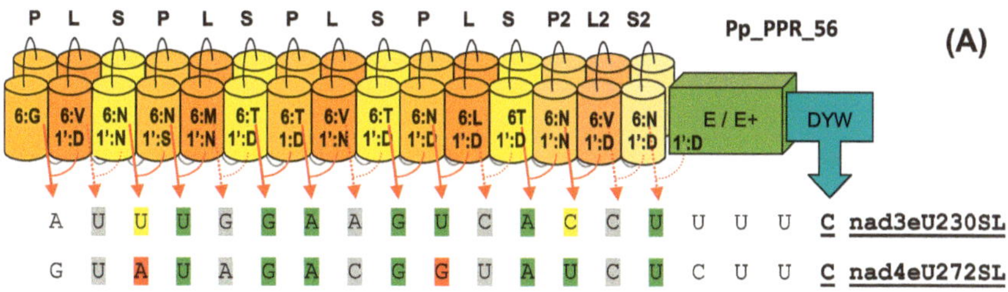

6	1′	nt
N	D	U>C
N	N	C/U
N	S	C>U
T/S	N/S	A>G
T/S	D	G>A

(B)

CLB19/ At1g05750 (C)

PLSSPLSPLS
TITNTINNTG 6
TNHNDDNTDR 1′

AGAAGCCCAAGCUC clpPeU559HY
ACACGUGCAAAAUC rpoAeU200SF
AUAUGUUCCUCCGC petBeU85HY
AAAUGCCCGGGCGC atpFeU250PC
*AUACGACCCUAAUC rpl2eU253P**
AAACGAUCCAAUUC cemAeU285FF
AUAUGGUCGAAAGC ycf2eU300SS
ACAUGUCCUGUAAC rpoC1eU341TI
AUAUGCCCUUAGGC rpl2eU408GG
ACACGACCUAUUGC ndhFeU512AV
AAACGGCCGAUUCC atpAeU693SS
AUAUGGUCUUCAAC rpoBeU2565NN

M. Schallenberg-Rüdinger and V. Knoop, Figure 3 (A) Key amino acid positions 6 and 1′ of DYW-type PPR protein PPR_56 of *Physcomitrella patens* are aligned to the upstream target sequences of their assigned RNA editing sites nad3eU230SL and nad4eU272SL with the last S repeat opposite the 4th nucleotide upstream of the editing sites (C). Green shading indicates matches of amino acids in positions 6/1′ with nucleotides according to the rules described in (B) Mismatches are shaded in red. Pyrimidines (C, U) are less well distinguished than purines (A, G). Hence, pyrimidine transition mismatches are shaded in yellow. Nucleotides at positions assigned to L motifs are shaded in grey, as their role in recognition is minor regarding the rules proposed in Barkan et al. (2012). (B) Combinations of amino acids at positions 6 and 1′ in PPR motifs that specify binding to specific bases proposed in Barkan et al. (2012). (C) Alignment of E-type PPR protein CLB19 of *Arabidopsis thaliana* with the upstream target sequences of its assigned RNA editing sites clpPeU559HY and rpoAeU200SF and 10 putative target sequences in plastid coding sequences. Shading rules are listed in (A).

D. Lang *et al.*, Figure 1 ***Habitus and natural habitat of the moss* P. patens.** Photographs of the natural habitus and habitat of *Physcomitrella* populations on a fallow acre near Freiburg, Germany discovered by Michael Lüth, that is forked by a small stream, visible as a green/red grass line in Fig. 1A. Fig. 1A–E demonstrate the habitat at different resolutions. Photographs (A–D) were taken by Daniel Lang in October 2008. *Arrows* indicate the location of a *Physcomitrella* plant with green-yellow, immature capsules. (E) was taken at the same location in 2006 by Mark von Stackelberg. The drawing in the lower right depicts the schematic habitus of an adult *Physcomitrella* gametophore with a mature sporophyte. Diploid sporophytic [2n] tissues are annotated in orange while haploid [1n] tissues are shown in green colour.

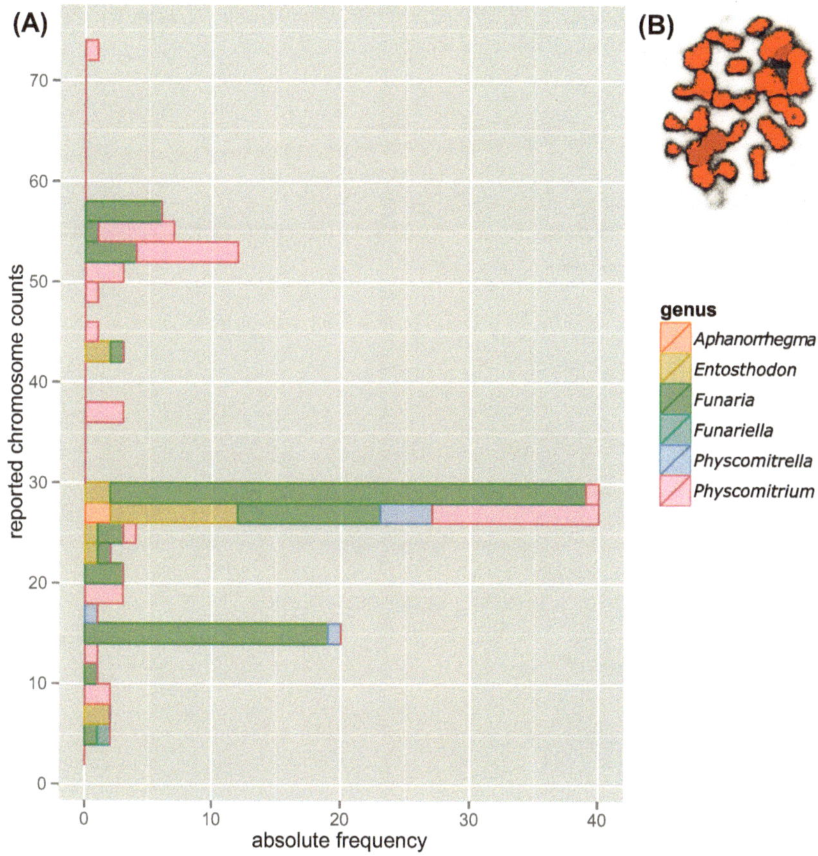

D. Lang *et al.*, Figure 2 ***Chromosome numbers of selected Funariaceae.*** Fig. 2A depicts a stacked histogram (bin width = 2) of Funariaceae chromosome counts collected from Rensing et al. (2007) and the Chromosome Count Database (CCDB; http://ccdb.tau.ac.il/Bryophytes/Funariaceae/); 158 data points for 40 species were mapped to synonyms, curated and aggregated at genus level. Fig. 2B depicts an edited digital scan of an original meiotic chromosome squash published by Reski et al. (1994). Image processing was applied to select chromosomal boundaries and to highlight them in red. Putatively overlapping chromosomes are depicted in distinct red tones.

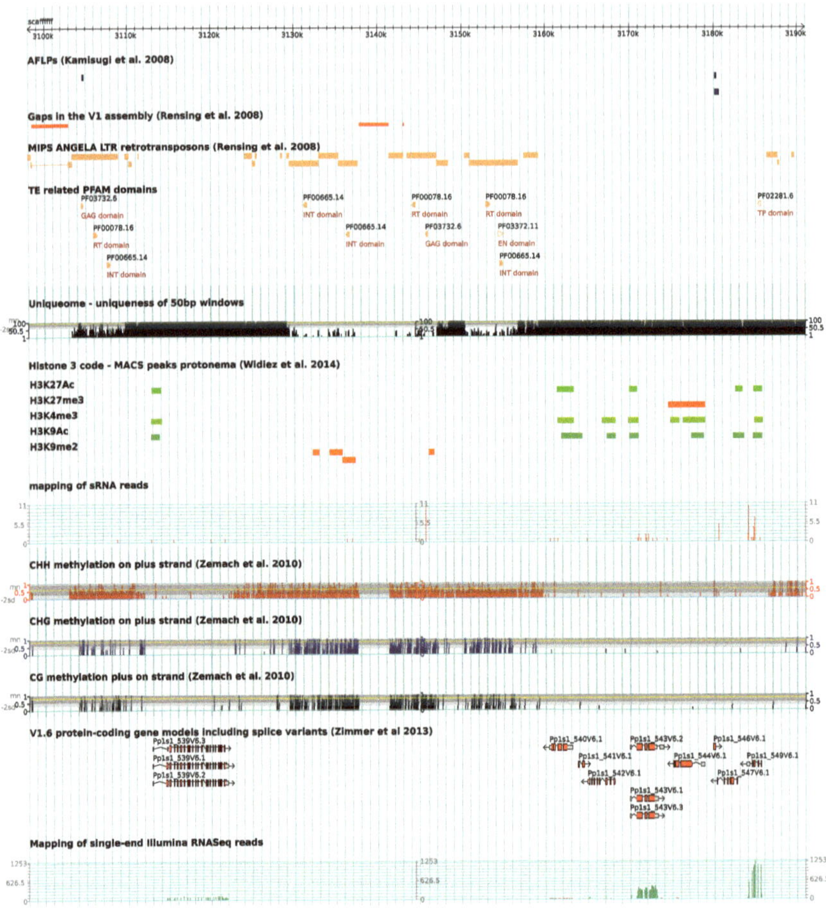

D. Lang *et al.*, Figure 3 ***Genomic region of* Physcomitrella *in the cosmoss.org genome browser.*** Snapshot of the cosmoss.org genome browser displaying available V1.6 annotation as tracks comprising chromosomally localized features. Tracks represent (from top to bottom): (1) Mappings of AFLP markers used for genetic map (blue segments). (2) Assembly gaps (red segments). (3) LTR-Rs (brown segments). (4) PFAM protein domains specific for LTR-Rs (brown segments). (5) Uniquenome: histogram of sliding-window analysis of uniqueness of 50 bp windows (black). (6) Peak regions of histone 3 ChIP-Seq data derived from protonema cultures (red = repressive mark; green = activating mark). (7) Histogram of spliced alignments of sRNA Illumina reads. (8–10) Histograms of CHH (red), CHG (blue) and CG (black) methylation on the plus strand. (11) V1.6 gene models for splice variants (red segments = coding sequence; light-grey segments = UTR). (12) Mapping of RNA-Seq reads derived by Trueseq single-end Illumina sequencing.

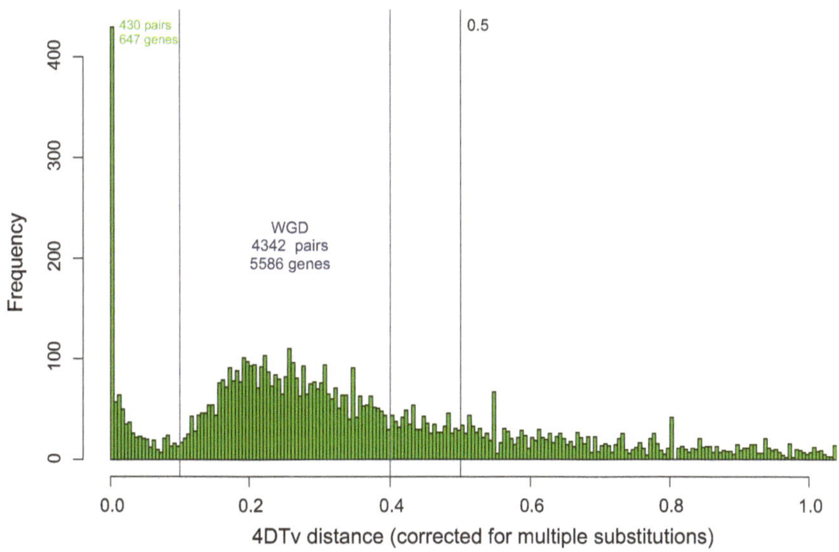

D. Lang *et al.*, Figure 6 ***The analysis of paralogs in the* Physcomitrella *genome provides evidence for at least one large-scale genome duplication event.*** Histogram of pairwise synonymous substitution rates among pairs of paralogs in the V1.2 genome measured using the 4DTv method demonstrates at least one broad peak comprising 4342 paralog pairs involving 5586 unique loci. This peak is consistent with what was observed using assembled EST data (Rensing et al., 2007). A higher, sharper peak is observed at bin 0, comprising 647 loci. Further, possibly older peaks cannot reliably be detected by this method due to the problem of multiple substitution (4DTv $>$ 0.5).

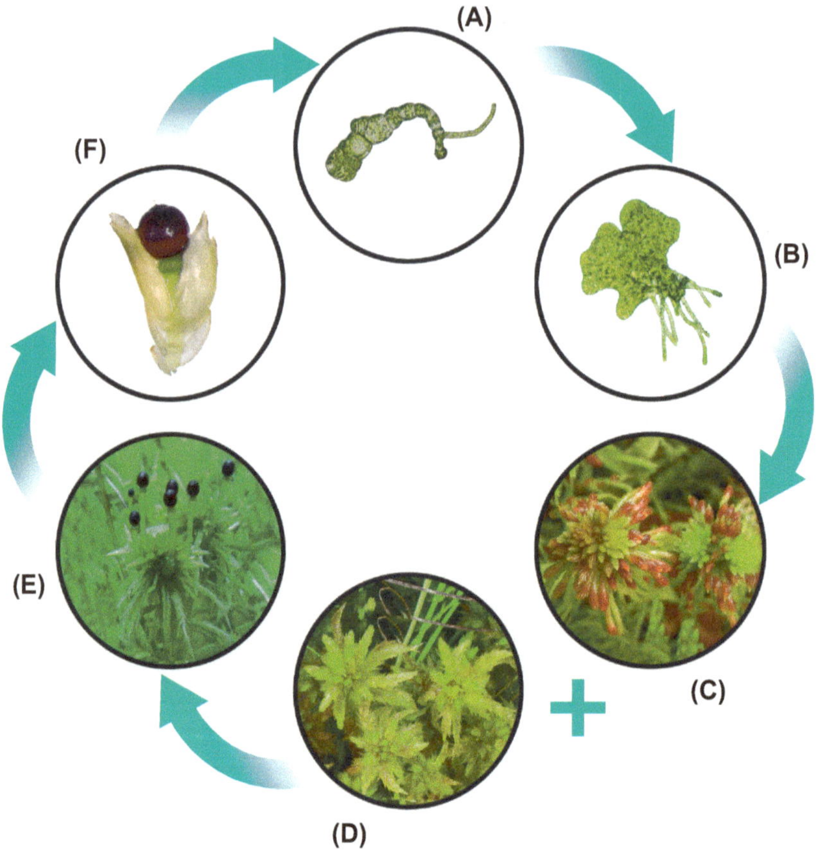

A.J. Shaw *et al.*, Figure 2 *Sphagnum* life cycle. (A) Young protonema with a single rhizoid. (B) Mature thalloid protonemata with rhizoids. (C) Male gametophore. (D) Female gametophore. (E) Female gametophyte bearing multiple sporophytes. (F) Mature sporophyte (round, brown capsule, bearing meiotically derived spores) borne on a short pseudopodium (mostly surrounded by perichaetial leaves).

A.J. Shaw *et al.*, Figure 3 Representative cultivation techniques for *Sphagnum*. Gametophytes can be cultivated on solid medium in (A) liquid Knop medium in a photobioreactor, (B) magenta vessels, (C) in petri dishes or (D) populations in a series of multi-well tissue culture plates. *In photo credit for panel A: Anna Beike, Reski Lab, University of Freiburg, Germany.*

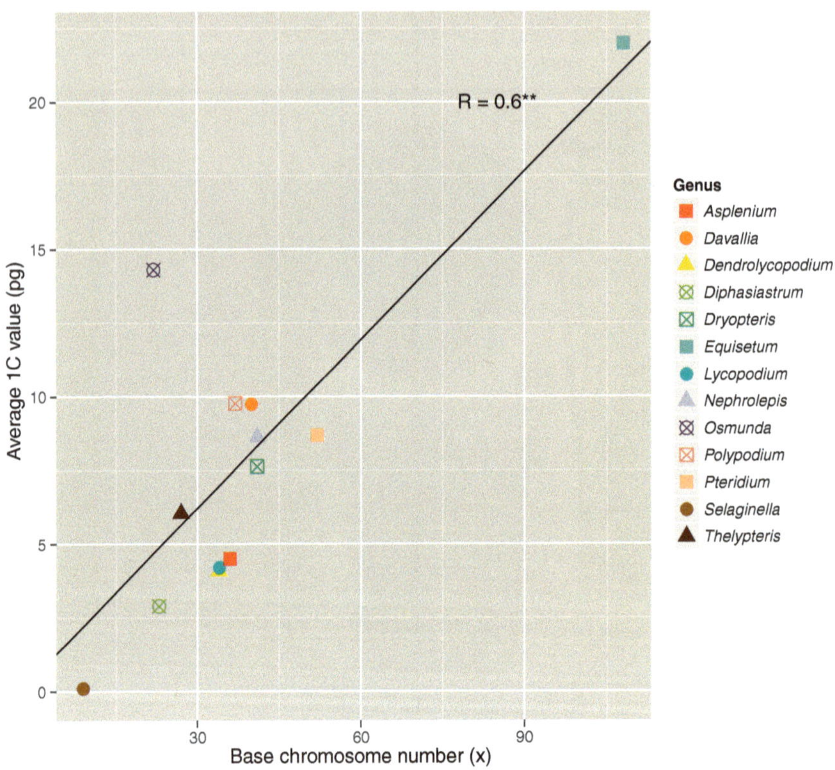

E.B. Sessa and J.P. Der, Figure 2 Average genome size (1C, pg) versus base chromosome number for 13 genera of ferns and lycophytes. There is a significant correlation ($R^2 = 0.6$, $DF = 12$, $P = 0.001$**). Homosporous lycophytes include *Lycopodium*, *Dendrolycopodium*, and *Diphasiastrum*; *Selaginella* is a heterosporous lycophyte. All other taxa are homosporous ferns.

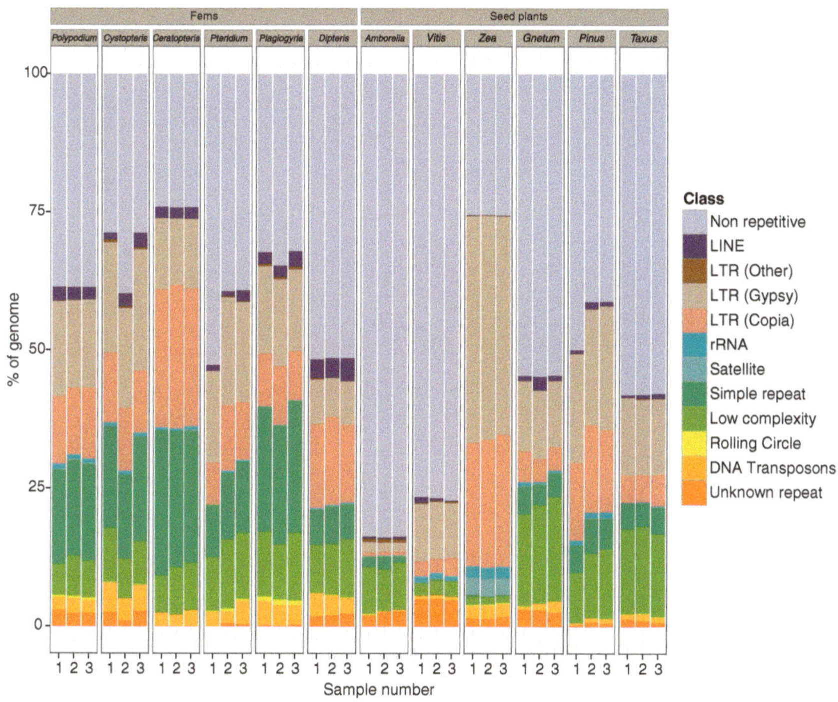

E.B. Sessa and J.P. Der, Figure 3 Genome proportions represented by ten sequence-based repeat classes, plus unknown repeats and nonrepetitive sequences, in six fern and six seed plant taxa, with three samples per taxon. *Reprinted from Wolf, P.G., Sessa, E.B., Marchant, D.B., Li, F.-W., Rothfels, C.J., Sigel, E.M. … Der, J.P. (2015). An exploration into fern genome space.* Genome Biology and Evolution, *7(9), 2533–2544; under a Creative Commons license.*

FSC
www.fsc.org
MIX
Papier | Fördert gute Waldnutzung
FSC® C083411

Kolibri 360 GmbH
Ferdinand-Jühlke-Straße 7
99095 Erfurt, Deutschland
produktsicherheit@kolibri360.de